Estimation Problems in Hybrid Systems

Recent developments in sensor and processor sophistication have created a need for effective estimation and control algorithms for hybrid, nonlinear systems. This book develops and illustrates a highly effective, computationally efficient, and flexible family of algorithms that can be used for the design of state estimators and feedback controllers for a variety of nonlinear plants. Several applications are studied, including tracking a maneuvering aircraft, automatic target recognition, and the decoding of signals transmitted across a wireless communications link.

The authors begin by setting out the necessary theoretical background, discussing infinite-dimensional algorithms and methods of nonlinear estimation. They develop a practical, finite-dimensional approximation to an optimal estimator and demonstrate its application to such problems as target tracking (including the jammed radar case), control of hybrid systems, warhead impact prediction, and an innovative signal demodulator. Throughout the book they illustrate theoretical results by simulation of real-world hybrid systems, drawn from a variety of engineering fields.

The book will be of great interest to graduate students and researchers in electrical and computer engineering. It will also be a useful reference for practicing engineers involved in the design of estimation, tracking, or wireless communications systems.

David Sworder received his Ph.D. from the University of California, Los Angeles and is Professor of Electrical and Computer Engineering at the University of California, San Diego. He has served as a consultant for many corporations, including Hughes Aircraft Company and Rockwell International. He is a Fellow of the IEEE.

John Boyd received his Ph.D. from the University of California, San Diego. He is a systems scientist at Cubic Defense Systems Inc., in San Diego, working on the design of novel communications and navigation systems.

Estimation Problems in Hybrid Systems

DAVID D. SWORDER
University of California, San Diego

JOHN E. BOYD
Cubic Defense Systems, Inc.

CAMBRIDGE
UNIVERSITY PRESS

CAMBRIDGE UNIVERSITY PRESS
Cambridge, New York, Melbourne, Madrid, Cape Town, Singapore, São Paulo

Cambridge University Press
The Edinburgh Building, Cambridge CB2 2RU, UK

Published in the United States of America by Cambridge University Press, New York

www.cambridge.org
Information on this title: www.cambridge.org/9780521623209

First published 1999
This digitally printed first paperback version 2006

A catalogue record for this publication is available from the British Library

Library of Congress Cataloguing in Publication data

Sworder, David D.

Estimation problems in hybrid systems / David D. Sworder, John E. Boyd.

p. cm.

includes bibliographical references.
ISBN 0-521-62320-0
1. Feedback control systems. 2. Estimation theory. 3. Nonlinear theories. I. Boyd, John E., 1946– . II. Title.
TJ216. S96 1999
629.8′3 – dc21 99-12281
CIP

ISBN-13 978-0-521-62320-9 hardback
ISBN-10 0-521-62320-0 hardback

ISBN-13 978-0-521-02452-5 paperback
ISBN-10 0-521-02452-8 paperback

CONTENTS

List of Illustrations

Preface

Who Should Read This Book

This book is intended for engineers and designers who seek to develop effective estimation and control algorithms for nonlinear systems. The reader is assumed to have some background in random processes and estimation (see, for example, [Pap91]) along with familiarity with concepts of feedback control phrased within the context of linear state space models (see, for example, [DB95] or [Wol94]). This background should include knowledge of

- random variables and processes,
- probability density and distribution functions for random variables, both continuous and discrete,
- moments and cross moments, including correlation functions,
- second-order properties of stationary processes, including power spectral densities,
- conditional expectations with respect to an observation process,
- fundamental properties of feedback systems, including stability and controllability of a system model.

Both time-continuous and time-discrete processes will be encountered. Some familiarity with mean-square estimation is useful. For example, in the development of the Kalman filter [BW92, Chapter 7], linear state space models are integrated with Gaussian white noise. The Kalman filter will form a basis of comparison for many of the estimators that follow.

Our treatment is applications oriented, but the reader will find that nonlinear systems require more detailed analysis than is necessary in the study of linear systems. For example, a common approach to linear estimation uses a system model phrased as a set of ordinary differential equations with continuous white noise excitation [May79, Chapter 4]:

$$\dot{x}_t = Ax_t + Bu_t + Cw_t, \tag{0.1}$$

where x_t is the state, u_t is the actuating signal, and w_t is an exogenous, Gaussian white noise excitation. This model, or its time-discrete analogue, suffices to represent the dynamics of system evolution. The Kalman filter is based, in part, upon such a model.

Equation (0.1) is an adequate expression of system dynamics in many situations. But because of the pathological properties of white noise, it is sometimes preferable to replace the differential equation (0.1) with an integral equation:

$$x_t = x_0 + \int_0^t (Ax_s + Bu_s)\, ds + \int_0^t C\, dw_s. \tag{0.2}$$

It is easier to give consistent meaning to the integrals in Equation (0.2) than it is interpret the path properties of white noise [WH85, Chapter 3, Section 8]. Stochastic integral equations such as (0.2) are often written in a differential form using increments [Ell82]; for example, (0.2) would be written

$$dx_t = (Ax_t + Bu_t)\, dt + C\, dw_t, \tag{0.3}$$

where (0.3) is taken to be shorthand for (0.2). System analysis can then be carried out in terms of the increments, retaining only terms of order one or less in dt. This formal calculus of increments permits a coordinated treatment of processes, both continuous and piecewise continuous.

When the system is nonlinear, additional difficulties arise because nonlinear operations on white noise paths are difficult to interpret, whether in the form (0.1) or (0.2). Suppose the dynamic evolution of the system is represented by a vector stochastic differential equation:

$$dx_t = \mathbf{f}(x_t, u_t)\, dt + \mathbf{g}(x_t, u_t)\, d\eta_t, \tag{0.4}$$

where $\{\eta_t\}$ is a random process and represents the unpredictable disturbances that influence state evolution. Equation (0.4) is interpreted to say that, from state x_t at time t, the plant has a deterministic drift in the direction $\mathbf{f}(x_t, u_t)$. About this extrapolation, there is a random perturbation $(d\eta_t)$ with multiplier $\mathbf{g}(x_t, u_t)$.

In an introductory analysis, we might divide both sides of (0.4) by dt to arrive at a model that has a more conventional appearance:

$$\dot{x}_t = \mathbf{f}(x_t, u_t) + \mathbf{g}(x_t, u_t)\dot{\eta}_t.$$

If $\{\eta_t\}$ were Brownian motion, $\{\dot{\eta}_t\}$ would be *Gaussian white noise*. However, $\{\eta_t\}$ is not necessarily Brownian and may indeed have discontinuous sample paths. Equation (0.4) is better written

$$x_t = x_0 + \int_0^t \mathbf{f}(x_s, u_s)\, ds + \int_0^t \mathbf{g}(x_s, u_s)\, d\eta_s. \tag{0.5}$$

Because the integrands in (0.5) are random, care must be exercised in their explication.

Preliminaries

First, we present some basic notational conventions. An integer index set $\{1, \ldots, S\}$ will be designated $\mathbf{S}$. Boldface vectors labeled $\mathbf{e}$ are canonical unit vectors whose dimension is always clear from the context: $\mathbf{e}_i$ is the ith canonical unit vector in $\mathbb{R}^k$. We shall encounter $\mathbf{E}_{ij} = \mathbf{e}_i\mathbf{e}_j'$ and $\mathbf{E}_i = \mathbf{E}_{ii}$; $\mathbf{1}$, a vector of "ones;" and $\mathbf{I}$, the identity matrix. The dimension of these matrices will be determined by context. The statement "x is $\mathbf{N}(m, P)$" (or "$x \in \mathbf{N}(m, P)$") means that x is a Gaussian random variable with (mean, covariance) equal to (m, P), though sometimes $\mathbf{N}(m, P)$ will represent the probability density itself. The Hadamard product "$*$" is defined by $(x * y)_i = x_i y_i$. If λ is a vector, none of whose components is zero, we shall refer to the vector of inverses as λ^{-1}: $\lambda * \lambda^{-1} = \mathbf{1}$. An integral over the whole space is written $\int_\Omega$; for example, $\int_\Omega f(u)\, du$ indicates an integration over the full range of the variable labeled u.

In what follows, subscripts are used in a variety of ways. We wish to avoid iterated subscripts because such forms make the equations harder to read. Suppose we are dealing with the time interval [0, T], and at time t, ν_t is the value of the vector process $\{\nu_t; t \in [0, \mathrm{T}]\}$ (written $\{\nu_t\}$). Where no confusion will arise, a subscript may identify time, the component of the vector, or a particular set of components of the vector. For the process $\{\nu_t\}$, $\{\nu_1\}$ denotes the scalar process that is the first component of $\{\nu_t\}$, while $\{\nu_x\}$ is a subvector of processes in $\{\nu_t\}$ that is associated in some way with another process $\{x_t\}$.

This notation becomes ambiguous when the process is time discrete: $\{\nu_t; t = kT, k \in \mathbf{N}\}$. The sequence $\{\nu_{kT}; k \in \mathbf{N}\}$ will be written $\{\nu[k]\}$. A component sequence from $\{\nu[k]\}$ would be $\{\nu_i[k]\}$. This notational convention becomes complicated when $\{\nu[k]\}$ is a sequence of functions of some spatial variable "z." When the spatial variable needs to be made explicit, the sequence is written $\{\nu[k](z)\}$ with components $\{\nu_i[k](z)\}$. If there is concern that the meaning of the subscript is hard to determine from context, the more explicit notation will be used (e.g., $(\nu_i[k] = \mathbf{e}_i'\nu[k])$.

Subscripts also appear as identifiers. To make explicit the variables involved in correlation and covariance matrices, they are sometimes identified with subscripts (e.g., $E[xy'] = R_{xy}$). The time dependence of the moment may be written as a direct argument (e.g., $E[x_t y_t'] = R_{xy}(t)$ or, alternatively, $E[x[k]y[k]'] = R_{xy}[k]$). Subvectors and submatrices will be denoted in different ways depending on context: If P_{xx} is a matrix related to a vector $\boldsymbol{x}$, $P_{xx}(r : s, t : v)$ is the submatrix formed from rows r through s of columns t through v; $P_{x_i x}$ is the ith row of the matrix (and is usually associated with the ith component of the vector x) and P_{xx_i} is the ith column. A matrix $S_{yy}(t)$ is the square root of the positive symmetric matrix $P_{yy}(t)$ if $S_{yy}(t)'S_{yy}(t) = P_{yy}(t)$. There are many square roots of a positive matrix and their differences are important in computation. We are concerned only with representation of matrices and thus any of the square roots will do for our purposes. Given the multiplicity of uses, when confusion regarding the interpretation of a subscript may exist, multiple subscripts will perforce be used.

Random Processes

In this book we will look at the properties of random processes defined on a probability space $(\Omega, \mathcal{F}, \mathcal{P})$ (see, [Ell82]) on a time interval [0, T] (alternatively $t = kT; k \in \mathbf{N}$). The set of events $\mathcal{F}$ is a σ-field. A *random variable* is a (real-valued) $\mathcal{F}$-measurable function. A random vector has components that are random variables, and a random process is a time-indexed set of random vectors [Pap91, Chapter 10].

In estimation and control the notion of conditional expectation is important. Our definition of conditional expectation differs from that found in introductory engineering texts. Suppose x and y are random variables (understood to be on $(\Omega, \mathcal{F}, \mathcal{P})$). Another σ-field, $\mathcal{Y}$, on Ω is said to be coarser than $\mathcal{F}$ if every element of $\mathcal{Y}$ is necessarily an element of $\mathcal{F}$: $\mathcal{Y}$ is *coarser* than $\mathcal{F}$ (i.e., $\mathcal{Y} \subset \mathcal{F}$); $\mathcal{F}$ is *finer* than $\mathcal{Y}$. The coarsest σ-field (necessarily within $\mathcal{F}$) with respect to which y is a random variable is said to be the σ-field generated by y and is possibly labeled descriptively (e.g., $\mathcal{Y}$). Clearly y is a $\mathcal{Y}$-random variable (a random variable on $(\Omega, \mathcal{Y}, \mathcal{P})$). The expectation of x given y (denoted $E[x \mid \mathcal{Y}]$) is the $\mathcal{Y}$-random variable with all of the orthodox properties of conditional expectation given in [Pap91, Chapter 7]. Idiomatically, we would say that $E[x \mid \mathcal{Y}]$ is a random variable expressible as a function of y.

We will deal with conditioning, not just on random variables, but on the sample paths of random processes. On $(\Omega, \mathcal{F}, \mathcal{P})$, there are several elemental random processes. All of them are piecewise *right continuous* (or continuous); that is, if $\{y_t\}$ is a random process, $y_t = y_{t+}$. Let $\{x_t\}$ and $\{y_t\}$ be random processes and consider $\{y_t\}$ on the interval $[0, s]$. There is a coarsest σ-field within $\mathcal{F}$ with respect to which events determined by $\{y_u; u \in [0, s]\}$ (the past and present of $\{y_t\}$) are measurable. This σ-field will be called $\mathcal{Y}_s$. The indexed family of σ-fields, $\{\mathcal{Y}_t\}$, is the *filtration* generated by $\{y_t\}$ (see, [Ell82, p. 332]). This filtration is right continuous because $\{y_t\}$ is right continuous and, moreover, is such that if $s \leq t$ then $\mathcal{Y}_s \subset \mathcal{Y}_t$. There is also a left continuous filtration generated by $\{y_u; u \in [0, s)\}$ and labeled $\{\mathcal{Y}_{t-}\}$. A process $\{x_t\}$ is *$\mathcal{Y}_t$-adapted* if x_t is a $\mathcal{Y}_t$-random variable for every t. A process $\{x_t\}$ is *$\mathcal{Y}_t$-predictable* if x_t is $\mathcal{Y}_{t-}$-measurable for every t. Indeed, the predictable version† of a right continuous process is given by its left continuous modification. So, if $\{x_t\}$ is a right continuous random process and generates $\{\mathcal{X}_t\}$, then $\{x_{t-}\}$, the left continuous modification of $\{x_t\}$, generates $\{\mathcal{X}_{t-}\}$, the filtration of "past events."

There may be different filtrations on $(\Omega, \mathcal{F}, \mathcal{P})$ relevant to the application, and if we wish to distinguish the filtration of interest we will write the probability space and filtration as $(\Omega, \mathcal{F}, \mathcal{P}; \mathcal{F}_t)$. All of the elemental processes in this book are $\mathcal{F}_t$-adapted.

If $\{x_t\}$ is a state process and $\{y_t\}$ is an observation process, an engineer may seek the expectation of x_t conditioned on the past of $\{y_t\}$: Idiomatically this estimate is said to be a function of the observations up to time t or to be *causal*. Write this conditional expectation $E[x_t \mid \mathcal{Y}_t]$. The conditional mean $E[x_t \mid \mathcal{Y}_t]$ is a $\mathcal{Y}_t$-random variable, which we will write as $\hat{x}_t$ if the conditioning filtration is apparent from context. The random process $\{\hat{x}_t\}$ is a $\mathcal{Y}_t$-adapted random process (also called a $\mathcal{Y}_t$-random process). If confusion might arise as to the conditioning filtration, the conditional mean process would be written $\{\hat{x}_t; \mathcal{Y}_t\}$. Since $\mathcal{Y}_t$ is coarser than $\mathcal{F}_t$ the estimation error is an $\mathcal{F}_t$-random variable. For example, x_t is $\mathcal{F}_t$-adapted and $\hat{x}_t$ is $\mathcal{Y}_t$-adapted. Hence, the error $\tilde{x}_t = x_t - \tilde{x}_t$ must be an $\mathcal{F}_t$-random variable but likely not a $\mathcal{Y}_t$-random variable.

The structure of random processes on $(\Omega, \mathcal{F}, \mathcal{P}; \mathcal{F}_t)$ can be quite complex. Fortunately we will not have to face processes of a general type. Instead, only two circumscribed classes of elemental $\mathcal{F}_t$-random processes will appear in the applications that follow. The first is composed of $\mathcal{F}_t$-Brownian motions: A (vector) random process $\{w_t\}$ is a *Brownian motion* if $w_0 = 0$, and when $s \le t$, $w_t - w_s$ is $\mathbf{N}(0, W(t-s))$ and independent of $\mathcal{F}_s$ (see, [Ell82, Definition 12.27]). We will refer to W as the *intensity* of the Brownian motion. It is easily seen that $E[w_t w_t'] = Wt$. Brownian motion is an $\mathcal{F}_t$-martingale process: If $s \le t$, $E[w_t \mid \mathcal{F}_s] = w_s$.

It is useful to develop a formal calculus of increments. Increments are defined in the forward direction (e.g., $dw_t = w_{t+dt} - w_t$ with $dt > 0$). Associated with $\{w_t\}$ is the $\mathcal{F}_t$-predictable quadratic variation process, $\langle w, w; \mathcal{F}_t\rangle_t$. The *$\mathcal{F}_t$-predictable quadratic variation process* is the integral of its increments, where $d\langle w, w; \mathcal{F}_t\rangle_t = E[dw_t dw_t' \mid \mathcal{F}_t]$. Since $dw_t dw_t' = W\,dt$ [WH85, Proposition 3.4], it follows that $\langle w, w; \mathcal{G}_t\rangle_t = Wt$ for any filtration $\{\mathcal{G}_t\}$ $(d\langle w, w; \mathcal{G}_t\rangle_t = Wdt)$.

The second class of elemental processes contains $\mathcal{F}_t$-Markov processes on the canonical unit vectors in $\mathbb{R}^S$. Such a process, $\{\phi_t\}$, is characterized by its initial probability distribution, $(\hat{\phi}_0)$, and its transition rates. Let the $S \times S$-matrix Q have as its elements $Q_{ij} = \mathcal{P}(\phi_{t+dt} = \mathbf{e}_j \mid \phi_t = \mathbf{e}_i)/dt$ if $i \neq j$, with $Q_{ii} = -\sum_{j \neq i} Q_{ij}$. The *generator* of the Markov process $\{\phi_t\}$ is Q', a matrix with nonnegative elements

† If x_t and y_t are random variables on the same probability space and $\mathcal{P}(x_t = y_t) = 1$ for all t, then the variables are said to be versions or modifications of each other.

off the diagonal and column sums equal to zero. Because the state space of $\{\phi_t\}$ is the canonical unit vectors, $\mathcal{P}(\phi_t = \mathbf{e}_i) = E[\mathbf{e}_i{}'\phi_t]$. Let us add the discontinuities to the forward increment. Define $d\phi_t$ as $\phi_{t+dt} - \phi_{t-}$. If $\{\phi_t\}$ has a discontinuity at time t, this will be denoted $\Delta\phi_t = \phi_t - \phi_{t-}$. It is easily shown that $E[d\phi_t \mid \mathcal{F}_t] = Q'\phi_t dt$.

Define $dm_t = d\phi_t - Q'\phi_t\, dt$. Then $\{m_t\}$ has discontinuities where $\{\phi_t\}$ does:

$$\Delta m_t = m_t - m_{t-} = \phi_t - \phi_{t-}.$$

Clearly, $E[dm_t \mid \mathcal{F}_t] = 0$: $\{m_t\}$ is an $\mathcal{F}_t$-martingale (see, [EAM95, Section 7.2]). The $\mathcal{F}_t$-predictable quadratic variation of $\{m_t\}$ is defined as with $\{w_t\}$, but $\{m_t\}$ has a fundamentally different character. First, if $\{\phi_t\}$ makes no transitions in the interval $[t, t+dt]$, then $dm_t dm_t' \approx 0$: $\{m_t\}$ is called a purely discontinuous $\mathcal{F}_t$-martingale because, excluding jumps, its quadratic variation is zero. Alternatively, if $\{\phi_t\}$ makes the transition $\mathbf{e}_i \mapsto \mathbf{e}_j$ in $[t, t+dt]$, then $\Delta m_t \Delta m_t' = (\mathbf{e}_j - \mathbf{e}_i)(\mathbf{e}_j - \mathbf{e}_i)' = \mathbf{E}_i + \mathbf{E}_j - \mathbf{E}_{i,j} - \mathbf{E}_{j,i}$. If $\phi_t = \mathbf{e}_i$,

$$d\langle m, m; \mathcal{F}_t\rangle_t = \sum_j (\mathbf{E}_i + \mathbf{E}_j - \mathbf{E}_{ij} - \mathbf{E}_{ji})\mathcal{P}(\phi_{t+dt} = \mathbf{e}_j \mid \phi_t = \mathbf{e}_i).$$

The general expression for $d\langle m, m; \mathcal{F}_t\rangle_t = E[\Delta m_t \Delta m_t' \mid \mathcal{F}_{t-}]$ is given as a function of Q in the Appendix 1.

Sometimes martingales with continuous paths (e.g., $\{w_t\}$) appear in combination with martingales with discontinuous paths (e.g., $\{m_t\}$) to form a *composite* martingale $\{\eta_t\}$. In fact, any $\mathcal{F}_t$-martingale can be separated into its continuous and discontinuous parts: $\eta_t = \eta_t^c + \eta_t^d$ where $\{\eta_t^c\}$ is a continuous process and $\{\eta_t^d\}$ is purely discontinuous. The two are mutually orthogonal: $d\langle \eta_t^c, \eta_t^d; \mathcal{F}_t\rangle_t = 0$ [Ell82, Chapter 9]. For example, $d\langle w, m; \mathcal{F}_t\rangle_t = E[dw_t dm_t' \mid \mathcal{F}_t] = 0$. Additionally, two purely discontinuous processes without common jump times are orthogonal: If $\{\Delta\phi_t \Delta\psi_t'\}$ is essentially the zero process, $d\langle \phi, \psi; \mathcal{F}_t\rangle_t = 0$.

The composite martingale $\{\eta_t\}$ has associated with it another quadratic process. The *optional quadratic variation*, $[\eta, \eta]_t$, is determined from its increments: $d[\eta, \eta]_t = d\eta_t d\eta_t'$. It can also be found by adding the outer product of the jumps in $\{\eta_t\}$ to the predictable quadratic variation:

$$d[\eta, \eta]_t = d\langle \eta_t^c, \eta_t^c; \mathcal{F}_t\rangle_t + \Delta\eta_t \Delta\eta_t'.$$

The *optional cross quadratic variation* of two martingales is similarly defined [Kri84, Chapter 4].

Stochastic Differential Equations

Equation (0.5) relates the actuating signal to the system state. This is an integral equation with differential embodiment given in (0.4). For this model to be useful, each of the terms on the right side of (0.5) must be given clear meaning. The first integral, $\int_{[0,t]} \mathbf{f}(x_s, u_s)ds$, is of a conventional sort if the sample functions of $\{\mathbf{f}(x_t, u_t)\}$ are well behaved. The second integral, $\int_{[0,t]} \mathbf{g}(x_s, u_s)d\eta_s$, is more problematic. The integrand $\{\mathbf{g}(x_t, u_t)\}$ is an $\mathcal{F}_t$-random process and in what follows $\{\eta_t\}$ is an $\mathcal{F}_t$-martingale. It is advantageous to define the integral using the predictable version of the integrand; that is, $\int_{[0,t]} \mathbf{g}(x_s, u_s)d\eta_s$ is better written $\int_{[0,t]} \mathbf{g}(x_{s-}, u_{s-})d\eta_s$ [Ell82, Theorem 11.44]. For consistency, the stochastic differential equation could be written

$$dx_t = \mathbf{f}(x_{t-}, u_{t-})\,dt + \mathbf{g}(x_{t-}, u_{t-})\,d\eta_t$$

since the increment in $\{x_t\}$ depends upon the antecedent values of the arguments rather than their current values. For simplicity, we will not distinguish the predictable versions of the random processes in the differential equations even though the left continuous version of the integrands will appear in the integrals.

The output of the system is represented with a stochastic differential equation too:

$$dg_t = \mathbf{r}(x_t, u_t)\,dt + \mathbf{s}(x_t, u_t)\,dn_t, \tag{0.6}$$

where $\{g_t\}$ is the output process or the observation process as appropriate. The components of the $\mathcal{F}_t$-martingale $\{n_t\}$ that appear in (0.6) would be called *observation noise* or the equivalent. It is through $\{g_t\}$ that the value of $\{x_t\}$ can be determined. Let the filtration generated by $\{g_t\}$ be labeled $\{\mathcal{G}_t\}$. This output filtration is a subfiltration of $\mathcal{F}_t$. For any $\mathcal{F}_t$-random process $\{\zeta_t\}$, denote the $\mathcal{G}_t$-conditional expectation with a circumflex and the $\mathcal{F}_t$-conditional error with a tilde: For example, $\hat{\zeta}_t = E[\zeta_t \,|\, \mathcal{G}_t]$; $\tilde{\zeta}_t = \zeta_t - \hat{\zeta}_t$.

An important process related to the $\mathcal{G}_t$-mean is the *innovation process*. The innovation process, labeled $\{\nu_t\}$, is generated from its increments:

$$d\nu_t = dg_t - E[dg_t \,|\, \mathcal{G}_t].$$

This terminology is "motivated by the observation that, formally, $\nu_{t+h} - \nu_t$ represents the 'new' information about (the system state) obtained from observations between t and $t + h$" [Ell82, Definition 18.6].

Sometimes the observations are not time continuous but instead have a natural sampling interval. An example of this is a radar tracking an aircraft. The aircraft path is continuous (i.e., modeled as in (0.4)), but the observations occur every T seconds beginning at $t = 0$. In this case, we would replace (0.6) with

$$g[k] = \mathbf{r}(x[k], u[k]) + \mathbf{s}(x[k], u[k])n[k], \tag{0.7}$$

where $g[k]$ is the output (or observation) at time $t = kT$, and similarly for $x[k]$ and $u[k]$. In this model, $\{n[k]\}$ is not typically an $\mathcal{F}_t$-martingale but may be a sequence of martingale increments ($E[n[k] \mid \mathcal{F}_{kT}] = 0$). The output sequence, $\{g[k]\}$, generates a filtration, $\{\mathcal{G}_t\}$, which is defined for all $t \in [0, \mathrm{T}]$, not just $t \in \{0, T, 2T, ...\}$. However, since new information appears at the output at distinct times, it is true that $\{\hat{x}_t\}$ tends to have discontinuities at sample times.

In some circumstances, the system state is time discrete:

$$x[k+1] = \mathbf{f}(x[k], u[k]) + \mathbf{g}(x[k], u[k])\eta[k+1]. \tag{0.8}$$

If the state and measurement grid are the same, a time-discrete system with time-discrete measurements has a structure like that given above with natural changes in terminology.

Some Useful Results from Martingale Theory

This section lists some useful results from martingale theory. The statements do not include certain qualifications to be found in the references [Ell82].

Definition 1 *The process $\{X_t\}$ is* ***corlol*** *(for continuous on the right, limits on the left) if there is a modification of $\{X_t\}$ such that*

$$X_t(\omega) = \lim_{s \to t^+} X_s(\omega)$$

and

$$X_{t^-}(\omega) = \lim_{s \to t^-} X_s(\omega).$$

Definition 2 *Given any process $\{X_t\}$ adapted to $\mathcal{F}_t$, if there exists a process $\{A_t\}$ such that $A_0 = 0$, $\{A_t\}$ is $\mathcal{F}_t$-predictable, $\{A_t\}$ has corlol sample paths of locally finite variation, and $\{X_t - A_t\}$ is an $\mathcal{F}_t$-martingale, then $\{A_t\}$ is called the predictable compensator of $\{X_t\}$ relative to $\mathcal{F}_t$*

Theorem 1 (Doob–Meyer Decomposition Theorem) *If the random process $\{X_t\}$ has a predictable compensator, then it is unique in the sense that any two predictable compensators are equal to each other for all t.*

This statement of the Doob–Meyer Decomposition Theorem is Proposition 3.2 in [WH85]. See [DM82] for a proof.

Theorem 2 (Martingale Representation Theorem) *Suppose the filtration $\{\mathcal{F}_t\}$ is generated by the local semimartingale $X_t = B_t + W_t$, where $\{B_t\}$ is of bounded variation and $\{W_t\}$ is a Brownian motion or a point process. Then any $\mathcal{F}_t$-local semimartingale $\{Z_t\}$ can be written as a stochastic integral against $\{W_t\}$. That is, there exists an $\mathcal{F}_t$-predictable function $\{\gamma_t\}$*

such that

$$Z_t = Z_0 + \int_0^t \gamma_s \, dW_s.$$

The Brownian motion version is due to a generalization by Fujisaki et al. [FKK72] of the works of Itô [Itô51], Kunita and Watanabe [KW67], and Clark [Cla70]. The extension to point processes is due to Brèmaud [Brè72].

Theorem 3 *Let $\{S_t\}$ be a process, not necessarily adapted to $\mathcal{G}_t$, a filtration generated by the continuous or purely discontinuous martingale $\{\nu_t\}$. Let $\hat{S}_t = E[S_t|\mathcal{G}_t]$, and define a process $\{B_t\}$ by $dB_t = E(dS_t|\mathcal{G}_t)$. Then $\{\hat{S}_t - B_t\}$ is a $\mathcal{G}_t$-martingale, and there exists a $\mathcal{G}_t$-predictable function γ_t such that*

$$d\hat{S}_t = E(dS_t|\mathcal{F}_t) + \gamma_t \, d\nu_t.$$

Proof ([WH85]): Since $\mathcal{G}_t$ is increasing,

$$\begin{aligned} E(d\hat{S}_t \,|\, \mathcal{G}_t) &= E\{[E(S_{t+dt} \,|\, \mathcal{G}_{t+dt}) - E(S_t \,|\, \mathcal{G}_t)]\} \\ &= E(S_{t+dt} \,|\, \mathcal{G}_t) - E(S_t) \,|\, \mathcal{G}_t) \\ &= dB_t. \end{aligned}$$

Therefore, $\hat{S}_t - B_t = M_t$ is a $\mathcal{G}_t$-martingale. By Theorem 2, M_t can be represented as a stochastic integral against $\{\nu_t\}$.

Theorem 3 provides a cornerstone for system estimation theory. It implies that under modest conditions, the estimator of S_t is the solution to a stochastic differential equation driven by the innovation process.

1

Hybrid Estimation

1.1 Introduction

Common problems in design require that an engineer devise a control or decision algorithm that converts measurements of system and environmental variables into signals that aid in system regulation. For example, a control node converts sensor outputs into an actuating signal that moves the system toward the desired operating point and keeps it there. At this foundational level, the engineer must formulate a mapping from the system observables into an action or report; for example, a feedback regulator converts the measured outputs of the system to be controlled (the *plant*) into an input that stabilizes the system.

Design is made difficult by disturbances internal to the system and by noise at its output. For example, there may be no sensors that measure those plant variables most useful for regulation, or, if measured, the variables may be masked by noise in the sensor-to-regulator link. Lacking omniscience, an engineer must process the available measurements to produce a good approximation to relevant but "hidden" variables. And this inference must be done on-line. The processing algorithm must not only be adapted to the incoming data stream, it must be of a form that can be implemented: An implementable estimation algorithm is an explicit mapping of the sensor output process (the *measurements*) into a (nearly) concurrent estimate of the required variables. In the applications studied here, the need for contemporaneous response limits consideration to finite-dimensional recursive algorithms; new observations are integrated into an estimate in an accretive manner.

Analytical design in estimation and control begins with a formal mathematical description of the system to be controlled (the *plant model*). The model delineates the response of the plant to endogenous actuating signals as well as representing the influence of exogenous disturbances common to the application. The system designer selects a control policy or a state estimation algorithm based in large part upon the behaviors predicted by the model. The practicality of analytic procedures is linked closely to the realism of the plant model. However, realism must be tempered by the need to have a model that is simultaneously flexible and tractable.

One useful paradigm phrases the plant model in terms of a set of nonlinear stochastic differential equations. Let us start with a probability space $(\Omega, \mathcal{F}, \mathcal{P})$ and a time interval of interest, $[0, \mathrm{T}]$. On this space there is a right-continuous filtration $\{\mathcal{F}_t; 0 \leq t \leq \mathrm{T}\}$ and right-continuous, $\mathcal{F}_t$-adapted random processes, $\{\Phi_t\}$, $\{w_t\}$, and $\{n_t\}$. Subject to initial conditions χ_0 and g_0, the plant model is written:

plant model

$$d\chi_t = \mathbf{f}(\chi_t, \upsilon_t, \Phi_t)\, dt + \mathbf{g}(\chi_t, \upsilon_t, \Phi_t)\, dw_t, \tag{1.1}$$

$$dg_t = \mathbf{r}(\chi_t, \upsilon_t, \Phi_t)\, dt + \mathbf{s}(\chi_t, \upsilon_t, \Phi_t)\, dn_t, \tag{1.2}$$

where $\{\upsilon_t\}$ is an s-dimensional actuating process (the *plant input*), $\{g_t\}$ is an r-dimensional observation process (the *plant output*), and $\{\chi_t\}$ is an n-dimensional internal process (the *plant state*). Equation (1.1) describes the temporal evolution of the internal variables within the plant, and (1.2) describes the sensor outputs available for estimation and/or control.

This plant model is more complicated than that encountered in introductory studies of feedback control. In applications, even when the actuating process is specified, the realizations of the state and output paths are unpredictable – there are many effects not well captured in a deterministic model. Chance influences in the plant and sensor are represented by the stochastic processes in (1.1) and (1.2). Various accretive effects are represented by $\{w_t\}$ and $\{n_t\}$; for example, $\{w_t\}$ could describe the high frequency modes ignored in a low-dimensional plant model, and $\{n_t\}$ could describe noise at the sensor output. The *environmental process*, $\{\Phi_t\}$, denotes external conditions of a more global sort that affect plant operation. The value of $\{\Phi_t\}$ might indicate the operational status of a subelement within the plant, external conditions that influence the plant dynamics (e.g., temperature), the level of loads placed upon the system by linked elements, etc. In contrast to $\{w_t\}$ and $\{n_t\}$, which tend to be aggregations of small increments, Φ_t may symbolize temporally distinct events. (Friedland called Φ_t the *metastate* when used in the context of adaptive control; see [Fri96, Chapter 10].) All of these disturbance processes are viewed by the designer as exogenous.

In both estimation and the control, the output signal, $\{g_t\}$, is processed to create causal estimates of important system variables. A *filter* provides estimates of the current values of both the plant state vector and the environmental process. A *predictor* estimates future values of the same variables. Often, the environmental process has a character fundamentally different from the plant state. The value of Φ_t

may be a symbolic variable (e.g., $\Phi_t \in \{$*normal operation, degraded operation*$\}$). In this event, the average value of Φ_t has no meaning. Rather, the probability distribution of Φ_t is required to properly assess the status of the plant. Denote the filtration generated by $\{g_t\}$ by $\{\mathcal{G}_t\}$. If mean square error is used as a performance index, the estimation problem can be posed as follows:

> Find an explicit processing algorithm to generate (or approximate) the mean plant state $\hat{\chi}_t = E[\chi_t \mid \mathcal{G}_t]$ and the $\mathcal{G}_t$-probability distribution of Φ_t.

There are applications in which even this will not suffice and more comprehensive statistical properties of the plant processes are required.

Unfortunately, even when formal descriptions of the exogenous processes are integrated with (1.1) and (1.2), an elementary solution to this estimation problem does not currently exist. There is, however, one special case in which astounding success has been achieved. So much so that the solution thus derived is used in circumstances far removed from those in which it was developed. Specifically, suppose that the system has "smooth" nonlinearities, that the plant noise, $\{w_t\}$, is a Brownian motion, and that the environmental process, $\{\Phi_t\}$, is constant with known value Φ_c. Associated with Φ_c there is a nominal operating condition, both in the state and in the actuating signal labeled (χ_n, υ_n). Frequently (χ_n, υ_n) is a condition of plant stasis: $f(\chi_n, \upsilon_n, \Phi_c) = 0$. The operating condition (or *regime*) is known by different names: in the process control industry, (χ_n, υ_n) is referred to as the set point or the operating point; in aircraft flight control, (χ_n, υ_n) is referred to as the trim condition; in other applications, (χ_n, υ_n) is simply the reference point. We will use these terms interchangeably and note in this context that Φ_t simply points to the operating mode or regime with its value having no intrinsic meaning.

For a particular regime, there is a local description of the plant phrased in terms of a set of perturbation variables. These are defined as the (usually small) deviations in state and excitation from the set point: $x_t = \chi_t - \chi_n$; $u_t = \upsilon_t - \upsilon_n$. Using orthodox methods and neglecting higher order terms, the perturbation processes are commonly represented by a linear stochastic differential equation with initial condition taken to be Gaussian: x_0 is $\mathbf{N}(\hat{x}_0, P_{xx}(0))$, and

$$dx_t = (Ax_t + Bu_t)\,dt + C\,dw_t, \tag{1.3}$$

where $\{w_t\}$ is a Brownian motion with intensity $W(d\langle w, w\rangle_t = W dt)$. Call $\{x_t\}$ the *base-state process* to distinguish it from the plant state process, $\{\chi_t\}$; call $\{u_t\}$ the *regulation signal* to distinguish it from the plant input, $\{\upsilon_t\}$. Equation (1.3) relates the base-state to the inputs $\{u_t\}$ (endogenous) and $\{w_t\}$ (exogenous). The base-state excitation is a Brownian motion with intensity $CWC' = R_\chi$. Of course, if the plant is linear over a large region of the state space, (1.3) is valid without consideration of the set point. In such applications, it is understood that χ_n and υ_n are both zero.

The set point is known ($\mathcal{P}[\Phi_t \equiv \Phi_c] = 1$) and need not be estimated, but the plant state is frequently not known and must be inferred from sensor outputs. Suppose a sensor provides a noisy but linear plant state measurement,

plant state measurement: time-continuous

$$dy_t = H\chi_t\, dt + dn_t, \tag{1.4}$$

where $\{n_t\}$ is a Brownian motion independent of $\{w_t\}$, with intensity $R_x > 0$ ($d\langle n, n\rangle_t = R_x\, dt$), and $y_0 = 0$. By subtracting the contribution of the set point from the output, (1.4) can be written as a noisy, linear measurement of the base-state: $dy_t - H\chi_n dt = Hx_t\, dt + dn_t$. The innovation increment $d\nu_t = dy_t - d\hat{y}_t$ can be written $H\tilde{x}_t\, dt + dn_t$, where $\tilde{x}_t = x_t - \hat{x}_t$. When there is only one sensor, $g_t \equiv y_t$. To differentiate this case from others that follow, denote the filtration generated by $\{y_t\}$ by $\{\mathcal{Y}_t\}$ ($= \mathcal{G}_t$ in this case), where a circumflex may be used to denote $\mathcal{Y}_t$-expectation if no confusion will result. Equations (1.3) and (1.4) will be called a *linear–Gauss–Markov* (LGM) model even when x_0 is not Gaussian. Although the observation is unconventional, the regime offset is known and is accommodated in a direct fashion. The base-state estimator is known for the LGM problem: the Kalman filter. The Kalman filter generates $\{\hat{x}_t\}$ using a simple recursive algorithm. The plant state estimator is $\hat{\chi}_t = \chi_n + \hat{x}_t$.

In the systems we will study, $\{\Phi_t\}$ is not nearly so obliging. Instead of a single operating point, $\{\Phi_t\}$ may move about in its range space in response to the macroevents that influence the plant. The temporal structure of the regime process has a fundamental impact on system analysis. If, for example, $\{\Phi_t\}$ has sample paths that are well described by a diffusion process, then $\{\Phi_t\}$ can be integrated into (1.1) as an additional plant state. This is an attractive option when the time constants of $\{\Phi_t\}$ are comparable with those of the plant, though this inclusion compounds the plant nonlinearity.

In other applications, $\{\Phi_t\}$ has a distinguishing feature that precludes orthodox state augmentation. Suppose the plant has S possible operating regimes, and at any particular time, Φ_t takes on a value selected from a set of size S: $\Phi_t \in \{\Phi_i; i \in \mathbf{S}\}$. The plant now has S possible reference points (or set points, etc.), and these are identified with the S possible values of $\{\Phi_t\}$; that is, there are S vector pairs, $\{(\chi_i, \upsilon_i); i \in \mathbf{S}\}$, which designate the S relevant stasis conditions for the plant. For example, the kth nominal operating point for the plant is (χ_k, υ_k), and if $\Phi_t = \Phi_k$ the plant input and state should be near (χ_k, υ_k).

For simplicity, array the nominal states (respectively nominal actuating signals) as an $n \times S$ matrix $\boldsymbol{\chi}$ (respectively an $s \times \mathrm{S}$ matrix $\boldsymbol{\upsilon}$): $\boldsymbol{\chi} = [\chi_i]$ (respectively

$\upsilon = [\upsilon_i]$). During operation, the system will operate in one regime for a time ($\Phi_t = \Phi_i$ for $t \in [a, b)$) and then suddenly shift ($\Phi_b = \Phi_j$) to another in response to an external event or change in the surrounding environment. In most applications, the discontinuous sample paths of $\{\Phi_t\}$ are an approximation to the continuous though abrupt modal transitions that actually occur. Nevertheless, the representation of $\{\Phi_t\}$ with a process of piecewise constant paths is a useful abstraction when the interval over which the modal transition takes place is short as compared to the important time constants of the plant.

Since the environmental process has a finite state space, $\{\Phi_t\}$ can be represented using a more illuminating notation. Let ϕ_t be a pointer to the current regime: The state space of ϕ_t consists of the S canonical unit vectors in $\mathbb{R}^S$ ($\phi_t \in \{\mathbf{e}_1, \ldots, \mathbf{e}_S\}$). The component in ϕ_t with value one marks the current mode of operation: If $\Phi_t = \Phi_k$ then $\phi_t = \mathbf{e}_k$. The $\{\phi_t\}$ process is called the *modal-state process* to differentiate it from the base-state process. The base-state variables are deviations from the current set point: $x_t = \chi_t - \chi\phi_t$; $u_t = \upsilon_t - \upsilon\phi_t$. The comprehensive state of the system is the composition of the base- and modal-states: The *zygostate* is the pair (x_t, ϕ_t). Since ϕ_t is an indicator vector, the expectation of the modal-state is actually the conditional probability vector $\hat{\phi}_t = [\mathcal{P}\{\phi_i = \mathbf{e}_i \,|\, \mathcal{G}_t\}]$.

Control in a multiregime environment presents some subtle challenges. When the regime is known and constant (e.g., $\phi_t \equiv \mathbf{e}_i$), the actuating signal has a natural decomposition ($\upsilon_t = u_t + \upsilon\phi_t$) into a feedforward component associated with the set point ($\upsilon\mathbf{e}_i = \upsilon_i$) and a feedback component (u_t) that maintains the plant state near the set point ($\chi_t \approx \chi_i$). When the modal-state is neither known nor measured, this implementation is not possible because proper feedforward control cannot be generated. In applications, a variety of replacements for $\{\upsilon\phi_t\}$ have been proposed. We will not explore issues of feedforward control in any depth here. We will simply employ $\{\upsilon\hat{\phi}_t\}$ as the "feedforward" component of the actuating signal: Ideal set point actuation will be replaced with its expectation. Note, however, that a failure to generate the proper feedforward actuating signal has an influence that must be included in the base-state dynamics.

A comprehensive plant model requires a representation of evolution, both intramodal and intermodal. Consider the former first. During an extended (known) modal sojourn, proper control will place and maintain the plant state vector near the correct set point. The natural plant model in this circumstance would be that local model, selected from a family of regime-specific, linear models, associated with the present mode of operation. The modal-state is a pointer, and the intrasojourn model can be written:

$$dx_t = \sum_i ((A_i x_t + B_i(u_t + \upsilon(\hat{\phi}_t - \mathbf{e}_i)))\,dt + C_i\,dw_t)\phi_i, \qquad (1.5)$$

where $\{A_i, B_i, C_i; i \in \mathbf{S}\}$ are determined from (1.1) in precisely the way (1.3) was in the unimodal (or unimorphic†) case.

Suppose the plant is in the ith mode ($\phi_t = \mathbf{e}_i$) and the modal estimate is a good one ($\hat{\phi}_t \approx \mathbf{e}_i$). The base-state dynamic equation is the ith selection from the family of models: $(A_i x_t + B_i u_t)\,dt + C_i\,dw_t$. The exogenous excitation is a Brownian motion with intensity $R_\chi(i) = C_i W C_i'$. There is an atypical term in (1.5) that is connected with failure to implement the proper feedforward excitation $(-B_i \upsilon \tilde{\phi}_t \phi_i dt)$. When the estimate of ϕ_t is good, this last term is negligible, and the intramodal dynamics are LGM.

The intramodal representation is but a part of the model of plant evolution. When the regime changes, many things can happen to the plant state. There will be no attempt to be exhaustive in this list, but we will encounter situations in which the plant state translates, rotates, and/or is scaled. More specifically, suppose $\{\Phi_t\}$ makes the transition $\mathbf{e}_i \mapsto \mathbf{e}_l$ at time t. Then $\{\chi_t\}$ may experience:

Translation: $\Delta\chi_t = \rho(i,l);\ i \neq l.$

Rotation and/or scaling: $\Delta\chi_t = M(i,l)\chi_{t-};\ i \neq l,$

where $\Delta\chi_t = \chi_t - \chi_{t-}$.

When the mode changes, the plant state may be transformed in a way that creates a path discontinuity. This abrupt change in plant state is an approximation in most cases. But, if the interval over which a change takes place is small, a discontinuous path model may provide a far simpler representation of the state variation than would a continuous path model created from an intricate diffusion process. To fill out the list of transformation matrices, let $\rho(i,i) = 0$, $M(i,i) = 0;\ i \in \mathbf{S}$. The indicator vector of the discontinuity event $\mathbf{e}_i \mapsto \mathbf{e}_l$ at time t can be written as $\phi_i \mathbf{e}_l' \Delta\phi_t$. The plant state discontinuity can be written explicitly as

$$\Delta\chi_t = \sum_{i,l} (M(i,l)\chi_{t-} + \rho(i,l))\phi_i \mathbf{e}_l' \Delta\phi_t.$$

Discontinuities in $\{\chi_t\}$ are reflected directly in $\{x_t\}$, but there is an additional source of base-state discontinuity. When the mode changes $\mathbf{e}_i \mapsto \mathbf{e}_l$, the base-state reference level changes from χ_i to χ_l. Even if the plant state were continuous, the base-state would experience a discontinuity:

$$\Delta x_t = -\chi \Delta\phi_t.$$

These intermodal transition conditions can be combined to yield the base-state

† We say a system may have one or several modes or, equivalently, forms. Hence a single-mode plant is called unimodal (or unimorphic) to distinguish it from a polymodal (polymorphic) system.

discontinuity model:

$$\Delta x_t = \sum_{i,l} (M(i,l)x_{t-} + (\chi_i - \chi_l) + M(i,l)\chi_i + \rho(i,l))\phi_i \mathbf{e}_l' \Delta\phi_t. \quad (1.6)$$

Now combine the intermodal discontinuity with the intramodal dynamics to yield:

base-state model

$$dx_t = \sum_i ((A_i x_t + B_i(u_t - \upsilon\phi_t))\, dt + C_i\, dw_t)\phi_i + \sum_{i,l} (M(i,l)x_t + (\chi_i - \chi_l) + M(i,l)\chi_i + \rho(i,l))\phi_i \mathbf{e}_l' \Delta\phi_t. \quad (1.7)$$

Equation (1.7) is the fundamental model of time-continuous base-state evolution. Its appearance is formidable. Be assured that while the various discontinuity and set point conditions will appear in what follows, in no application will all occur simultaneously! In many cases, (1.7) takes on a strikingly simpler form. It is advantageous to set apart some special instances of (1.7) because they are easier to interpret.

LJS: The most often studied specialization of (1.7) is called a linear jump system (LJS). In an LJS there is no regime-specific set point reference ($\chi = 0$, $\upsilon = 0$), nor are there plant state discontinuities at modal transition [Mar90]. The LJS model is simply

$$dx_t = \sum_i ((A_i x_t + B_i u_t)\, dt + C_i\, dw_t)\phi_i. \quad (1.8)$$

Often the intensity of the Brownian excitation is constant across regimes and there is no feedback control:

$$dx_t = \sum_i A_i x_t \phi_i\, dt + C\, dw_t. \quad (1.9)$$

We will find this simpler model to be useful in certain tracking applications.

JTS: In some applications, the plant state discontinuity has a particular structure. There is neither rotation nor scaling. The plant state discontinuity is a translation in the form of a difference between mode-specific levels: $\rho(i,l) = \rho_l - \rho_i$. Array these levels as rows of an $s \times n$ matrix $\rho = [\rho_i]$. The base-state dynamic equation of a jump translating system

(JTS) can be written

$$dx_t = \sum_i ((A_i x_t + B_i(u_t - \upsilon\phi_t))\, dt + C_i\, dw_t)\phi_i + (\rho' - \chi)\, d\phi_t. \tag{1.10}$$

If the plant state is a continuous process and there is no control, the JTS-model becomes even simpler:

$$dx_t = \sum_i A_i x_t \phi_i\, dt - \chi\, d\phi_t + C\, dw_t, \tag{1.11}$$

where the model is shown with constant Brownian intensity.

In interpreting the results derived on the basis of (1.7), we should recognize the approximations inherent in the model. If we ignore the drift identified with the feedforward implementation, the intrasojourn base-state dynamics are LGM. This model is easily justified in a region about the set point where higher order deviation variables are negligible. Exactly this kind of linearization procedure is accepted practice in applications involving unimodal plants, and during quiescent periods, Equation (1.5) – the intermodal restriction of (1.7) – is reasonable. If the set point changes, the magnitude of the base-state vector will increase abruptly. The state of a well-regulated plant will move expeditiously toward the new set point. In (1.7) the evolution model uses the dynamics of the successor regime. There are systems for which this concatenation of local models would be inappropriate (e.g., an unstable system moves away from the new set point). We will not pursue this issue further and will accept (1.7) as an adequate for our purposes.

The comprehensive plant state (base, mode) is a combination of continuous and discrete elements. The base-state moves within $\mathbb{R}^n$, and though $\phi_t \in \mathbb{R}^s$, the modal-state has a finite range space. The modal process is usually thought to be exogenous: The path of $\{\Phi_t\}$ is indifferent to $\{x_t\}$. Because it modulates the base-state motion, $\{\Phi_t\}$ is not, however, independent of $\{x_t\}$. With this heterogeneous state space structure, such plants are called *hybrid*. Heterogeneity of various kinds is becoming more common in applications, and the adjective "hybrid" is applied quite broadly. Nevertheless, because it is so descriptive, we will use hybrid to refer to plants and systems with this state space decomposition.

To complete the plant model, the temporal evolution of the modal-state must be quantified. In much of what follows, $\{\phi_t\}$ will be represented by an $\mathcal{F}_t$-Markov process satisfying the stochastic differential equation:

modal-state model

$$d\phi_t = Q'\phi_t\, dt + dm_t \tag{1.12}$$

with initial condition ϕ_0. The $S \times S$ matrix Q is called the modal transition rate matrix: If $i \neq j$, $\mathcal{P}(\phi_{t+dt} = \mathbf{e}_j \mid \phi_t = \mathbf{e}_i) = Q_{ij}\, dt$ with $Q_{ii} = -\sum_{l \neq i} Q_{il}$. The off-diagonal elements of the Q-matrix are nonnegative. The diagonal elements are such as to make the row sums of Q equal zero. It is known that the mean sojourn time in state $\phi_t = \mathbf{e}_i$ is $-1/Q_{ii}$, and if $\phi_t = \mathbf{e}_i$, the probability that the next modal transition will be $\mathbf{e}_i \mapsto \mathbf{e}_j$ is $-Q_{ij}/Q_{ii}$. Consequently, Q can be particularized from observations of the modal process. The second term in (1.12) is a purely discontinuous $\mathcal{F}_t$-martingale increment: $E[dm_t \mid \mathcal{F}_t] = 0$.

Equation (1.12) can be integrated into (1.7). Note that $\phi_i \mathbf{e}_l' \, d\phi_t = (Q_{il}\, dt + dm_l)\phi_i$. So

$$dx_t = \sum_i ((A_i x_t + B_i(u_t - \upsilon\phi_t))\, dt + C_i\, dw_t)\phi_i + \sum_{i,l} (M(i,l)x_t + (\chi_i - \chi_l) + M(i,l)\chi_i + \rho(i,l))(Q_{il}\, dt + dm_l)\phi_i. \tag{1.13}$$

Though not a particularly appealing relation, (1.13) can be made easier to interpret if we collect some of the terms that have a common influence. Let

$$\begin{aligned}\mathbf{A}_i &= A_i + \sum_l Q_{il} M(i,l),\\ \Theta(i,l) &= \chi_i - \chi_l + M(i,l)\chi_i + \rho(i,l),\\ \rho_{i\cdot} &= \sum_l \Theta(i,l) Q_{il}.\end{aligned} \tag{1.14}$$

In these terms, the base-state model can be written

$$dx_t = \sum_i ((\mathbf{A}_i x_t + B_i(u_t - \upsilon\phi_t))\, dt + C_i\, dw_t)\phi_i + \sum_{i,l} (M(i,l)x_t + \Theta(i,l))\phi_i\, dm_l + \rho'\phi_t\, dt. \tag{1.15}$$

The equation of base-state dynamics has the general appearance of an LGM model but it differs in important particulars. The state matrix, $\mathbf{A}_i$, of $\{x_t\}$ is composed of the intramodal component (A_i) plus a component determined by both the direction of the linear, intermodal discontinuity and its likelihood ($\sum_l Q_{il} M(i,l)$). The control matrix, B_i, is that of the intramodal model. The translational discontinuity in the plant state is reflected in $\rho'\phi_t\, dt$. There is a collection of terms in the drift of $\{x_t\}$ not found in the classical models of control and estimation. The model is highly nonlinear with the modal-state a multiplier throughout.

The increment in $\{x_t\}$ also contains exogenous forcing terms. One is a wideband noise term ($C_i\, dw_t$) also found in LGM models. The other is neither linear nor Gaussian. The plant state discontinuity term, $\sum_{i,l}(M(i,l)x_t + \Theta(i,l))\phi_i\, dm_l$, is an increment of a purely discontinuous martingale. The coefficient, $(M(i,l)x_t + \Theta(i,l))\phi_i$, contains base- and modal-state products.

The specialized dynamics of an LJS are not changed when the modal process is Markovian because the modal dynamics do not enter the base-state equation. The base-state evolution of the JTS can be written:

$$dx_t = \sum_i ((A_i x_t + B_i(u_t - v\phi_t) + (\rho' - \chi)Q'\mathbf{e}_i)\,dt + C_i\,dw_t)\phi_i + (\rho' - \chi)\,dm_t. \quad (1.16)$$

Equation (1.16) contains the same types of excitation found in the more comprehensive model, (1.15), but the simpler structure of (1.16) will be reflected in the estimation algorithm; compare $(\rho' - \chi)\,dm_t$ with $\sum_{i,l}(M(i,l)x_t + \Theta(i,l))\phi_i dm_l$.

In this book, we will present algorithms for generating (or approximating) $\{\hat{x}_t\}$ and $\{\hat{\phi}_t\}$. The accuracy of the estimates depends upon the quality and kind of sensors available in the application. A model for one kind of sensor is displayed in (1.4). The measurement is time continuous, linear in plant state, and the noise is additive and Gaussian. We will refer to (1.4) as the model of the *plant state sensor* even though $\{y_t\}$ may be generated by a collection of individual devices arrayed in a suite. For example, there may be radars aboard a set of geographically diverse platforms (shipboard, land-based, and air-based) with all tracking the same target. It is this aggregate that is called the plant state sensor. The noise in the observation is determined by both the sensor and the geometry (e.g., range), after linearization if necessary.

When the measurement frequency is too slow to justify using (1.4), the plant state sensor outputs are more accurately viewed as a time-discrete sequence. Suppose observations occur with intersample period T. A linear, time-discrete measurement of the plant state at time $t = kT$ is a direct analogue of (1.4):

plant state measurement: time-discrete

$$y[k] = H\chi[k] + n[k], \quad (1.17)$$

where $\{n[k]\}$ is a Gaussian white noise process with covariance R_x ($R_x > 0$), independent of the exogenous processes in (1.13). As is the case when the measurements are time continuous, if $\{\phi_t\}$ is known, $\{y[k]\}$ can be recast as a measurement of the base-state uncontaminated by the mode: $y[k] - H\chi\phi[k] = Hx[k] + n[k]$, and the *measurement residual* is defined to be the difference between what the output is and what it is predicted to be:

$$r[k] = y[k] - E[y[k]\,|\,\mathcal{G}[k-1]] = H\tilde{x}[k]^- + n[k].$$

In this case, $r[k]$ is equal to the innovations increment $\Delta\nu[k]$. With imperfect knowledge of $\{\phi[k]\}$, the measurement residual (and the innovations increment) is

$$r[k] = H\tilde{x}[k]^- + H\chi\tilde{\phi}[k]^- + n[k].$$

There is thus a mixing of base-state and modal-state errors in which the base-state error, $\tilde{x}[k]^-$, is conflated with a *base-state equivalent error*, $\chi\tilde{\phi}[k]^-$. Of course, during long sojourns in a regime, the modal-state is probably identified rather well ($\tilde{\phi}[k] \approx 0$), and the observation reverts to its orthodox form.

For LJS with known modal path, the Kalman filter generates the conditional mean of the base-state for either time-continuous or time-discrete measurements. Look at the time-continuous case, and denote the filtration generated by $g_t = \text{vec}(y_t, \phi_t)$ by $\mathcal{G}_t^\phi$. The ϕ superscript is used to distinguish this filtration (perfect modal knowledge) from those that follow (noisy modal measurements or none). The Kalman filter generates two base-state moments: the conditional mean, $\hat{x}_t = E[x_t \mid \mathcal{G}_t^\phi]$, and the conditional error covariance, $P_{xx}(t) = E[\tilde{x}_t\tilde{x}_t' \mid \mathcal{G}_t^\phi]$. If x_0 is $\mathbf{N}(\hat{x}_0, P_{xx}(0))$, the estimate and the error are Gaussian. The Kalman filter generates the $\mathcal{G}_t^\phi$-conditional distribution of the base-state (x_t is $\mathbf{N}(\hat{x}_t, P_{xx}(t))$ from which other statistical properties of the estimate can be derived. One form of the Kalman filter is [BW92, Figure 7.1]

Kalman filter: time-continuous state, time-continuous measurement

$$d\hat{x}_t = \sum_i A_i\hat{x}_t\phi_i dt + \gamma_x\, d\nu_t \tag{1.18}$$

subject to

$$\frac{d}{dt}P_{xx} = \sum_i (A_i P_{xx} + P_{xx}A_i' + R_\chi(i))\phi_i - \gamma_x R_x \gamma_x'. \tag{1.19}$$

The factor $\gamma_x = P_{xx}H'R_x^{-1}$ is the Kalman gain and

$$d\nu_t = dy_t - H(\chi\phi + \hat{x}_t)\, dt$$

is a $\mathcal{G}_t^\phi$-innovation increment.

The Kalman filter is familiar to engineers, and comprehensive studies of its properties are available. There are features of the Kalman filter that warrant comment. The base-state estimate prescribes concurrent extrapolation ($\sum_i A_i\hat{x}_t\phi_i\, dt$) and correction ($\gamma_x\, d\nu_t$). The direction of extrapolation is determined by the current A-matrix (selected by ϕ_t). Correction is achieved by weighting the innovations increment with γ_x. The Kalman gain, $\gamma_x = P_{xx} H' R_x^{-1}$, increases with observation quality (H) and decreases with sensor noise intensity (R_x). The gain also increases as the estimation uncertainty increases. (The base-state error covariance, P_{xx}, is sometimes called the *uncertainty matrix*.) An increase in P_{xx} makes the Kalman filter more data-driven whereas a decrease in P_{xx} makes the Kalman filter more model-driven.

Modal dependence enters the Kalman filter in a direct manner. The $\{\hat{x}_t\}$ and $\{P_{xx}\}$ dynamics change in concert with $\{\phi_t\}$. Although $\{\phi_t\}$ is a random process, $\{P_{xx}\}$ is random only because the coefficients in (1.19) are random: If the modal process were known a priori, P_{xx} could be precomputed. In any case, the error covariance is independent of the base-state observation $\{y_t\}$.

Equation (1.18) has an intuitively appealing form. Note that

$$E[dx_t \,|\, \mathcal{G}_t^\phi] = \sum_i A_i \hat{x}_t \phi_i \, dt$$

and that the innovation process is a $\mathcal{G}_t^\phi$-martingale. The equation of evolution of the base-state estimate is

$$d\hat{x}_t = E[dx_t \,|\, \mathcal{G}_t^\phi] + d\mu_t,$$

where μ_t is a $\mathcal{G}_t^\phi$-martingale [Kri84]. The increment in the mean is the mean of the increment plus a correction that is a martingale increment. For the system under study, all $\mathcal{G}_t^\phi$-martingales are integrals with respect to the innovations process: All $\mathcal{G}_t^\phi$-martingale increments are $\mathcal{G}_t^\phi$-predictable multiples of the innovation increment. The last term above must be of the form of a gain multiplying $d\nu_t$ [Ell82].

In applications in which the measurement is time discrete, the Kalman filter can be deduced formally from (1.18) and (1.19). Begin at time kT with the filter in state $(\hat{x}[k], P_{xx}[k])$. For the discrete-time case, it is convenient to distinguish the *pre-update* version of the base-state estimate from the *post-update* estimate. Denote the extrapolated state vector estimate at time $(k+1)T$ by $\hat{x}[k+1]^- = \hat{x}_{(k+1)T^-}$, and similarly denote the covariance by $P_{xx}[k+1]^- = P_{xx}((k+1)T^-)$. Integration of the measurement at time $(k+1)T$ gives rise to a correction to the pre-update estimate: $\Delta\hat{x}[k+1] = \hat{x}[k+1] - \hat{x}[k+1]^-$ and similarly $\Delta P_{xx}[k+1] = P_{xx}[k+1] - P_{xx}[k+1]^-$. The filter residual $r[k+1] = y[k+1] - H(\chi\phi[k+1] - \hat{x}[k+1]^-)$ is the innovations increment. The residual process is a white Gaussian process with covariance $R_{yy}[k]$ (with inverse $D_{yy}[k] = S_{yy}[k]' S_{yy}[k]$):

$$R_{yy} = E[r[k]\, r[k]' \,|\, \mathcal{G}^\phi[k-1]] = H P_{xx}[k]^- H' + R_x = D_{yy}[k]^{-1}.$$

The discrete form of the $\mathcal{G}^\phi[k]$-filter is given in [BW92, Figure 5.9].

Kalman filter: time-continuous state; time-discrete measurements

Between observations:

$$\frac{d}{dt}\hat{x}_t = \sum_i A_i \hat{x}_t \phi_i, \tag{1.20}$$

$$\frac{d}{dt}P_{xx} = \sum_i (A_i P_{xx} + P_{xx} A_i' + R_\chi(i))\phi_i. \tag{1.21}$$

At an observation:

$$\Delta\hat{x}[k+1] = \gamma_x r[k+1], \tag{1.22}$$

$$\Delta P_{xx}[k+1] = -\gamma_x R_{yy}[k+1]\gamma_x', \tag{1.23}$$

where $\gamma_x = P_{xx}[k+1]^- H' D_{yy}[k+1]$. Equations (1.20)–(1.23) follow from (1.18) and (1.19) by making the replacements

(i) in extrapolation, $R_x^{-1} \to 0$,

(ii) in correction, $E[dv_t\, dv_t' \,|\, \mathcal{G}_t^\phi]/dt \to R_{yy}[k]$.

Like the continuous Kalman filter, its discrete sibling is a predictor–corrector, but the prediction and correction are not concurrent. Correction takes place at the observation times with a difference in the gain: For time-discrete measurements, $\gamma_x = P_{xx} H' R_{yy}^{-1}$; for time-continuous measurements, $\gamma_x = P_{xx} H' R_{yy}^{-1}/dt$, where R_{yy}^{-1}/dt $(= E[dv_t\, dv_t' \,|\, \mathcal{G}_t^\phi]/dt)$ is the intensity of the residual process. As in the time-continuous case, the observation gain increases with improved sensor quality. The time-discrete residual process is a white, Gaussian process: $r_k \in \mathbf{N}(0, R_{yy}[k])$. If the residual is scaled by $S_{yy}[k]$, a unit Gaussian white sequence is obtained: $S_{yy}[k]\, r[k] \in \mathbf{N}(0, I)$.

The Kalman filter is a complete solution to the estimation problem as posed, but most applications do not fall neatly within the modeling paradigm. Nonlinearities and discontinuities neglected in the model cause the performance of the Kalman filter to degrade. The influence of mismodeling is seen frequently in simulation exercises where the size of the estimation error can be contrasted with the computed error covariance. In an actual system, the true error is not known. But the residuals can be measured, and if $\{S_{yy}[k] r[k]\}$ is not a unit white noise process, the model of the plant and sensor may need to be refined. When $\{r[k]\, r[k]'\}$ consistently exceeds $R_{yy}[k]$, the filter is said to exhibit *excess error*; if P_{xx} is small, the filter residual may exceed the standard deviation of the noise in a single measurement.

1.2 A Tracking Example

To illustrate some of the issues that arise in hybrid estimation within the context of a concrete example, consider a tracking problem in which we wish to determine the position and velocity of an evasive aircraft moving in the X–Y plane. (The altitude is essentially constant.) Targets with limited thrust control maneuver by *jinking*: The turn rate tends to be nearly constant over intervals with sudden changes at unpredictable times. Suppose the aircraft is detected at a range of 36 km ($t = 0$) traveling at a speed of 300 m/s. The aircraft coasts (nearly constant velocity flight) for three seconds ($t \in [0, 3)$), makes a 7 g turn to the right for six seconds ($t \in [3, 9)$), coasts for two seconds ($t \in [9, 11)$), makes a 7 g turn to the left for five seconds ($t \in [11, 16)$), and then returns to coast. Increased drag during a turn causes the aircraft to slow to 60% of the speed that it had entering the turn with a 40% increase in speed when a turn transitions to coast. During an interval of constant turn rate (including coast), the speed is fairly constant.

A rudimentary motion model for the aircraft between changes in turn mode is

$$d\begin{bmatrix} X \\ Y \\ V_x \\ V_y \end{bmatrix} = \begin{bmatrix} 0 & 0 & 1 & 0 \\ 0 & 0 & 0 & 1 \\ 0 & 0 & 0 & -\Phi \\ 0 & 0 & \Phi & 0 \end{bmatrix} \begin{bmatrix} X \\ Y \\ V_x \\ V_y \end{bmatrix} dt + \begin{bmatrix} 0 & 0 \\ 0 & 0 \\ 1 & 0 \\ 0 & 1 \end{bmatrix} d\begin{bmatrix} w_x \\ w_y \end{bmatrix}. \tag{1.24}$$

In this tracking problem, there is no plant state reference point; that is, the plant dynamics are linear over $\mathbb{R}^4$ ($\chi = \mathbf{0}$), and the plant state is the base-state. Moreover, there is no endogenous actuating signal; that is, the tracker has no control over target motion ($\upsilon_t \equiv 0$). The base-state consists of $\{X, Y\}$, the position coordinates, and $\{V_x, V_y\}$, the associated velocities. The target is subject to two types of acceleration: (i) a wide band, omnidirectional acceleration described by the Brownian motion $\{w_x, w_y\}$ with intensity W and (ii) a maneuver acceleration represented by the turn rate process $\{\Phi_t\}$. The speed is slowly varying when the turn rate is constant, and so the omnidirectional acceleration is small: Let $C_i = \mathbf{e}_2 \otimes \mathbf{I}_2$ for all i and $W = \mathbf{I}_2$. The intensity of the acceleration is about 0.1 g.

The jinking behavior can be captured by partitioning the range of possible turn rates into three levels:

$$\Phi_t \in \{a_1 = 0.2r/s,\ \phi_t = \mathbf{e}_1;\ a_2 = 0r/s,\ \phi_t = \mathbf{e}_2;\ a_3 = -0.2r/s,\ \phi_t = \mathbf{e}_3\}.$$

The turn rate is given by $\Phi_t = a'\phi_t$. A change in motion mode causes a change in speed, but no rotation:

- At the beginning or end of a turn, the position process is continuous: $[X_{t+}, Y_{t+}] = [X_{t-}, Y_{t-}]$.
- At the beginning of a turn ($\Phi \mapsto a_1$ or $\Phi \mapsto a_3$) the target slows by 40%: $[V_x(t+), V_y(t+)] = 0.6[V_x(t-), V_y(t-)]$.

- At the end of a turn ($\Phi \mapsto \mathbf{e}_2$), the target speed increases by 40%, but not enough to attain the pre-turn velocity: $[V_x(t+), V_y(t+)] = 1.4\,[V_x(t-), V_y(t-)]$

In this example, a turn-to-turn transition is not allowed. The intraregime model of the aircraft can be written as the stochastic differential equation:

$$dx_t = \sum_i A_i x_t \phi_i dt + dw_t, \tag{1.25}$$

where

$$A_i = \begin{bmatrix} 0 & 0 & 1 & 0 \\ 0 & 0 & 0 & 1 \\ 0 & 0 & 0 & -a_i \\ 0 & 0 & a_i & 0 \end{bmatrix}.$$

At the origin of the coordinate system, (0, 0), there is a sensor. A radar measures the position of the target every second with Gaussian errors of 40 m in range and 1.75 mr in bearing (approximately 63 m at 35 km). This measurement is not linear in the coordinate system selected for the motion model: $y[k] = \mathbf{r}(x[k]) + n[k]$ instead of $y[k] = Hx[k] + n[k]$. The measurement relation can be linearized, not about the set point, but about the computed state estimate, $\hat{x}_t$, itself. A replacement for the measurement residual is $r[k] = y[k] - \mathbf{r}(\hat{x}[k]^-)$. The covariance update is computed using the χ-gradient of $\mathbf{r}$ evaluated at $\hat{x}[k]$ in place of H in (1.22) and (1.23). This ancillary linearization is commonly done when the sensor nonlinearities are smooth and the estimation errors reasonably small, and it leads to an instance of the extended Kalman filter (EKF) [GA93, Table 5.4]. Similar output linearization will be performed in what follows wherever required without further comment.

The most rudimentary approach to the tracking problem would be to ignore the turn process and design an EKF based upon the specification of radar quality given above. Suppose $\hat{x}_0 = x_0$ and the initial covariance is taken to be diagonal with standard deviation in position (100 m) and velocity (20 m/s): The tracker is initialized at the true state of the aircraft and the initial uncertainty is larger than the single-measurement sensor error. The Brownian disturbances on the path are small: Set $W = 1$. Figure 1.1 shows a sample path of the nominal EKF as a *feather plot* referenced to a target path generated with $W = 0$. (A feather plot connects the estimates of location after a measurement to the true location. A point is shown every 0.1 s for clarity. The speed changes are not visible on the target path.) With the advantageous initialization, EKF(W=1) begins well. The target model ignores turns and none occur at first.

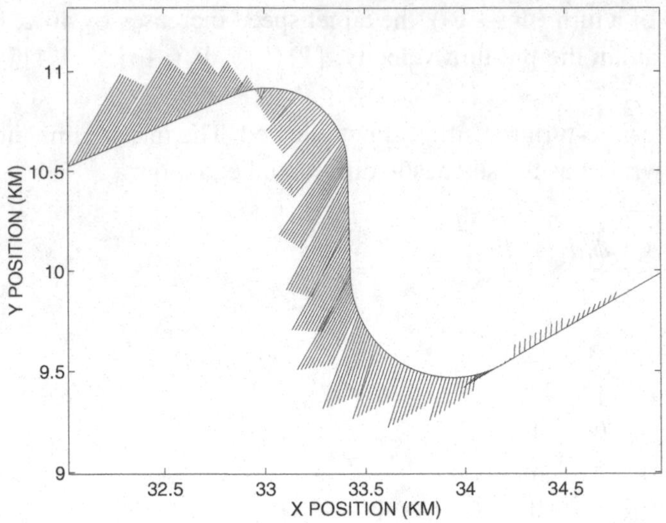

Figure 1.1. The path of a target with estimates from EKF(W=1).

When the target turns and slows, the performance of EKF(W=1) degrades. As the target turns to the right, EKF(W=1) fails to follow and extrapolates between radar measurements in the direction of the initial velocity. The position error is corrected in part when a radar measurement is received, but the gain is too small to bring $\hat{x}_t$ back to x_t. The velocity correction is also far too small. With no direct velocity measurement, EKF(W=1) misinterprets $\{y[k]\}$. This creates tracking errors far in excess of the raw radar noise (about 60 m). It is not until the reverse turn has begun that EKF(W=1) identifies the velocity, but this is an artifact of the path. The error going into the final coast is quite large and the velocity estimate is abysmal.

The EKF, in contrast to less structured estimators (e.g., the $\alpha - \beta$ tracker), not only generates an estimate of the base-state, but it also provides an assessment of its own performance. The upper left submatrix $P_{xx}(1:2, 1:2)$ gives the error covariance in position. A one-σ region of target location is found by centering an ellipse determined by $P_{xx}(1:2, 1:2)$ about $\hat{x}_t$. In some adaptive estimators, the radar pulse shape (and the signal-to-noise ratio (SNR) of the sensor) and the tracking window are dependent on the size, shape, and location of this error ellipse.

Figure 1.2 displays the target path along with the one-σ error ellipses (shown every 0.2 s for clarity) centered at the location estimates. The ellipses are near circles in this case because of the symmetry in the measurement. On the first coast, when the dynamic hypotheses of the EKF match the motion, tracking uncertainty

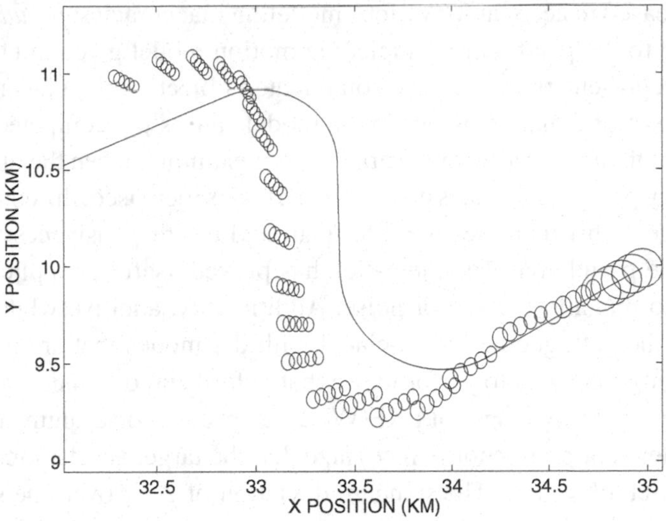

Figure 1.2. The path of a target with error ellipses generated by EKF(W=1).

is reduced (the ellipses shrink) with each radar measurement. The true path lies within or next to the envelope of the one-σ error circles. When the target turns to the right and slows, EKF(W=1) fails to react. Although EKF(W=1) tacks away from the true path, the error ellipses evidence no sensitivity to the growth in the size of the measurement residuals. The residuals may exceed 10 σ, a near impossibility if the errors were truly Gaussian. After completing the first turn, the target path lies several standard deviations away from $\{\hat{x}_t\}$ except when the target turns back into the estimate: The estimates are bad but the filter fails to acknowledge just how bad they are. A Gaussian density has thin tails, and the persistent presence of excess error as shown in Figure 1.1 is highly unlikely. The EKF's sanguine attitude would lead to loss-of-lock if the radar energy were focused in a three-σ window about $\{\hat{x}_t\}$.

With the approximations we have made in the design of this EKF, it is not surprising that it may need to be adjusted or tuned for this application. The nominal EKF is too sluggish to follow an agile target. In principle, any of the coefficients in EKF(W=1) could be changed to make it more responsive. However, the aircraft dynamics and the observation equation are constrained by the physics of the path (e.g., the $\{A_i, i \in \mathbf{S}\}$) or the geometry of the sensors (e.g., H). Tuning in the EKF usually concentrates on the intensities of the exogenous disturbances: $W = E[dw_t dw_t']/dt$ (the plant noise) or $R_x = E[n_k n_k']$ (the sensor noise). In fact, the focus is more commonly on the former because there are stronger empirical restrictions on the latter.

When W is increased to account for various modeling inaccuracies, *pseudonoise* is said to be added to the plant. For example, the motion model given in (1.24) is a low dimension representation of a very complicated object. An engineer could argue that the neglect of dynamic modes in the model causes the computed value of P_{xx} to be smaller than the true error covariance. For example, when the turns are ignored, the primary plant excitation is disregarded. If W is increased, the computed $\{P_{xx}\}$ is made larger. This increases the filter gain and the responsiveness of the EKF as well. While pseudonoise augmentation has proved useful in applications, the higher gains do magnify the sensor noise. Additionally, additive white noise does not preserve the path geometry associated with the modes that are ignored, and this mismodeling may lead to performance that is far from optimal.

Let us try to improve the response of EKF(W=1) by pseudonoise augmentation. To rationalize the level of pseudonoise, recognize that the target accelerations also include the turns. Set $W = 100$. The standard deviation of ΔV_x over one second is 10 m/s, which is equivalent to a 1 g constant acceleration. Over six seconds or so this would be roughly the *white-noise equivalent* of the turn process involving a 7 g turn over six seconds and intervening coasts. Of course, this equivalence is crude: The Brownian motion is continuous whereas the turn rate process is not; the Brownian motion acts throughout the tracking interval, whereas the turn rate changes at isolated times; the Brownian motion is omnidirectional, whereas the turn places specific geometric constraints on the target path.

Figure 1.3 shows the feather plot of a sample of an EKF with this pseudonoise augmentation. The effect of pseudonoise is beneficial for the most part. After the first

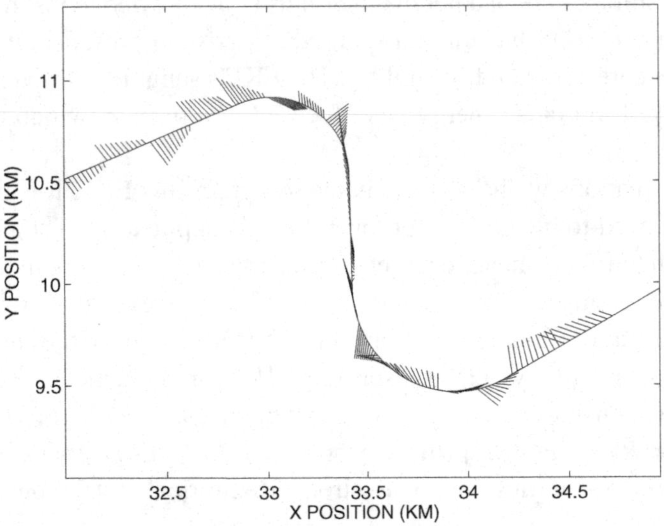

Figure 1.3. The feather plot for EKF(W=100).

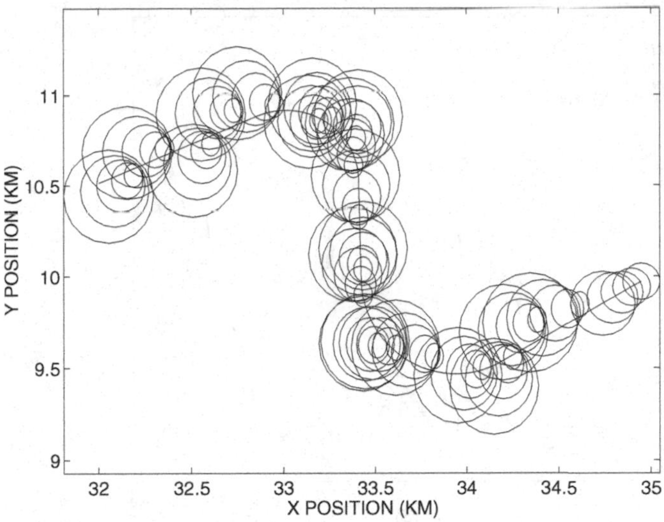

Figure 1.4. The ellipse plot for EKF(W=100).

turn begins, the estimation accuracy is improved significantly over that displayed in Figure 1.1. Velocity estimates are also improved during intrasojourn intervals. As expected, the increased gain makes the estimate more responsive to the peculiarities of the target path.

Unfortunately, pseudonoise carries with it a concomitant shortcoming. The one-σ error ellipses generated by EKF(W=100) are shown in Figure 1.4. The envelope of the ellipses does contain the target path (in almost all cases), but this is achieved by making the the axes of the ellipses overly long. The ellipses appear to be reasonable in size and location after an update in that, with few exceptions, the target path lies in or near the ellipse. However, they grow excessively between radar measurements. The predictive capability of EKF(W=100) is not good: This EKF does not sharply delineate the region in which the aircraft is likely to be. If the SNR of the radar were made inversely proportional to the size of the error ellipse, even position tracking would degrade significantly.

Since W is a tuning parameter, an engineer might ignore our superficial rationalization and simply select W to yield the best performance based upon trade-offs gleaned from the sample behavior. Figure1.5 shows the mean radial estimation error for three filters: EKF(W=1); EKF(W=10); EKF(W=100). This figure is formed from the sample average of 50 independent experiments in which each filter saw the same observations and the aircraft flew the path shown in Figure 1.1 ($W = 0$). We can extract three important subintervals from this experiment: $t \in [0, 3]$, initialization and opening coast; $t \in [4, 9]$, slowing and turning; $t \in [18, 20]$, return to coast. With the advantageous initialization, the EKF with the smallest

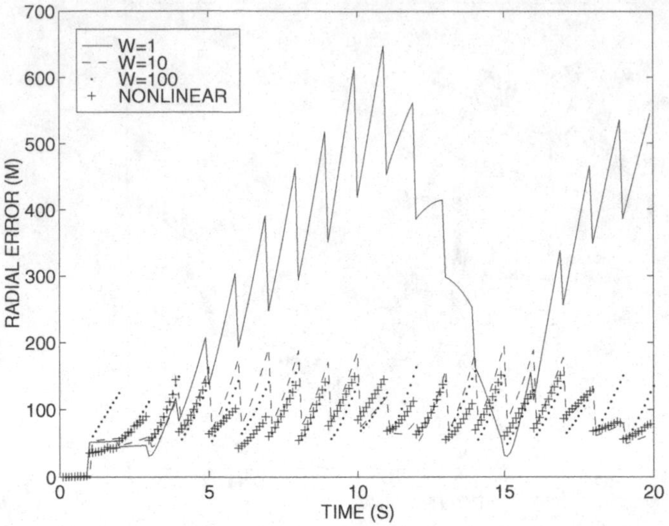

Figure 1.5. Mean radial estimation error for three EKFs and a nonlinear estimator.

gain is best on $t \in [0, 3]$. Let us use "$\prec$" to indicate preference: On $t \in [0, 3]$, EKF(W=1) $\prec$ EKF(W=10) $\prec$ EKF(W=100). In the first turn ($t \in [4, 9]$) the preferences are reversed with preference ordering in accord with the gain: EKF(W=100) $\prec$ EKF(W=10) $\prec$ EKF(W=100). The radial error in the interval after $t = 10$ is small because the target is turning toward the estimated position.

The interval [18, 20] is in the interior of a coast, and it would be expected that the initial preference ordering would again prevail since the model upon which EKF(W=1) is based is most representative of the target path for $t \in [16, 20]$. Actually the ordering is contrary: EKF(W=10) $\prec$ EKF(W=100) $\prec$ EKF(W=1). The reason behind this anomaly is related to the way in which the Kalman gain influences *memory* in the estimator. After the second turn ends, both EKF(W=10) and EKF(W=100) return quickly to their premaneuver error condition; they have a short memory. Alternatively, EKF(W=1) is disoriented at the end of the turn, particularly with regards to velocity – Figure 1.1 shows EKF(W=1) tacking in the positive Y direction when the true Y velocity is clearly negative. None of the EKFs have a velocity measurement, and when the accelerations in the model are small, motion aberrations are seen as coming from velocity errors. During a maneuver, there is an accumulation of velocity errors that must be dissipated before the EKF can return to proper operation. When the innovations gain is small, as it is in EKF(W=1), this takes considerable time. The deviant ordering shown in the figure would eventually be corrected, but the time required could be significant.

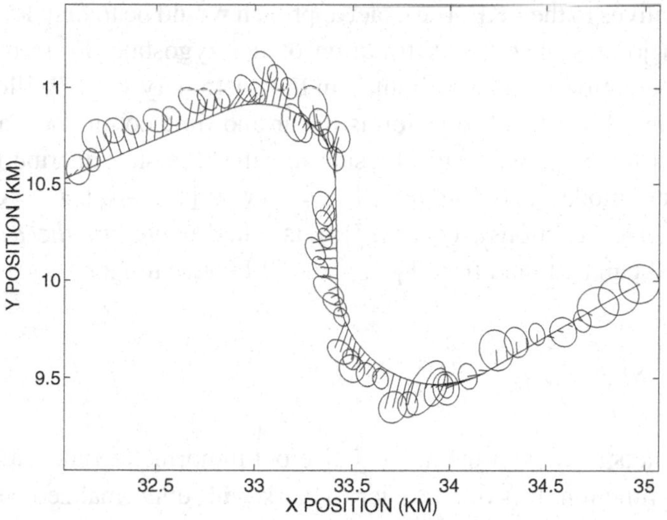

Figure 1.6. The path of the target along with estimates from a nonlinear filter.

This example illustrates the trade-offs that arise in hybrid estimation. Algorithms with high gains magnify the sensor noise but are responsive to modal changes; algorithms with low gains average out the sensor noise but are slow to adapt themselves to changing modal conditions. More attractive would be a response like that shown in Figure 1.6. The same target path leads to generally smaller radial errors than any of the EKFs (as shown in Figure 1.5) with error ellipses more faithful in shape and smaller in size. The ellipses are more variable, growing on that part of the trajectory where the uncertainty is greatest and with major axis closer to the direction of greatest uncertainty (along the estimated velocity vector). The development of algorithms that provide this balance between inter- and intrasojourn performance will be the topic of what follows.

1.3 Infinite-Dimensional Algorithms

The previous example illustrates some of the difficulties associated with hybrid estimation. The modal process modulates the base-state dynamics. If the modal-state is measured, the estimation problem can be solved using classical methods. Lacking measurement of $\{\phi_t\}$, the problem is intractable. Because the composite estimation problem arises in so many applications, engineers have been forced to create algorithms for hybrid systems using the analytical tools available. Thus the EKF-tracker used the radar measurements, $\{y[k]\}$, to obtain a $\mathcal{Y}_t$-estimate of the kinematic state while ignoring the turn state.

There are alternatives to the EKF. The exact approach would be to first determine the $\mathcal{Y}_t$-conditional joint probability distribution of the zygostate. From this, the $\mathcal{Y}_t$-conditional expectations can be computed in the customary way. To illustrate, consider the case in which the observation is linear and time continuous, and the plant is a LJS (see (1.8)). Suppose the plant is such that the base-state distribution has a $\mathcal{Y}_t$-density for every modal condition: $p_t^i(z)\,dz = \mathcal{P}(x_t \in [z, z+dz], \phi_t = \mathbf{e}_i \mid \mathcal{Y}_t)$. A vector of nonnegative functions $q_t(z) = [q_t^i(z)]$ is called an *unnormalized density* if the $\{p_t^i(z); i \in \mathbf{S}\}$ can be found from $\{q_t^i(z); i \in \mathbf{S}\}$ by rescaling:

$$p_t^i(z) = q_t^i(z) \Big/ \sum_i \int_\Omega q_t^i(u)\,du. \tag{1.26}$$

Division in (1.26) acts to convert the nonnegative but unnormalized densities into a joint probability function. It is often easier to work with unnormalized densities than their normalized counterparts because their equations of evolution are simpler. Let $\{\mathcal{L}_{(i)}; i \in \mathbf{S}\}$ be a set of differential operators:

$$\mathcal{L}_{(i)} = \frac{1}{2}\sum_{k,j}(R_\chi(i))_{jk}(\partial^2/\partial z_k \partial z_j) - \sum_k (\partial/\partial z_k)(A_i z + B_i u)_k. \tag{1.27}$$

It is shown in [LRE86] that $q_t(z)$ satisfies the stochastic partial differential equation:

$$dq_t(z) = (\mathcal{L}^* q_t(z) + Q' q_t(z))\,dt + (dy_t' H z) q_t(z). \tag{1.28}$$

Equation (1.28) is intimidating. It consists of a set of nonlinear, second-order, stochastic partial differential equations in which each of the S unnormalized density functions (the components of $q_t(z)$) have n spatial variables. All S equations must be solved simultaneously because of the coupling among the densities in Q.

To utilize (1.28) or its analogues, restrictions and approximations are necessary. For example, if the modal process has a "direction of evolution" (e.g., when in regime $\mathbf{e}_i$, $\{\phi_t\}$ can only transition to an $\mathbf{e}_j$ for which $j < i$), the solution becomes more manageable (see [LRE86, Theorem 3.1] and [LBV91]). Look first at all modes that are absorbing: the set of $\mathbf{e}_i$ for which $Q_{ii} = 0$. The solution for $\{q_t^i(z)\}$ for these modes can be done in parallel and each such subproblem has a Kalman filter–like solution. Look next at those regimes that transition to an absorbing state. They are linked to the $\{q_t^i(z)\}$ for the absorbing states in (1.28), but these densities are known from the previous calculation. The solution to (1.28) for the single-transition regimes is not Gaussian, but only one jump time need be considered. This process can be continued backward, but the complexity of the calculation is such as to preclude real-time implementation.

1.4 Multiple Model Algorithms

Implementable algorithms for calculating the unnormalized conditional densities of the zygostate are seen most often in the context of time-discrete plant models. Time-discrete models arise in applications in which the engineer is only concerned with the behavior of the plant state at a set of sample times. The sample times do not have to be evenly spaced nor do they have to be simultaneous with the observations. It is, however, simpler to look at all of the system variables on an evenly spaced time grid contemporaneous with the output samples. The zygostate process for a time-discrete plant is a random sequence in which $(x[k], \phi[k])$ is identified with the state of the time-continuous plant sampled every T seconds; (x_{kT}, ϕ_{kT}). In contrast with the time-continuous plant, the modal transitions will occur at sample times and are therefore coincident with an observation. Let $\{\mathcal{F}[k]\}$ be the filtration generated by $\{x[k], y[k], \phi[k]\}$. The time-discrete modal process will be represented by an $\mathcal{F}[k]$-Markov process:

modal-state model: time-discrete

$$\phi[k] = \Pi\phi[k-1] + m[k], \tag{1.29}$$

where $\{m[k]\}$ is a time-discrete, $\mathcal{F}[k]$-martingale difference:

$$E[m[k+1] \,|\, \mathcal{F}[k]] = 0,$$

and $\Pi_{ij} = \mathcal{P}(\phi[k] = \mathbf{e}_i \,|\, \phi[k-1] = \mathbf{e}_j)$ is the modal transition matrix (compare (1.12)).

In the time-discrete plant, it is assumed that there are no modal transitions interior to an intrasample interval. The time-discrete, base-state model can be deduced by integrating (1.5) over a single period. In the examples of time-discrete systems that follow, issues of feedforward control will not be addressed ($v = \mathbf{0}$), and the regulating signal will be assumed to be constant between samples:

$$x[k] = \sum_i \Big(\Phi_i(T,0)x[k-1] + \int_{(k-1)T}^{kT} \Phi_i(kT,s)(B_i u[k-1]\,ds + C_i\,dw_s)\Big)\mathbf{e}_i'\phi[k-1], \tag{1.30}$$

where Φ_i is the transition matrix associated with the ith mode:

$$\Phi_i(t,s) = \exp(A_i(t-s)).$$

We will assume that for every $i \in \mathbf{S}$, $\{(A_i, C_i W^{1/2})\}$ is controllable. Regime-specific controllability assures that exogenous excitation is a Gaussian white sequence with positive covariance:

$$\int_{(k-1)T}^{kT} \Phi_i(kT, s)C_i\, dw_s \in \mathbf{N}(0, R_\chi(i));\; R_\chi(i) > 0.$$

Let us rewrite (1.30) using the following replacements:

$$\Phi_i(T, 0) \to A_i,$$

$$\int_0^T \Phi_i(T, s)B_i\, ds \to B_i,$$

$$\int_0^T \Phi_i(T, s)C_i W^{1/2} \to C_i.$$

Then C_i is a square root of $R_\chi(i)$: $C_i'C_i = R_\chi(i)$. The time-discrete base-state model uses a parameter set identical to that of (1.5). Despite the labeling, the values of the coefficient matrices are not the same for the time-continuous and the time-discrete base-state models. Matching the labels permits the identification of similar operations across model categories. Care must be exercised in interpreting specific algorithms because a statement about A_i, for example, in a time-discrete application may not be true about A_i in a time-continuous application. With this caveat, (1.30) can be written:

$$x[k] = \sum_i (A_i x[k-1] + B_i u[k-1] + C_i w[k])\mathbf{e}_i'\phi[k-1], \tag{1.31}$$

where $\{w[k]\}$ is a Gaussian white sequence of unit covariance: $w[k] \in \mathbf{N}(0, I)$.

The complete time-discrete model is found by integrating the modal transition conditions into (1.31). The applications to be considered here do not involve the full range of discontinuities given in (1.6). Suppose only that the plant state set point changes with regime. The equation of base-state evolution can then be written:

base-state model: time-discrete

$$x[k] = \sum_i (A_i x[k-1] + B_i u[k-1] + C_i w[k])\mathbf{e}_i'\phi[k-1] - \chi\Delta\phi[k-1], \tag{1.32}$$

where $\Delta\phi[k-1] = \phi[k] - \phi[k-1]$. Because they derive from a time-continuous, regime-controllable plant, the coefficient matrices in (1.32) have special properties that will prove useful (e.g., all of the matrices A_i and C_i; $i \in \mathbf{S}$ are nonsingular).

The $\mathcal{Y}[k]$-estimation problem where the $\{B_i; i \in \mathbf{S}\}$ and χ are zero has been studied by several investigators. Let us see how this special case has been approached for it will give us insight into the more general problem. The reduced base-state (or plant) dynamics are

$$x[k] = \sum_i (A_i x[k-1] + C_i w[k])\mathbf{e}_i'\phi[k-1]. \tag{1.33}$$

Suppose first that $\{\phi[k]\}$ is known, and denote the filtration generated by $g[k] = \text{vec}(y[k], \phi[k-1])$ by $\mathcal{G}^\phi[k]$. If the initial condition on (1.33) is $\mathbf{N}(\hat{x}_t[0], P_{xx}[0])$, then the $\mathcal{G}^\phi[k]$-mean of $\{x[k]\}$ is generated by the time-discrete Kalman Filter (see [GA93, Table 4.3] and (1.20)–(1.23)):

Kalman filter: time-discrete state, time-discrete measurement

Extrapolation:

$$\hat{x}[k+1]^- = \sum_i A_i \hat{x}[k]\mathbf{e}_i'\phi[k], \tag{1.34}$$

$$P_{xx}[k+1]^- = \sum_i (A_i P_{xx}[k]A_i' + R_\chi(i))\mathbf{e}_i'\phi[k]. \tag{1.35}$$

Update:

$$\Delta\hat{x}[k+1] = \gamma_x r[k+1], \tag{1.36}$$

$$\Delta P_{xx}[k+1] = -\gamma_x R_{yy}[k+1]\gamma_x'. \tag{1.37}$$

The innovations process $\{\nu[k]\}$ (with increments $r[k] = y[k] - H\hat{x}[k]^-$) is a time-discrete $\mathcal{G}^\phi[k]$-martingale ($E[r[k]|\mathcal{G}^\phi[k]] = 0$). As was the case of a time-continuous plant with discrete measurements, the covariance of $r[k]$ is given by $R_{yy}[k] = HP_{xx}[k]^-H' + R_x$ with inverse $D_{yy}[k]$. The gain $\gamma_x = P_{xx}[k+1]^-H'D_{yy}[k+1]$ is identical to that for the time-continuous plant with time-discrete observations. The $\mathcal{G}^\phi[k]$-conditional distribution of $x[k]$ is $\mathbf{N}(\hat{x}[k], P_{xx}[k])$. If $\{\phi[k]\}$ is a deterministic sequence, $\{\hat{x}[k]\}$ is a Gaussian process and $\{P_{xx}[k]\}$ can be precomputed. If $\{\phi[k]\}$ is a random process, $\{\hat{x}[k]\}$ is only conditionally Gaussian and $\{P_{xx}[k]\}$ is random.

The observation, $\{y[k]\}$, appears only in the base-state update; the error covariance is independent of $\{y[k]\}$. To make clearer the dependence of the base-state estimate on the model process, denote the realized modal sequence over the interval $[iT, \ldots, jT]$ by $\phi(i : j)$. We could write the $\mathcal{G}^\phi[k]$-mean of $\{x[k]\}$ and the $\mathcal{G}^\phi[k]$-error covariance more descriptively as $\{\hat{x}([k]; \phi[0 : k-1])\}$ and $\{P_{xx}([k]; \phi[0 : k-1])\}$.

Let us make this problem more realistic and assume that $\{\phi[k]\}$ is not known. With only $\{y[k]\}$ as the observation, the relevant filtration is $\{\mathcal{Y}[k]\}$. It is true that

$$\begin{aligned} E[x[k] \mid \mathcal{Y}[k]] &= E[E[x[k] \mid \mathcal{G}^{\phi}[k]] \mid \mathcal{Y}[k]] \\ &= E[E[\hat{x}([k]; \phi[0:k-1])] \mid \mathcal{Y}[k]]. \end{aligned}$$

The $\mathcal{Y}[k]$-mean of the base-state is found by taking the elementary estimates based upon $\mathcal{G}^{\phi}[k]$ and averaging them over the set of possible modal paths. This is a conceptually attractive representation of the base-state estimator since each $\hat{x}([k]; \phi[0:k-1])$ is found from a finite-dimensional recursion.

The actual calculation of $\{\hat{x}[k]\}$ would proceed as follows. Order the allowable modal sequences. There are no more than S^k of them on $[0, k-1]$, and let $\phi([\kappa]; [0:k-1])$ be the κth such sequence. Let $\hat{\phi}([\kappa]; [0:k-1]) = \mathcal{P}[\phi[0:k-1] = \phi([\kappa]; [0:k-1]) \mid \mathcal{Y}[k]]$. Then the $\mathcal{Y}[k]$-mean of the base-state is simply

$$\hat{x}[k] = \sum_{\kappa} \hat{x}([\kappa]; \phi([\kappa]; [0:k-1]))\hat{\phi}([\kappa]; [0:k-1]). \tag{1.38}$$

This multiple model (MM) base-state estimator is composed of a finite number of Kalman filters, and the evaluation of $\hat{\phi}([\kappa]; [0:k-1])$ is a separate estimation problem. From $\{\hat{\phi}([\kappa]; [0:k-1]); \kappa \in S^k\}$, $\hat{\phi}[k-1]$ can be found by summing across sequences with the same $(k-1)$th element.

Equation (1.38) is an exact solution to the time-discrete LJS filtering problem, but there are insurmountable difficulties associated with its implementation. Observe that $\hat{x}([k]; \phi([\kappa]; [0:k-1]))$ is not known as an explicit $\mathcal{G}_t^{\phi}$-adapted function. Each elementary estimate is given implicitly by a path-individuated recursion. If each modal-state can transition to any other modal-state in one sample time, there are S^k such recursions, and there is simply no way to accommodate this geometric growth. Also, there exists no tractable algorithm for finding the $\mathcal{Y}$[k]-probability of a modal path. Not only do the number of elements in $\{\hat{\phi}([\kappa]; [0:k-1])\}$ grow geometrically, but their computation involves sophisticated data smoothing.

Failing to find a solution that is both exact and implementable, we must turn to an approximation. The simplest way to reduce the algorithmic complexity would be to ignore the modal transitions (set $\Pi = \mathbf{I}$). Instead of S^k possible modal paths there are now S of them: $\phi([\kappa]; [0:k-1]) \in \{\mathbf{1}_k' \otimes \mathbf{e}_1, \ldots, \mathbf{1}_k' \otimes \mathbf{e}_s\}$. There are only S regime-specific subfilters required to generate $\{\hat{x}([k]; \mathbf{1}_k' \otimes \mathbf{e}_i) = \hat{x}^i[k]; i \in \mathbf{S}\}$. Equation (1.38) reduces to [May82a, Section 10.8]:

multiple model filter: path-length-one

$$\hat{x}[k] = \sum_{i} \hat{x}^i[k]\mathbf{e}_i'\hat{\phi}[k-1]. \tag{1.39}$$

Equation (1.39) is called the path-length-one multiple model (PL1-MM) filter. The number of subfilters is fixed. The path probabilities required in (1.38) reduce to the modal probabilities, $\mathbf{e}_i'\hat{\phi}[k-1] = \mathcal{P}[\phi[k-1] = \mathbf{e}_i \mid \mathcal{G}[k]]$, in (1.39). The $\mathcal{G}$[k]-conditional distribution of the zygostate is the S-vector $p_t = [\mathbf{e}_i'\hat{\phi}[k-1]\mathbf{N}(\hat{x}_k^i, P_{xx}^i[k])]$. In implementation, the residuals generated by the individual subfilters are used to compute the modal probabilities [May82a, Equation (10-107)]. This MM filter works well when the regime process is invariant. Of more interest here is the fact that the MM filter has been used with success in applications in which $\{\phi[k]\}$ is not a constant process. Equation (1.39) is only an approximation in this circumstance. However, if the SNR is high, $\{\hat{\phi}[k]\}$ follows $\{\phi[k]\}$ with small delay and selects the "correct" subfilter in (1.39).

The path-length-one MM algorithm may require adjustment when the modes communicate. Because the PL1-MM filter is framed on the basis of regime invariance, it can generate inaccurate estimates after a modal transition. Suppose that over an interval preceding $t = NT$, $\{\phi[k]\} \equiv \mathbf{e}_j$. If the SNR ratio is high, the modal estimate will be good when $t = (N-1)T$: The modal estimator will select the correct regime ($\hat{\phi}[N-2] \approx \mathbf{e}_j$), and the error covariance will be small ($P_{\phi\phi}[N-2] = E[\tilde{\phi}[N-2]\tilde{\phi}[N-2]' \mid \mathcal{Y}[N-1]] \approx 0$). If $R_\chi(i)$ is small for all $i \in \mathbf{S}$, the computed error covariance in each subfilter will be small as will the subfilter gains. If now $\{\phi[k]\}$ makes the transition $\mathbf{e}_j \rightarrow \mathbf{e}_l$ at $t = (N-1)T$, the proper subfilter is suddenly the lth. The lth subfilter, unmatched to the observations prior to $t = NT$, may have accumulated a significant amount of error. It was noted by Maybeck that the "mismatched filter estimates can drift significantly from the true state values. When the (regime) changes and one of these 'mismatched' filters becomes the 'correct' one, a very long transient is required to achieve proper identification" [May82a, p. 135]. The large residuals in the lth subfilter were of little consequence when $\mathbf{e}_l'\hat{\phi}[N-2]$ was small, but as $\{\hat{\phi}[k]\}$ moves toward $\mathbf{e}_l$, the drift error in the lth filter will manifest itself in $\{\hat{x}[k]\}$.

To reduce these transfer transients, two alterations are made in the implementation of (1.39). All components of $\{\hat{\phi}[k]\}$ are kept above a threshold to shorten the delay in identifying the new regime. To mitigate the influence of filter drift, pseudonoise is added to each of the filters.

The earlier tracking example illustrates some of the untoward effects created by regime transfers. The MM formalism was not used in the filter design, but the nominal estimator, EKF(W=1), is the EKF identified with the coast mode. On the time set $t \in [0,3) \cup [9,11) \cup [16,20]$, EKF(W=1) would be the proper subfilter for the PL1-MM algorithm to select. Compare, therefore, the response of EKF(W=1) on the interval after detection with that on an interval in final coast (e.g., around $t = 18$ s). The initialization is uncomplicated: The error covariance is reduced quickly from its initial value, and the estimation error stays well within the one-σ error envelope. Thus, EKF(W=1) provides both an accurate estimate of position and a valid measure of tracking uncertainty. This interval shows the proper functioning of a MM filter.

On the final coast, the performance of EKF(W=1) is poor even though EKF(W =1) is again the correct subfilter. This difference is attributable to the large errors at $t = 16$ s, particularly in velocity. Furthermore, the computed error covariance EKF(W=1) is far too small, and the subfilter is slow to correct the tracking error.

Generalizations of the MM approach are useful when the plant has a short memory for modal changes. In this circumstance, the conditioning on the full modal path could be replaced with conditioning on a shorter segment. For example, order the modal paths of length s and let

$$\begin{aligned}\hat{\phi}([\kappa]; [k-s:k-1]) &\approx \mathcal{P}[\phi(k-s:k-1)\\ &= \phi([\kappa]; 0:s-1), \ \text{(the } \kappa\text{th such sequence)} \,|\, \mathcal{Y}[k]].\end{aligned}$$

Now the number of models that must be considered for a path-length-s MM filter is bounded. Unfortunately, the number of required subfilters still grows so rapidly (of order S^s) that only short spans have ever been proposed (perhaps two). Further, because the modal paths are truncated, subfilter (re)initialization must be handled in an ad hoc manner.

In many applications, constraints on the computational complexity of the estimation algorithm severely limit the number of elemental filters that can be maintained. The PL1-MM filter utilizes only S subfilters, but it suffers from carryover errors at modal transitions. Without pseudonoise, the gains of the subfilters are too small, and transition transients take a long time to decay. An algorithm that uses S subfilters but achieves performance closer to that attained with a path-length-two MM filter is the *interacting-multiple-model* filter (IMM).

Suppose the observation cycle at time $t = kT$ is complete. The observation reflects the modal path up to time $t = (k-1)T$. Write $\hat{\phi}[k-1] = E[\phi[k-1] \,|\, \mathcal{Y}[k]]$. Suppose the $\mathcal{Y}[k]$-conditional probability density of the zygostate is $p_t = [\mathbf{e}_i'\hat{\phi}[k-1]\mathbf{N}(\hat{x}^i[k], P_{xx}^i[k])]$. The $\mathcal{Y}[k]$-density of $x[k]$ is $\sum_i \mathbf{e}_i'\hat{\phi}[k-1]\mathbf{N}(\hat{x}^i[k], P_{xx}^i[k])$. This is an instance of a Gaussian wavelet distribution in which the individual wavelets are basis functions, translated and shaped by the mean ($\hat{x}^i[k]$) and the covariance ($P_{xx}^i[k]$), and weighted by the coefficients $\mathbf{e}_i'\hat{\phi}[k-1]$. This wavelet distribution would be formally interpreted to mean that the conditional density of $x[k]$ given both $\mathcal{Y}[k]$ and $\phi[k-1] = \mathbf{e}_i$ is $\mathbf{N}(\hat{x}_k^i, P_{xx}^i[k])$.

The IMM propagates the conditional distribution forward to time $t = (k+1)T$. As does the PL1-MM filter, the IMM employs S subfilters. Initialize the individual subfilters at $t = kT$ with states $\{\hat{x}^i[k], P_{xx}^i[k]; i \in \mathbf{S}\}$. For each subfilter, the equations for extrapolation, update, residual, and gain are those of the associated Kalman filter.

To present the IMM most concisely, let us rewrite the subfilter algorithms in a different form. Equations (1.34)–(1.37) give the Kalman filter in *covariance form.* The filter state as displayed is the $\mathcal{Y}[k]$-mean of the base-state, $\hat{x}[k]$, and the $\mathcal{Y}[k]$-covariance of the estimation error, $P_{xx}[k]$. The latter is indicative of the uncertainty that remains in the estimate of $x[k]$ after all of the information in $\mathcal{Y}[k]$ is exploited. The state space of a linear system is arbitrary to some degree. The *information* implementation of the Kalman filter utilizes an alternative (but equivalent as long as $\{P_{xx}[k]\}$ is a positive sequence) state space representation. Instead of the error covariance, $P_{xx}[k]$, the filter propagates its inverse $D_{xx}[k] = P_{xx}[k]^{-1}$. Instead of the $\mathcal{Y}[k]$-conditional mean of the base-state, the filter propagates the product $d[k] = D_{xx}[k]\hat{x}[k]$. To contrast it with the *uncertainty matrix* $P_{xx}[k]$, $D_{xx}[k]$ is called the *information matrix*, and the Kalman filter in this transformed state space is called the *information filter* [AM79]. The relative computational advantage of these alternative implementations is explored in textbooks on filtering. Here we will mix these state spaces to clarify the organization of the subfilters rather than to suggest the best way to implement the IMM.

Write the inverse of the observation noise covariances as $D_x = R_x^{-1}$. The extrapolation and update relations in the subfilters in the IMM can be written [LBS93]:

the IMM base-state estimator

Extrapolation:

$$\hat{x}^j[k+1]^- = A_j\hat{x}^j[k];\ j \in \mathbf{S}, \tag{1.40}$$

$$P^j_{xx}[k+1]^- = A_j P^j_{xx}[k]A'_j + R_X(j);\ j \in \mathbf{S}. \tag{1.41}$$

Update:

$$\Delta d^j[k+1]^+ = H'D_x y[k+1];\ j \in \mathbf{S}, \tag{1.42}$$

$$\Delta D^j_{xx}[k+1]^+ = H'D_x H;\ j \in \mathbf{S}. \tag{1.43}$$

As in (1.36) and (1.37), $\Delta d^j[k+1]^+$ (respectively $\Delta D^j_{xx}[k+1]^+$) is the increment in the estimate at an observation, for example,

$$d^j[k+1]^+ = d^j[k+1]^- + \Delta d^j[k+1]^+.$$

The updated mean and covariance, $(\hat{x}^j[k+1]^+, P^j_{xx}[k+1]^+)$, are directly produced from $(\hat{d}^j[k+1]^+, D^j_{xx}[k+1]^+)$. Instead of the single filter with random coefficients displayed in (1.34)–(1.37), Equations (1.40)–(1.43) describe S constant coefficient subfilters.

The algorithm for each of the elemental filters has an intuitively appealing form. The plant noise ($R_X(j)$ in (1.41)) adds uncertainty to the estimate during each extrapolation, and each update adds information ($H'D_xH$ in (1.43)). The extrapolation equation is linear in $\hat{x}^j[k]$, and the base-state update is linear in $y[k+1]$, with the latter weighted proportionately to the sensor gain (H) and inversely to the sensor noise (D_x).

To update the modal probabilities, look at the measured residual at the output of each of the subfilters,

$$r^j[k+1] = y[k+1] - H\hat{x}^j[k+1]^-;\ j \in \mathbf{S}.$$

In a regime-invariant system in which $\phi[k] \equiv \mathbf{e}_j$, the residual of the jth filter is actually the innovations increment and has covariance

$$R^j_{yy}[k+1] = HP^j_{xx}[k+1]^- H' + R_x.$$

In a hybrid system, the modes communicate, and none of the residuals is truly the innovations increment. Still, let us retain the name $R^j_{yy}[k+1]$ for the functional form with $D^j_{yy}[k+1]$ its inverse and $S^j_{yy}[k+1]$ the square root of the inverse:

$$R^j_{yy}[k+1]^{-1} = D^j_{yy}[k+1] = S^j_{yy}[k+1]' S^j_{yy}[k+1].$$

The modal estimate is produced in the IMM as follows. Let $\Phi(u)$ be the unit Gaussian density function, $\mathbf{N}_u(0, I)$. Let

$$\gamma^j[k+1] = \Phi(S^j_{yy}[k+1] r^j[k+1]).$$

The regime probabilities are given by:

the IMM modal-state estimator

Extrapolation:

$$\hat{\phi}[k]^- = \Pi\hat{\phi}[k-1]. \tag{1.44}$$

Update:

$$q[k] = \hat{\phi}[k]^- * \gamma[k+1]. \tag{1.45}$$

The quantity after the update, $q[k] = [q^i[k]]$, is a vector of unnormalized probabilities and, when normalized, is accepted as a replacement for the regime probabilities: $q[k]/\mathbf{1}'q[k] \mapsto \hat{\phi}[k]$. The rationale for this algorithm is that the smallest residuals (when scaled by $S^j_{yy}[k+1]$) should be associated with that subfilter identified with the actual regime. After propagating $\hat{\phi}[k-1]$ forward using the matrix of modal transition rates Π, the likelihoods are corrected with factors that are (relatively) large for the current regime and small for the others. The calculation of $\hat{\phi}[k]$ is inexact, and there is no claim that the true $\mathcal{Y}[k+1]$-modal probabilities have been computed.

We now have a new $\mathcal{Y}[k+1]$-distribution of $x[k+1]$ as a Gaussian sum with S terms:

$$x[k+1] \in \sum_j \mathbf{e}'_j \hat{\phi}[k] \mathbf{N}(\hat{x}^j[k+1])^+, P^j_{xx}[k+1]^+).$$

The base-state mean and covariance can be obtained using the formulas that apply to Gaussian sums [AM79, Section 8.4, Equation (4.5)]:

$$\hat{x}[k+1] \approx \sum_j \hat{x}^j[k+1]^+ \mathbf{e}'_i \hat{\phi}[k], \tag{1.46}$$

$$P_{xx}[k+1] \approx \sum_j (P^j_{xx}[k+1]^+ + (\hat{x}^j[k+1]^+ - \hat{x}[k+1])(\hat{x}^j[k+1]^+ - \hat{x}[k+1])')\mathbf{e}'_i \hat{\phi}[k]. \tag{1.47}$$

To this point, the implementation of the IMM is close to that of the PL1-MM algorithm. Both use S subfilters of the same form and both update $\{\hat{\phi}[k]\}$ using the residuals from the subfilters. The IMM includes the regime transition rates and the MM-filter ignores them. The calculation of the mean of the base-state is a natural consequence of the S-fold Gaussian wavelet representation of the base-state density. If modal transitions were not permitted (e.g., in the PL1-MM filter), then

$$\{\hat{x}^j[k+1]^+, P^j_{xx}[k+1]^+; j \in \mathbf{S}\}$$

would provide the initial conditions for the next cycle of the subfilters.

The singular properties of the IMM follow from a sophisticated merging step. The Gaussian sum used to compute the base-state estimates is not used as a starting point for the next cycle of extrapolation and update. To approximate the $\mathcal{Y}[k+1]$-probability of a length-two modal path, a mixing probability $\alpha^i_j[k+1]$ is defined as follows:

$$\alpha^i_j[k+1] = \frac{\Pi_{ji}\hat{\phi}_j[k]}{(\Pi'\hat{\phi}[k])_i}. \tag{1.48}$$

For every $i \in \mathbf{S}$, $\{\alpha^i_j[k+1]\}$ is a probability in j and is accepted as a replacement for $\mathcal{P}[\phi[k-1] = \mathbf{e}_j \mid \phi[k] = \mathbf{e}_i \wedge \mathcal{Y}[k+1]]$; that is, $\alpha^i_j[k+1]$ is the conditional probability that regime $\mathbf{e}_i$ was immediately preceded by $\mathbf{e}_j$.

Let us focus on the ith subfilter at $t = (k+1)T$. The base-state mean, $\hat{x}^i[k+1]$, should involve the estimates of the subfilters, $\{\hat{x}^l[k+1]^+; l \in \mathbf{S}\}$, and the likelihood of a transition $\mathbf{e}_l \mapsto \mathbf{e}_i$ at time $(k+1)T$ (and similarly with the base-state error covariance, $P^i_{xx}[k+1]$). The updated outputs of the subfilters, $\{\hat{x}^j[k+1]^+, P^j_{xx}[k+1]^+; j \in \mathbf{S}\}$, can be combined using the merging formula of the Gaussian sum but this time restricted to the subfilter level:

$$\hat{x}^i[k+1] = \sum_j \hat{x}^j[k+1]^+ \alpha^i_j[k+1], \tag{1.49}$$

$$P^i_{xx}[k+1] = \sum_j (P^j_{xx}[k+1]^+ + (\hat{x}^i[k+1]^+ - \hat{x}^j[k+1])(\hat{x}^i[k+1]^+ - \hat{x}^j[k+1])')\alpha^i_j[k+1]. \tag{1.50}$$

The augmentation,

$$\sum_j \cdot + (\hat{x}^i[k+1]^+ - \hat{x}^j[k+1])(\hat{x}^i[k+1]^+ - \hat{x}^j[k+1])'\alpha^i_j[k+1],$$

is an adaptive analogue to adding pseudonoise. The next cycle begins with a $\mathcal{Y}[k+1]$-distribution of $x[k+1]$ given by the Gaussian sum:

$$x[k+1] \in \sum_j \mathbf{e}'_j \hat{\phi}[k] \mathbf{N}(\hat{x}^j[k+1], P^j_{xx}[k+1]).$$

If modal transitions are not allowed, $\alpha^i_j[k+1] = 0$ for $i \neq j$, and

$$(\hat{x}^j[k+1], P^j_{xx}[k+1]) \equiv (\hat{x}^j[k+1]^+, P^j_{xx}[k+1]^+).$$

When modal changes can occur, the IMM avoids subfilter drift by moving the means after mixing toward an intermediate value. For example, if an $\mathbf{e}_j \mapsto \mathbf{e}_i$ transition is likely, the initialization of ith subfilter would weight heavily the state of the jth subfilter. The IMM algorithm also augments the subfilter covariances through the use of a local version of the mean spreading term found in (1.50). The gain of "less-likely" filters is kept higher thereby. Simulation suggests that the IMM provides the performance found in path-length-two, multiple model filters [BBS88].

The motivation behind the IMM is the efficient computation of an approximation to $\{\hat{x}[k]\}$. The wavelet distribution

$$\sum_j \mathbf{e}'_j \hat{\phi}[k] \mathbf{N}(\hat{x}^j[k+1], P^j_{xx}[k+1])$$

is not claimed to be the $\mathcal{Y}[k]$-joint distribution of the zygostate, nor is it. Approximations induced by the merging of the subfilters and the update rule for $\{\hat{\phi}[k]\}$ preclude accurate calculation of the distribution function as well as relevant higher moments (e.g., $R_{x\phi}[k] = E[x[k]\phi[k]' \mid \mathcal{Y}[k]]$). Since only the mean of $\{x[k]\}$ is sought, the regime models can be fairly crude.

There are different ways in which the IMM can be configured. The modal decomposition is arbitrary to some degree as are the regime models selected for base-state evolution. To illustrate the manner of implementation of the IMM, return to the tracking problem studied earlier. In [LBS93] and [WKBS98] applications related to air traffic control were explored. In contrast to the earlier tracking example, the expected turn rates of the aircraft are moderate. The simplest IMM is framed with a base-state representation like that used in the earlier EKF analysis with the turns neglected. (This decouples the X and the Y motion.) The IMM uses two regimes instead of the single regime employed by the EKF:

(i) A high plant noise level during a turn ($\phi = \mathbf{e}_1$).
(ii) A low plant noise level during constant velocity flight ($\phi = \mathbf{e}_2$ with $R_\chi(1) \gg R_\chi(2)$).

It was observed in the references that the IMM acts as a "self-adjusting variable bandwidth filter" because the $\phi_t = \mathbf{e}_2$ filter has a low bandwidth while the $\phi_t = \mathbf{e}_1$ has a high bandwidth [WKBS98]. The SNR was good in the case studied with a 3°/s turn identified in three samples. Moreover, the "accuracy of the turn rate estimate is not important as far as the quality of the (base-state) estimates are concerned" [LBS93, p. 190].

This example illustrates the robustness of the IMM. The rudimentary regime models do not mimic the geometry of the motions, and this prevents the modal estimate from being particularly good. But good modal estimates are not required for good estimates of $\{x[k]\}$: Low quality estimates of the conditional distribution function can still be used to obtain high quality estimates of the mean.

Although the elemental IMM works well in a benign environment, it would not be acceptable for use with highly maneuverable targets (as noted in [LBS93]) because it completely neglects the geometry of the turning motion. The turn-submodel replaces a directional acceleration with one that is omnidirectional, albeit with a higher intensity. To represent the directionality of the acceleration more closely, adjustments to the regime models can be made. In [WKBS98] it was proposed that:

(i) During a turn ($\phi_t = \mathbf{e}_1$) the aircraft acceleration is a Gaussian white process with standard deviation 2 g in a direction perpendicular to the velocity and standard deviation 0.4 g along the velocity vector.
(ii) During coasting motion ($\phi_t = \mathbf{e}_2$), the aircraft acceleration is a Gaussian white process with standard deviation 0.01 g in both directions.

It is still true that $R_\chi(1) \gg R_\chi(2)$, but the X and Y motion are now coupled during a turn. Submodel 1 is nonlinear, but conventional methods can be used to linearize it. It was found by simulation that to achieve comparable performance from the EKF, the selection $R_\chi(1)$ is required throughout the experiment. The EKF has a concomitant deterioration in tracking accuracy during uniform motion.

As a further refinement of the IMM, the turn rate process, $\{\Phi[k]\}$, can be treated as a component of the base-state with first-order dynamics. This raises the dimension of the kinematic model and creates a multiplicative nonlinearity that must be accommodated. However, when the turn rate is integrated into the base-state vector, the intensity of the additive excitation in the IMM can be reduced considerably: During a turn the Gaussian-white acceleration has standard deviation 0.2 g in a direction perpendicular to the velocity and 0.1 g along the velocity vector (as compared with 2 g and 0.4 g in a submodel without turn rate as a base-state). For mild turns the inclusion of the turn rate yields performance superior to that of the Kalman filter and the lower order IMMs.

1.5 Modal Observations

Multiple model methods work quite well in a LJS when the output SNR is high and the accurate computation of higher (or cross-) moments is not required. High quality estimation in the IMM depends more on the celerity of detection of a modal transition than it does on accurate determination of the specific change. There are situations in which the process $\{y_t\}$ alone is not sufficiently informative to achieve the estimation quality required. In air traffic control, suppose that the radar is of low quality or that the sampling frequency is low. By the time $\{y_t\}$ has been processed to identify a change in the modal-state, the aircraft could have drifted outside the tracking window resulting in loss-of-lock (LOL). In this circumstance, a quicker and less ambiguous indication of changes in $\{\phi_t\}$ is needed.

When the modal process is not clearly distinguishable in $\{y_t\}$, a direct measurement of the mode is useful. If the regime is identified with a physical quantity (e.g., a temperature) the representation for the modal observation could be an analogue of (1.4) in which $\{\Phi_t\}$ is measured in a white noise channel. However, the plant regime is often a category variable: It may have a name rather than an intrinsic value, and the ordering of the regimes follow some prevailing convention. In this case, identification of the current regime is done by looking at global attributes of the plant. The information that goes into what we call the modal measurement may be compiled from a diverse collection of sensors. These sensors provide a raw data aggregate (a *data frame*) consisting of the measurements bearing on the operating regime of the system. These measurements may differ in kind and in form; for example a FLIR imager may provide a two-dimensional picture of a target, whereas another sensor may provide a multispectral decomposition of the exhaust plume of the same target. An online processor would be overwhelmed if forced to treat each element of the primitive data set (e.g., each pixel in the image) as an "output." To reduce the data handling demands on the estimator–controller, the data frame is first reduced to a category statement by a high-level preprocessor; for example, based upon its shape and emission spectrum, the target is classified as being an aircraft of type 1. After processing, the modal observation process comprises a sequence of classifications. Using various artifices, the observation categories can be identified with the set of regimes. In this event, the modal observation process, $\{z_t\}$, can be written as a vector counting process in which the ith component is the number of times the regime has been classified as $\mathbf{e}_i$: $\Delta z_t = \mathbf{e}_i$ if the ostensible regime at time t is the ith. In this way the estimator–controller throughput is reduced to a manageable level.

When the modal measurement is ambiguous and/or infrequent, a problem of considerable difficulty arises. Suppose that the full sensor suite consists of the conventional measurement, $\{y_t\}$, augmented with a direct modal measurement, $\{z_t\}$. How then should these disparate measurements processes be fused for optimal performance? Should $\{z_t\}$ be used to identify $\{\phi_t\}$, and $\{\phi_t\}$, so identified, be used in a path-specific Kalman filter (*track data fusion*)? Alternatively, should $\{z_t\}$ be merged with $\{y_t\}$ to form $\{g_t\}$, and $\{\mathcal{G}_t\}$ used to form the estimates? This would be *low-level* data fusion.

For example, suppose an aircraft of unknown type is in the field of view of an imaging sensor and a radar. The radar gives a line-of-sight measurement to the center of reflection of the aircraft (bearing) and there is perhaps also a distance measurement (range). The radar observations are measurements of a point target and easily accommodated in $\{y_t\}$. But the aircraft type (the modal-state) is a category variable. If the aircraft is close enough, a multipixel image can be created (the data frame). Shape analysis can be used to classify the target (the initial preprocessing). The output of the image generation/processing node is a statement about the aircraft (e.g., "the aircraft is an F-14"). The modal sensor would in this case be the composite image generation–preprocessing node associated with the imager–image processor. The modal sensor maps the true mode (F-14) into a statement of modal condition. (Image distortion might lead to the conclusion that the aircraft is a helicopter.)

In principle, the modal status could be determined from $\{y_t\}$ alone; for example, a helicopter has a different motion pattern than does an F-14. Modal estimation from a radar can be successfully done if $\{y_t\}$ is such as to distinguish between modal categories in an expeditious manner. This type of analysis is done in the multiple model algorithms described above. But a quicker and often more accurate approach is to take the aggregate contained in the data frame and use that to assist in motion estimation. The observation processes should be processed synergistically rather than in parallel.

There are two common methods for modeling the modal sensor. The first mimics (1.4). Suppose each regime has a characteristic signature in the measurement (e.g., $\phi_t = \mathbf{e}_i$ is distinguishable in the measurement by a slope h_i). If we wish to determine the current mode, we need only look at dz_t/dt (or more likely $\Delta z_t/\Delta t$ for some small Δ). Since $dz_t/dt \in \{h_i; i \in \mathbf{S}\}$, the value of ϕ_t can be recovered from $\{z_t\}$. Clearly, the signature vector $h = [h_i]$ could be a function of time with no essential change in the argument.

Unfortunately, the plant is not usually so cooperative. The modal measurement is less determinate. Specifically, suppose the modal signature is observed in an additive white-Gaussian channel:

modal-state measurement: time-continuous; additive noise

$$dz_t = h'\phi \, dt + d\eta_t, \tag{1.51}$$

where $\{\eta_t\}$ is a Brownian motion with intensity $R_\phi > 0$. Now $\{z_t\}$ has no "slope." The average slope can still be determined ($E[dz_t \,|\, \mathcal{F}_t] = \sum_i h_i \phi_i \, dt$), but over what interval should the average be calculated? This creates a dilemma: Long averaging intervals give better estimates of the slope of the measurement because the noise is white, but near a modal transition, the average slope is the signature of neither regime.

Equation (1.51) has an analogue when there are time-discrete modal measurements (see [LDB98]):

modal-state measurement: time-discrete; additive noise

$$z[k] = h'\phi[k] + \eta[k], \tag{1.52}$$

where $\{\eta[k]\}$ is a Gaussian white sequence with covariance $R_\phi > 0$. As was the case with time-continuous measurements, the noise-free ($R_\phi = 0$) case is easily solved; $\phi[k] = \mathbf{e}_i$ if $z[k] = h_i$. With noise, some form of averaging must be used to find the regime. With time-discrete measurements, not only is noise a problem, but also the intersample delay: Even if ϕ_{kT} is measured, $\phi_{kT+\delta}$ is random if $0 < \delta < T$.

Equations (1.51) and (1.52) have a conventional form, and many of the common approaches to estimation can be applied if the modal sensor can be so represented. For example, suppose Φ_t is the angular orientation of an aircraft in a video image. If target orientation is added as a base-state component, the output of an image classifier can be thought of as providing the true orientation plus some Gaussian noise. A high-resolution imager could be represented as a "data... sequence of 2-D images resulting from projection of the scene volume containing the targets onto the focal place of the imaging sensor. [MSG95]." After preprocessing, the measurement "is a nonzero-mean white Gaussian process with mean the projective transformation of the scene" [MSG95]. This approach fits nicely with (1.52).

In many applications, however, the classical white-noise-channel measurement models will not suffice. The modal sensor is both a data collector and a preprocessor; it collects data from diverse sources, and using suitable rules, it classifies each data frame into one of a predetermined set of categories. Suppose the 2-D projection of an F-14 is compared pairwise with a fixed set of templates formed by looking at an F-14 at different orientation angles. Figure 1.7 shows a sequence of images of an F-14 as it rolls. These data frames were created in infrared at a rate of 30 frames/s at the Point Mugu Naval Air Station and are displayed here as visible duplicates. The day was clear without foreground occlusions. Label the frames with an $X-Y$ ordering beginning in the lower left corner with frame (1,1). Even though the range is essentially constant, it is apparent that there is considerable variation in the number of pixels-on-target for different orientations (e.g., (3,7) is "clearer" than (3,1) say). Figure 1.7 superimposes upon the target the silhouette selected by the image processor from the replica data base. Except for frame (3,3), the classifier did a good job.

The experiment displayed in Figure 1.7 illustrates some of the potential problems that arise when a modal classifier is used. The limited replica data base and the need to minimize the latency interval during which image classification is performed combine to restrict the precision of the regime measurement. The mapping from

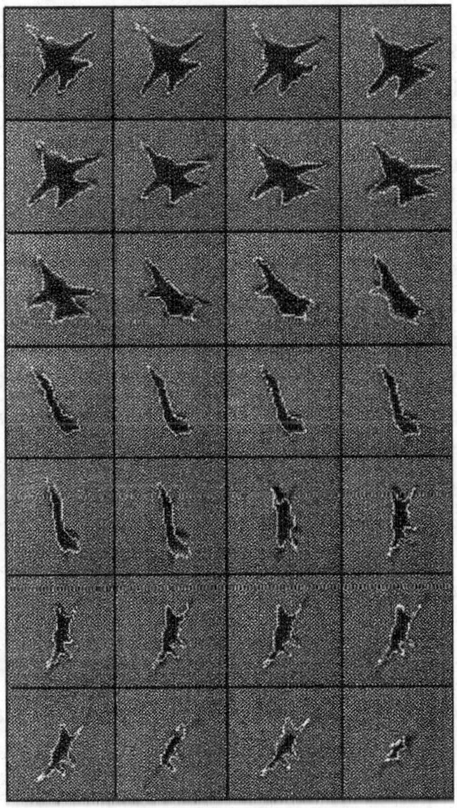

Figure 1.7. FLIR images of a maneuvering F-14 and the template selected by the image processor.

ϕ_t to z_t is not well modeled with an additive noise channel. Pattern classifiers are idiosyncratic. They make errors that are neither symmetric nor independent of the mode itself. Rather, if $\phi_t = \mathbf{e}_i$, a classifier may be biased toward certain modal groups to the exclusion of others.

To model the modal sensor–preprocessor suppose that the data frames are processed to yield an output statement with mean frequency λ samples/s and with intersample times that are exponentially distributed and independent of modal process. Each modal measurement is classified as coming from one of the S modal categories. Accumulate these output classifications to form a right-continuous, S-dimensional, counting process $\{z_t\}$. The increment in $\{z_t\}$ is the zero vector between measurements, and a unit vector at a measurement time. For example, $\Delta z_t = \mathbf{e}_i$ indicates that the modal sensor classified the data frame as coming from regime $\mathbf{e}_i$ at time t.

To illustrate, consider a simple situation in which the modal measurement is perfect and the plant is in regime $\mathbf{e}_1$ for an interval. In this case, the measurement sequence would be $\Delta z_t = \mathbf{e}_1$ with rate λ; that is, the observation string $\{\mathbf{e}_1, \mathbf{e}_1, \mathbf{e}_1, \mathbf{e}_1, \ldots\}$ is compatible with regime $\mathbf{e}_1$ and has mean reconfirmation time $1/\lambda$ s.

But suppose the sensor is not perfect and a different observation string is received (e.g., $\{\mathbf{e}_1, \mathbf{e}_1, \mathbf{e}_2, \mathbf{e}_1, \ldots\}$). The receipt of $\Delta z_t = \mathbf{e}_2$ has two consequences. First, the rate of occurrence of $\Delta z_t = \mathbf{e}_1$ is now less than λ, and this should be judged as a failure to reconfirm hypothesis $\phi_t = \mathbf{e}_1$. Secondly, the rate of $\Delta z_t = \mathbf{e}_2$ is now positive. This anomalous observation could be a false indication of modal change. ($\phi_t = \mathbf{e}_1$ but $\Delta z_t = \mathbf{e}_2$), or it could be symptomatic of a modal transition ($\{\phi_t\}$ transitions from $\mathbf{e}_1$ to $\mathbf{e}_2$).

Denote the filtration generated by $\{z_t\}$ by $\{\mathcal{Z}_t\}$. The output filtration is $\mathcal{G}_t = \mathcal{Y}_t \vee \mathcal{Z}_t$. Clearly, if the fidelity of the modal sensor is such that $\{\phi_t\}$ can be determined with good accuracy, estimation of $\{x_t\}$ given $\{\mathcal{G}_t\}$ reduces to the Kalman filter with $\{\phi_t\}$ replaced by its $\mathcal{Z}_t$-based estimate. The solution to the state estimation problem in a hybrid system requires at least the $\mathcal{G}_t$-expectation of $\{x_t, \phi\}$ with $\hat{\chi}_t = \hat{x}_t + \chi\hat{\phi}_t$.

To model a modal classifier, neglect delays in the forming and preprocessing of a data frame. Let quality of the modal sensor be represented by the $S \times S$ *discernibility* matrix $\mathbf{P} = [\mathbf{P}_{ij}]$, where $\mathbf{P}_{ij}$, is the probability that modal state $\mathbf{e}_i$ will be selected by the classifier if $\mathbf{e}_j$ is the true mode at time of measurement, that is,

$$\mathbf{P}_{ij} = \mathcal{P}(\Delta z_i = 1 \,|\, \phi = \mathbf{e}_j).$$

The columns of $\mathbf{P}$ are probability vectors: $\mathbf{P}_{ij} \geq 0$; $\mathbf{1}'\mathbf{P} = \mathbf{1}'$. Perfect modal measurements are implied by $\mathbf{P} = \mathbf{I}$. The form of $\mathbf{P}$ is flexible enough to include aliasing and bias.

For this measurement model, the quality of modal inference is clearly a function of the sample rate (λ) and the ability of the modal sensor to correctly classify a single measurement (**P**). Suppose $\phi_t = \mathbf{e}_i$. If a measurement is received at time t, its distribution across categories is given by $\mathbf{P}_{\cdot i}$. Denote the $\mathcal{F}_t$-conditional rate of individual observations by $\boldsymbol{\lambda}_t = \lambda \mathbf{P}\phi_t$: The ith component of $\boldsymbol{\lambda}_t\, dt$ is the $\mathcal{F}_t$-probability that an observation will be received and be classified in the ith modal bin in $[t, t+dt]$. Let $d\eta_t = dz_t - \lambda \mathbf{P}\phi_t\, dt = dz_t - \boldsymbol{\lambda}_t\, dt$. Then, since $E[dz_t - \boldsymbol{\lambda}_t\, dt \mid \mathcal{F}_t] = 0$, $\{\eta_t\}$ is an $\mathcal{F}_t$-martingale. The modal observation can thus be written:

modal-state measurement: time-continuous; classifier

$$dz_t = \lambda \mathbf{P}\phi_t\, dt + d\eta_t. \tag{1.53}$$

With $h = \lambda \mathbf{P}'$, Equation (1.53) looks exactly like (1.51). The observation increment is the sum of a term linear in modal-state ($h'\phi$) and a disturbance that is a martingale increment ($d\eta_t$). But $\{\eta_t\}$ in (1.53) is not a Gaussian process, is not continuous, is not centered around zero, and is not independent of the modal-state process. Rather, $\{\eta_t\}$ is a purely discontinuous martingale. If the discontinuities in $\{\eta_t\}$ are removed, the predictable quadratic variation of what remains is identically zero. Consequently, the EKF-like algorithms based upon the $\{z_t\}$ in (1.53) are not likely to be satisfactory.

There is a discrete analogue to (1.53). Let the sample rate again be $1/T$. Then the observation at $t = kT$ is

modal-state measurement: time-discrete; classifier

$$z[k] = \mathbf{P}\phi[k] + \eta[k], \tag{1.54}$$

where $\eta[k] = z[k] - \mathbf{P}\phi[k]$. Since $E[z[k] - \mathbf{P}\phi[k] \mid \mathcal{F}[k]] = 0$, $\{\eta[k]\}$ is an $\mathcal{F}[k]$-martingale difference sequence. The model given in (1.54) is useful in studies of time-discrete plants. Denote the filtration generated by $\{z[k]\}$ by $\{\mathcal{Z}[k]\}$. In discrete time, this model for $\{z[k]\}$ can be somewhat problematic. Equation (1.54) implies that the data frame upon which $z[k]$ is based can be formed and preprocessed to provide an observation with essentially no delay. This is reasonable if the plant is time continuous and measurement has a simple structure (e.g., preprocessing a radar return to form $\{y[k]\}$ requires little time). When the preprocessing delay is small and the estimator–controller is updated as soon as the preprocessing is complete, this measurement-state concurrence is justifiable. When the data frame requires a higher level of preprocessing, there is output latency: The observation lags the modal-state by the latency interval. By its nature latency is more common in $\{z[k]\}$ than it is in $\{y[k]\}$. The minimum time increment in a time-discrete system is T, and latency could cause any action to be delayed for a sample time or more. In this book, however, latency will be ignored.

The advantages of direct modal measurement in state estimation are apparent and were noted early in various contexts. Typically problems were studied within an EKF framework [AKG86]. Because of the nonlinear, discontinuous, non-Gaussian nature of the plant and the observation, the development of an exact, finite-dimensional, recursive algorithm for generating the $\mathcal{G}_t$-conditional mean of the comprehensive state is beyond reach. In subsequent chapters, approximations to $\{\hat{x}_t, \hat{\phi}_t\}$ are developed under various assumptions on the plant and measurement system.

2

The Polymorphic Estimator

2.1 Introduction

At its basic level, an estimation algorithm is a causal mapping from a spatiotemporal observation (the measurement) to an approximation of a primary process (the signal or state, depending on the context). Estimators have become increasingly sophisticated as more versatile online processors have become available. If the state process has a suitable structure, improved estimation and prediction is achieved through model-based synthesis procedures. These approaches, as their name suggests, use a comprehensive analytical model to delineate both the state dynamics and the precise relationship between the state and its measurement. Using this model as an intermediary, a problem in optimal inference is posed. The solution to this optimization problem is then said to be the *best* estimation algorithm in the application. Through the model, the estimator is tuned to subtle patterns in the measurement, thus enabling good performance to be achieved in the presence of significant measurement ambiguity. Of course, the model is only an abridgement of reality, and to the degree that the model fails to adequately portray the salient features of the actual state processes, there is justifiable concern that the algorithm may be tuned improperly and may *see* things in the observation that are not actually there.

Perhaps the most widely studied model-based algorithm is the Kalman filter and its lineal variants. As described in Chapter 1, the dynamic features of the plant are represented by a base-state process $\{x_t\}$. The observation process $\{y_t\}$ is a linear function of the base-state – albeit a noisy one. Although this summary description would suggest a modeling hierarchy in which the base-state vector gives the intrinsic description of plant evolution, and the observation is subordinate to it, frequently these two constituents of the plant model are interlinked in a horizontal manner. Measured properties of the plant are directly reflected in the equations of dynamic evolution, while unmeasured ones are ignored. This oft neglected linkage between sensor capabilities and the state space model is clearly seen in the motion model of an agile aircraft where the components of x_t would include positions, velocities, and perhaps accelerations in an appropriate reference frame. A radar provides measurements of the center-of-reflection attributes of the aircraft (e.g., range and bearing). Not coincidentally, the base-state space consists of the properties of a point aircraft. Other relevant but unmeasured variables (e.g., aircraft type and orientation) are not included in the target dynamics because they are not reflected clearly in the measurement. In this sense, the sensor suite has a major role in determining the components of the state space model; new sensors *create* new components in the state space.

Chapter 1 described a flexible model for quantifying the plant dynamics and observation. The plant model is phrased in terms of a set of nonlinear stochastic differential equations. Let us start with a probability space $(\Omega, \mathcal{F}, \mathcal{P}; \mathcal{F}_t)$ and a time interval [0, T]. On this space, the plant model was given by:

$$d\chi_t = \mathbf{f}(\chi_t, \upsilon_t, \Phi_t)\, dt + \mathbf{g}(\chi_t, \upsilon_t, \Phi_t)\, dw_t, \tag{1.1}$$

$$dg_t = \mathbf{r}(\chi_t, \upsilon_t, \Phi_t)\, dt + \mathbf{s}(\chi_t, \upsilon_t, \Phi_t)\, dn_t, \tag{1.2}$$

where $\{\upsilon_t\}$ is the plant input, $\{g_t\}$ is the plant output, and $\{\chi_t\}$ is the plant state. The plant model is subject to initial conditions χ_0 and g_0 and exogenous excitation $\{\Phi_t\}$, $\{w_t\}$, and $\{n_t\}$.

Estimation and control is difficult within the framework of the plant model. If there are a finite number of operating points and if the plant makes extended sojourns in these locations, a hybrid model is proposed as a replacement for (1.1) and (1.2). The modal-state, ϕ_t, is a pointer to the current regime: If $\Phi_t = \Phi_\kappa$, then $\phi_t = \mathbf{e}_\kappa$. The base-state variables are deviations from the current mode: $x_t = \chi_t - \boldsymbol{\chi}\phi_t$; $u_t = \upsilon_t - \boldsymbol{\upsilon}\phi_t$. The zygostate is the pair (x_t, ϕ_t). The plant model is replaced with a pair of equations:

base-state model

$$dx_t = \sum_i ((A_i x_t + B_i(u_t - \boldsymbol{\upsilon}\phi_t))\, dt + C_i\, dw_t)\phi_i + \sum_{i,l} (M(i,l)x_t$$

$$+ (\chi_i - \chi_l) + M(i,l)\chi_i + \rho(i,l))\phi_i \mathbf{e}_l' \Delta\phi_t, \tag{1.7}$$

modal-state model

$$d\phi_t = Q'\phi_t\, dt + dm_t. \tag{1.12}$$

The modal-state is represented by an $\mathcal{F}_t$-Markov process. Successive linearizations are used to delineate base-state evolution, and the exogenous processes in (1.7) and (1.12) are $\mathcal{F}_t$-martingales ($\{w_t\}$ is Brownian motion with intensity W and $\{m_t\}$ is a purely discontinuous martingale).

The sensor suite is composed of complementary elements. There is a base-state sensor with output $\{y_t\}$ generating the filtration $\{\mathcal{Y}_t\}$:

$$dy_t = H\chi_t\,dt + dn_t, \tag{2.1}$$

where $\{n_t\}$ is an $\mathcal{F}_t$-Brownian motion with intensity $R_x > 0$ ($d\langle n, n\rangle_t = R_x\,dt$) and $y_0 = 0$. There is a modal-state sensor with output $\{z_t\}$ generating the filtration $\{\mathcal{Z}_t\}$:

$$dz_t = h'\phi_t\,dt + d\eta_t, \tag{2.2}$$

where $\{\eta_t\}$ is an $\mathcal{F}_t$-martingale ($d\langle\eta, \eta\rangle_t = R_\phi\,dt$ with $R_\phi > 0$). In some applications $\{\eta_t\}$ is a Brownian motion and in others it is a discontinuous process. All of the exogenous processes are independent. Low level data fusion merges the data streams to create an observation $g_t = \mathrm{vec}(z_t, y_t)$ (sometimes treated as the array $g_t = (z_t, y_t)$) with filtration $\mathcal{G}_t = \mathcal{Y}_t \vee \mathcal{Z}_t$. An engineer seeks the $\mathcal{G}_t$-conditional distribution of the zygostate, or barring that, the $\mathcal{G}_t$-zygostate moments required for the task at hand.

The random processes generated by (1.7) and (1.12) are instances of a class of processes called *semimartingales*. An *$\mathcal{F}_t$-semimartingale* is a process $\{\varsigma_t\}$ that satisfies a stochastic differential equation:

$$d\varsigma_t = f_t\,dt + d\mu_t, \tag{2.3}$$

where $\{f_t\}$ is right continuous and $\mathcal{F}_t$-adapted, and $\{\mu_t\}$ is an $\mathcal{F}_t$-martingale. This is not the most general construction but will suffice for our purposes (see [Ell82, Definition 12.1] or [WH85, page 234]). In the integral equation associated with (2.3), recall that the integrands are replaced by their predictable modification. In a hybrid system then, both the state process and the observation process are semimartingales.

When working with semimartingales, it is expedient to generalize the notion of optional quadratic variation and define the *co-quadratic variation* of a pair of semimartingale processes as follows (see [Ell82, Definition 10.7]). Let $\{\varsigma_t\}$ and $\{\varphi_t\}$ be semimartingales. The co-quadratic variation $\{[\varsigma, \varphi]_t\}$ is generated from its increments: $d[\varsigma, \varphi]_t = (d\varsigma_t)d\varphi_t'$. A list of properties of the covariation process is presented in [WH85, Chapter 6, Proposition 6.1]:

- If $\{\varsigma_t\}$ is differentiable, its co-quadratic variation with any semimartingale is zero; for example, $(h'\phi_t\,dt)(h'\phi_t\,dt)' = 0$.†

† The process in view here is $\varsigma_t = \int_{[0,t]} h'\phi_\tau\,d\tau$. The significance of the observation is that, in semimartingale calculus, terms containing $(dt)^2$ vanish.

- If $\{\varsigma_t\}$ is a Brownian motion, its co-quadratic variation with any discrete martingale or any independent Brownian motion is zero; for example, $d[w, n]_t = 0$ and $d[w, m]_t = 0$, but $\{[w, w]_t\}$ is not the zero process.
- If $\{\varsigma_t\}$ is a purely discontinuous martingale, its co-quadratic variation with any other discontinuous martingale that shares no jumps is zero; for example, $d[m, \eta]_t = 0$, but $\{[m, m]_t\}$ is not a zero process.

In what follows, the *innovations process*, $\{\nu_t\}$, is particularly important. The innovations increment is the difference between the measurement increment and its $\mathcal{G}_t$-expectation: $d\nu_t = dg_t - E[dg_t \mid \mathcal{G}_t]$. The vector innovations process is easily partitioned into a composite vector with components readily identified with the sensor types: The modal-state innovation increment is $d\nu_\phi = dz_t - E[dz_t \mid \mathcal{G}_t]$, and the base-state innovation increment is $d\nu_x = dy_t - E[dy_t \mid \mathcal{G}_t]$. The innovations process is a $\mathcal{G}_t$-martingale: $E[d\nu_\phi \mid \mathcal{G}_t] = 0$ and $E[d\nu_x \mid \mathcal{G}_t] = 0$.

The observation processes can be written in two ways:

$$dy_t = H(x_t + \chi\phi_t)\,dt + dn_t, \tag{2.4}$$

$$dz_t = h'\phi_t\,dt + d\eta_t, \tag{2.5}$$

where $\{\mathrm{vec}(n_t, \eta_t)\}$ is an $\mathcal{F}_t$-martingale, and

$$dy_t = H(\hat{x}_t + \chi\hat{\phi}_t)\,dt + d\nu_x, \tag{2.6}$$

$$dz_t = h'\hat{\phi}_t\,dt + d\nu_\phi, \tag{2.7}$$

where $\nu_t = \mathrm{vec}(\nu_x, \nu_\phi)$ is a $\mathcal{G}_t$-martingale. The processes $\{\nu_x\}$ and $\{n_t\}$ are continuous, and their predictable quadratic variation is simply written

$$\begin{aligned} d\langle \nu_x, \nu_x; \mathcal{G}_t\rangle_t &= d\langle n, n; \mathcal{F}_t\rangle_t \\ &= R_x\,dt. \end{aligned} \tag{2.8}$$

The optional quadratic variation of both processes is $R_x t$, and similarly for $\{\nu_\phi\}$ if $\{\eta_t\}$ is a Brownian motion. In this event, the predictable quadratic variation of the noise in the modal sensor is

$$\begin{aligned} d\langle \eta, \eta; \mathcal{F}_t\rangle_t &= d\langle \nu_\phi, \nu_\phi; \mathcal{G}_t\rangle_t \\ &= R_\phi\,dt. \end{aligned} \tag{2.9}$$

We will assume that all of the $\mathcal{G}_t$-martingales we will encounter can be expressed as integrals with respect to $\{\nu_t\}$. (For related results see [Ell82, p. 279] and [WH85, Chapter 6, Proposition 7.3].)

2.2 Modal Estimation

Before we develop the full zygostate estimation algorithm, consider the simpler problem in which we seek only the regime probabilities. Two models have been proposed for the modal measurement, but let us focus on the case in which the measurement noise is white and Gaussian (see (1.51)). Observe that†

$$E[d\hat{\phi}_t \,|\, \mathcal{G}_t] = E[\hat{\phi}_{t+dt} - \hat{\phi}_t \,|\, \mathcal{G}_t]$$
$$= E[\phi_{t+dt} - \phi_t \,|\, \mathcal{G}_t] = E[d\phi_t \,|\, \mathcal{G}_t]. \tag{2.10}$$

Consequently, the increment in $\{\hat{\phi}_t\}$ can be expressed as

$$d\hat{\phi}_t = E[d\phi_t \,|\, \mathcal{G}_t] + d\mu_t, \tag{2.11}$$

where $\{\mu_t\}$ is a $\mathcal{G}_t$-martingale.

Any martingale increment can be expressed as a $\mathcal{G}_t$-multiple of the innovations increment:

$$d\hat{\phi}_t = Q'\hat{\phi}_t\,dt + \gamma_t d\nu_t, \tag{2.12}$$

where $\{\gamma_t\}$ is a $\mathcal{G}_t$-predictable gain process. Equation (2.12) is of the form we seek – a predictor $(Q'\hat{\phi}_t\,dt)$ corrector $(\gamma_t\,d\nu_t)$ with the innovations increment used for the latter (compare (1.18) of the Kalman filter). The correction term can be partitioned compatibly with the innovations process: $\gamma_t\,d\nu_t = \gamma_{\phi\phi}\,d\nu_\phi + \gamma_{\phi x}\,d\nu_x$ and

$$d\hat{\phi}_t = Q'\hat{\phi}_t\,dt + \gamma_{\phi\phi}\,d\nu_\phi + \gamma_{\phi x}\,d\nu_x. \tag{2.13}$$

The algorithm will be complete when the gain matrix, $\gamma_t = (\gamma_{\phi\phi}, \gamma_{\phi x})$, is determined. To find the gain, an approach used by [Ell82, Chapter 18] and [WH85, Chapter 7] is useful. Both $\{\phi_t\}$ and $\{g_t\}$ are semimartingales (see [WH85, Chapter 6, Equation (6.2)]):

$$d(\phi_t g_t') = (d\phi_t)g_t' + \phi_t(dg_t)' + (d\phi_t)dg_t'. \tag{2.14}$$

The modal process is purely discontinuous and the observation is continuous. Therefore $(d\phi_t)dg_t' = 0$. From (1.12), we have

$$(d\phi_t)g_t' - (Q'\phi_t\,dt + dm_t)g_t'$$
$$= Q'\phi_t g_t'\,dt + d\mu_t,$$

where here and in what follows $\{\mu_t\}$ is the matrix $\mathcal{F}_t$-martingale appropriate to the application. The second term in (2.14) can be written

$$\phi_t(dg_t)' = (\phi_t\phi_t'h\,dt + \phi_t\,d\eta_t', \phi_t\chi_t'H'\,dt + \phi_t\,dn_t').$$

Combining these equations, we obtain

$$d(\phi_t g_t') = (\phi_t\phi_t'h + Q'\phi_t z_t', \phi_t\chi_t'H' + Q'\phi_t y_t')\,dt + d\mu_t. \tag{2.15}$$

Taking the $\mathcal{G}_t$-expectation of (2.15) gives

$$E[d(\phi_t g_t') \,|\, \mathcal{G}_t]/\,dt = ((R_{\phi\phi}h + Q'\hat{\phi}_t)z_t', (R_{\phi\chi}H' + Q'\hat{\phi}_t)y_t'), \tag{2.16}$$

† Because $\mathcal{G}_t \subset \mathcal{G}_{t+dt}$, $E[\hat{\phi}_{t+dt} \,|\, \mathcal{G}_t] = E[E[\phi_{t+dt} \,|\, \mathcal{G}_{t+dt}] \,|\, \mathcal{G}_t] = E[\phi_{t+dt} \,|\, \mathcal{G}_t]$.

where $R_{\phi\phi}(t)$ is the $\mathcal{G}_t$-correlation of $\phi_t (R_{\phi\phi}(t) = E[\phi_t \phi_t' \mid \mathcal{G}_t])$, and $R_{\phi\chi}$ is the $\mathcal{G}_t$-cross correlation of ϕ_t and χ_t $(R_{\phi\chi}(t) = E[\phi_t \chi_t' \mid \mathcal{G}_t])$. The corresponding covariances are $P_{\phi\phi}$, the $\mathcal{G}_t$-covariance of ϕ_t $(P_{\phi\phi}(t) = E[\tilde{\phi}_t \tilde{\phi}_t' \,|\, \mathcal{G}_t])$, and $P_{\phi\chi}$, the $\mathcal{G}_t$-cross covariance of ϕ_t and χ_t $(P_{\phi\chi}(t) = E[\tilde{\phi}_t \tilde{\chi}_t' \,|\, \mathcal{G}_t])$.

We can express $E[d(\phi_t g_t') \mid \mathcal{G}_t]$ in another way. The measurement noises are independent. From (2.13) we have

$$(d\hat{\phi}_t)\, dg_t' = (\gamma_{\phi\phi}\, d\nu_\phi\, d\eta_\phi', \gamma_{\phi x}\, d\nu_x\, d\eta_x'),$$

which can be simplified to

$$(d\hat{\phi}_t)\, dg_t' = (\gamma_{\phi\phi}\, d\eta_\phi\, d\eta_\phi', \gamma_{\phi x}\, d\eta_x\, d\eta_x').$$

Also,

$$(d\hat{\phi}_t) g_t' = Q'\hat{\phi}_t g'\, dt + d\mu_t,$$
$$\hat{\phi}_t (dg_t') = (\hat{\phi}_t \phi_t' h, \hat{\phi}_t \chi_t' H')\, dt + d\mu_t.$$

Collecting the terms gives

$$d(\hat{\phi}_t g_t') = (\gamma_{\phi\phi} R_\phi + \hat{\phi}_t \phi_t' h + Q'\hat{\phi}_t z_t', \gamma_{\phi x} R_x + \hat{\phi}_t \chi_t' H' + Q'\hat{\phi}_t y_t')\, dt + d\mu_t.$$

Taking the $\mathcal{G}_t$-expectation of this, we obtain

$$E[d(\hat{\phi}_t g_t') \,|\, \mathcal{G}_t]/\, dt = (\gamma_{\phi\phi} R_\phi + \hat{\phi}_t \hat{\phi}_t' h + Q'\hat{\phi}_t z_t', \gamma_{\phi x} R_x + \hat{\phi}_t \hat{\chi}_t' H' + Q'\hat{\phi}_t y_t'). \tag{2.17}$$

The *predictable compensators*, (2.16) and (2.17), must be equal, and so

$$(\gamma_{\phi\phi} R_\phi + \hat{\phi}_t \hat{\phi}_t' h + Q'\hat{\phi}_t z_t', \gamma_{\phi x} R_x + \hat{\phi}_t \hat{\chi}_t' H' + Q'\hat{\phi}_t y_t') = ((R_{\phi\phi} h + Q'\hat{\phi}_t) z_t', (R_{\phi\chi} H' + Q'\hat{\phi}_t) y_t').$$

From this it follows that

$$\gamma_{\phi\phi} = P_{\phi\phi} h R_\phi^{-1}, \quad \text{and} \tag{2.18}$$

$$\gamma_{\phi x} = P_{\phi\chi} H' R_x^{-1}. \tag{2.19}$$

Substituting this into (2.13) yields

$$d\hat{\phi}_t = Q'\hat{\phi}_t\, dt + P_{\phi\phi} h R_\phi^{-1} d\nu_\phi + P_{\phi\chi} H' R_x^{-1}\, d\nu_x. \tag{2.20}$$

Equation (2.20) is the dynamic equation of the $\mathcal{G}_t$-conditional regime probabilities with the corrector given explicitly by

$$P_{\phi\phi} h R_\phi^{-1} d\nu_\phi + P_{\phi\chi} H' R_x^{-1} d\nu_x.$$

The probabilities propagate forward using the transition dynamics of the modal-state. The correction gain increases with uncertainty ($P_{\phi\phi}$), with observation gain (h), and inversely with observation noise (R_ϕ).

Equation (2.20) appears to be the algorithm we seek. It is a stochastic differential equation that generates the $\mathcal{G}_t$-regime probabilities (a semimartingale) as a function of the observation processes. It entails about the same level of complexity found in the Kalman filter. By analogy with the Kalman filter, the next step in deriving the modal filter would be to find the error covariances. In (2.20) there are two covariances required. The first is

$$P_{\phi\phi}(t) = E[(\phi_t - \hat{\phi}_t)(\phi_t - \hat{\phi}_t)' \mid \mathcal{G}_t] = \text{diag}(\hat{\phi}_t) - \hat{\phi}_t\hat{\phi}_t'. \tag{2.21}$$

The Kalman filter requires the solution to an auxiliary set of differential equations for $\{P_{xx}\}$ (see (1.19)). Here the modal-state error covariance matrix can be expressed as an algebraic function of the regime probabilities themselves, obviating the need to solve a matrix differential equation.

To implement (2.20), only $\{P_{\phi x}\}$ is needed. Unfortunately, we have no mechanism to compute it. The closure achieved in the Kalman filter depends upon the Gaussian structure of the problem. The presence of non-Gaussian elements in our problem precludes a similar closed-form solution here. Thus, (2.20) fails as a solution to the modal estimation problem.

The dual-sensor measurement architecture used in this book is motivated by the need for expeditious identification of the regime, an identification that is difficult to achieve from $\{y_t\}$. In many applications, the modal measurement is selected for the express purpose of tracking $\{\phi_t\}$. In such systems, high level data fusion algorithms use $\{z_t\}$ to determine $\{\phi_t\}$. The regime path so identified is then used in some sort of *certainty equivalence* algorithm to estimate $\{x_t\}$. Motivated by this partitioning of function, we can produce a good approximation to the modal estimator dynamics even if we ignore the base-state observation:

$$d\hat{\phi}_t \approx Q'\hat{\phi}_t\,dt + P_{\phi\phi}hR_\phi^{-1}\,d\nu_\phi. \tag{2.22}$$

Equation (2.22) is a finite-dimensional, albeit nonlinear, stochastic differential equation for the conditional modal probabilities and provides an easily implemented modal-state estimator [Ell82, Example 18.18].

2.3 The Polymorphic Estimator

The development of the modal estimator as given above illustrates the *moment formulation* of the estimation problem in which the complete algorithm includes the equations of evolution of both the principal $\mathcal{G}_t$-moments of the zygostate (e.g., $\{\hat{\phi}_t\}$) and those auxiliary moments required to compute the principal moments (e.g., $\{P_{\phi\phi}\}$

and $\{P_{\phi x}\}$). When the auxiliary moments can be displayed as an algebraic function of moments already calculated (e.g., $P_{\phi\phi}$ in (2.21)), the moment is called *sequent*. Those moments that must be found by integrating their equations of evolution are called *canonical* (e.g., $\hat{\phi}_t$ in (2.20)). Even though the observation model is linear in the relevant states, the equation of evolution of a specific $\mathcal{G}_t$-moment tends to involve higher order canonical $\mathcal{G}_t$-moments. And these in turn tend to involve yet higher order canonical $\mathcal{G}_t$-moments and so on.

There are situations in which this unending regress is avoided. In the LGM $\mathcal{G}_t^\phi$-estimation problem, the Kalman filter provides $d\hat{x}_t$ as a function of the second central $\mathcal{G}_t^\phi$-moment, P_{xx}(t). The stochastic equation for $\{E[\tilde{x}_t\tilde{x}_t'\tilde{x}_i \mid \mathcal{G}_t]; i \in \mathbf{n}\}$ would seemingly be required next. Fortuitously, this third central moment is known to be zero. From this it follows that $\{P_{xx}\}$ is a self-contained $\mathcal{G}_t^\phi$-semimartingale: The moment process terminates at the second step.

When there is no base-state measurement ($\mathcal{G}_t = \mathcal{Z}_t$), the analysis in the previous section can be extended to give a finite-dimensional equation for $\{\hat{x}_t\}$ [Bjo82]. Unfortunately, with no base-state updates, the base-state error covariance tends to grow large over time. The filter proposed in [Bjo82] can be modified to accept $\{y_t\}$, thereby reducing the growth of $\{P_{xx}\}$ [DB94, DB96].

In the $\mathcal{G}_t^\phi$-estimation problem, $\{P_{xx}\}$ is not a function of the base-state measurement at all. Let the filtration generated by $\{\phi_t\}$ be denoted by $\{\mathcal{O}_t\}$. Then $\{P_{xx}\}$ is independent of $\{\mathcal{Y}_t\}$ given $\{\mathcal{O}_t\}$. Specifically, dP_{xx}(t) in (1.19) has no term involving $d\nu_x$. Let us call moments for which the $\mathcal{G}_t^\phi$-increment is independent of the increment in the base-state innovations *ϕ-dominant* moments (PD-moment). Since $\{\phi_t\}$ is $\mathcal{O}_t$-adapted, the modal-state probabilities are trivially PD-moments. The $\mathcal{G}_t$-regime probabilities do depend upon both observation processes, but if the contribution from $\{d\nu_x\}$ is small, the only higher moment required for computing $\{\hat{\phi}_t\}$ is a sequent moment, and the estimation algorithm is closed. In what follows we will utilize the following approximation:

> **Approximation** *If $P(t)$ is a PD-moment, the contribution to $dP(t)$ from $d\nu_x$ can be ignored as compared to that of $d\nu_\phi$.*

The approximation does not imply that $\{P(t)\}$ is $\mathcal{Z}_t$-adapted. Nor does it imply that the quality of the modal sensor is particularly good. It does imply that information regarding the regime is more immediate in $\{z_t\}$ than it is in $\{y_t\}$. For example, consider an application in which we wish to track an aircraft (generate $\{\hat{x}_t\}$) and simultaneously identify the aircraft type (generate $\{\hat{\phi}_t\}$). A high quality radar ($\{y_t\}$) might suffice for tracking, and a long interval of $\{\hat{x}_t\}$ could be postprocessed to yield the value of the category variable. But how much better it would be to have a visual image of the aircraft. From an image, target identification might be immediate. Target identification from a sequence of distorted images is still faster than identification from the radar process alone. The fact that the radar sequence is neglected in modal estimation does not presuppose that the radar SNR ratio is low (a well designed tracker may follow the aircraft effectively even though the target type is uncertain) nor that the SNR of the imager is high (if high, fewer data frames will be require to classify the target). Neglecting the radar in target recognition acknowledges the fact that target type does not manifest itself clearly in $\{y_t\}$.

With the approximation, it is possible to derive a finite-dimensional equation for the zygostate estimate. This estimator is called the *polymorphic estimator* (PME). The details of the development are placed in the appendix. In this chapter we will present the algorithm and place it within the context of more conventional algorithms.

The PME can be developed for either a continuous or a classificational modal observation process. We will describe only the latter. The model given in (1.53) provides considerable flexibility and is particularly well suited to systems in which the mode is a category variable and the modal sensor is a sophisticated classifier. The PME utilizes the modal observations after first transforming them. Let $\{\vartheta_t\}$ be a process that is constant between modal observations and that has increments $\Delta\vartheta_t = h(\hat{\boldsymbol{\lambda}}_t^{-1} * \Delta z_t)$, where $\hat{\boldsymbol{\lambda}}_t^{-1}$ is understood componentwise. The jumps in $\{\vartheta_t\}$ have a natural interpretation. Suppose the modal observation is $\mathbf{e}_i$. Then $\Delta\vartheta_t$ is the ith row of $\mathbf{P}$ weighted by $\lambda/\hat{\boldsymbol{\lambda}}_i > 1$. If the observation is confirming ($\hat{\phi}_t \approx \Delta z_t = \mathbf{e}_i$), and if the discernibility matrix approximates the identity (good modal distinguishability), then $\lambda \approx \hat{\boldsymbol{\lambda}}_i$ and $\Delta\vartheta_t \approx \mathbf{e}_i$. If the observation is disconfirming ($\Delta z_t = \mathbf{e}_i$; $\hat{\phi}_t \approx \mathbf{e}_j$; $\lambda \gg \hat{\boldsymbol{\lambda}}_i$; $\Delta\vartheta_t \approx (\lambda/\hat{\boldsymbol{\lambda}}_i)\mathbf{e}_i$), the emphasis accorded to $\phi_t = \mathbf{e}_i$ is much greater than that given to $\phi_t = \mathbf{e}_i$ in the previous case because of the $\lambda/\hat{\boldsymbol{\lambda}}_i$ multiplier: Unanticipated events are given a higher weighting in $\{\vartheta_t\}$ than are confirming events.

A particular notation makes the PME algorithm more intuitive in what follows. Recall that x_t (respectively ϕ_t) is the base-state (respectively the modal-state) with $\mathcal{G}_t$-mean and error given by $\hat{x}_t$ and $\tilde{x}_t$ (respectively $\hat{\phi}_t$ and $\tilde{\phi}_t$). The second moments of these variables are

$$R_{x\phi}(t) = E[x_t\phi_t' \,|\, \mathcal{G}_t],$$

$$P_{x\phi}(t) = E[\tilde{x}_t\tilde{\phi}_t' \,|\, \mathcal{G}_t],$$

with similar definitions for P_{xx}, R_{xx}, etc. Let us extend this notational convention to higher moments as follows. Consider the two vectors x_t and ϕ_t and the scalar ϕ_i.

Let us display third moments as

$$R_{x\phi\phi_i}(t) = E[x_t\phi_t'\phi_i \,|\, \mathcal{G}_t],$$

$$P_{x\phi\phi_i}(t) = E[\tilde{x}_t\tilde{\phi}_t'\tilde{\phi}_i \,|\, \mathcal{G}_t],$$

with similar definitions for $P_{xx\phi_i}$, $R_{xx\phi_i}$, etc.

This notation can be extended to compound events; for example

$$P_{(xx\phi_i)\phi_m} = E[(\widetilde{x_t x_t'\phi_i})\tilde{\phi}_m \,|\, \mathcal{G}_t],$$

$$P_{(x\tilde{\phi}_i)x\phi_m} = E[(\widetilde{x_t\tilde{\phi}_i})\tilde{x}_t'\tilde{\phi}_m \,|\, \mathcal{G}_t].$$

The parentheses in the subscript act as delimiters. This can be extended to fourth moments; for example,

$$P_{xx\phi_i\phi_m} = E[\tilde{x}_t\tilde{x}_t'\tilde{\phi}_r\tilde{\phi}_m \,|\, \mathcal{G}_t].$$

The PME provides the dynamics of evolution for five canonical zygostate moments: the $\mathcal{G}_t$-mean of the zygostate, $\hat{x}_t$; $\hat{\phi}_t$; and three higher order central moments: $P_{x\phi}(t)$, $P_{xx}(t)$, and $P_{xx\phi_m}(t)$. Alternatively, the Kalman filter requires the calculation of only two canonical moments: $\hat{x}_t$ and $P_{xx}(t)$. In the Kalman filter, other error moments can be computed from these two because the base-state and error are jointly Gaussian. We are not so fortunate in the hybrid case because the $\mathcal{G}_t$-zygostate distribution is not a member of a common parametric family. However, it has already been noted that $\{P_{xx}\}$ is a PD-moment. Clearly $\{P_{x\phi}\}$ and $\{P_{xx\phi_m}\}$ are as well since both are identically zero under $\mathcal{G}_t^{\phi}$.

The PME is a finite-dimensional algorithm that generates the five canonical moments. The defining equations involve a potpourri of sequent moments. In Appendix 1 these sequent moments are tabulated. In this chapter, let us look at an abridged version of the PME in which there is neither plant state set point ($\chi = 0$) nor translation at modal transition ($\rho(i, j) \equiv 0$). There is a variable actuating signal and there may be rotation or scaling at a modal transition. The general case is presented in Appendix 1 and will be used where needed in the chapters that follow.

2.4 The Abridged PME: Time-Continuous Plant and Observation

The case of a time-continuous plant with continous plant state and time-continuous observation process is frequently treated in tutorials on Kalman filtering. Because of its familiarity and relative simplicity, it presents an attractive context in which to compare the Kalman filter and the PME . For this application, the PME comprises the following coupled stochastic differential equations:

modal estimation

Between modal measurements:

$$d\hat{\phi}_t = Q'\hat{\phi}_t\,dt. \tag{2.23}$$

At a modal measurement:

$$\hat{\phi}_t^+ = \hat{\phi}_t^- * \Delta\vartheta_t. \tag{2.24}$$

base-state estimation

Between modal measurements:

$$d\hat{x}_t = \sum_i \left(A_i R_{x\phi_i} + B_i\left(u_t\hat{\phi}_i - \upsilon P_{\phi\phi_i}\right)\right) dt + P_{x\chi} H' R_x^{-1} d\nu_x. \tag{2.25}$$

At a modal measurement:

$$\Delta\hat{x}_t = P_{x\phi}\Delta\vartheta_t. \tag{2.26}$$

base-state covariance

Between modal measurements:

$$\frac{d}{dt}P_{xx} = \left[\sum_i \left(A_i P_{(x\phi_i)x} + B_i\left(u_t P_{\phi_i x} - \upsilon P_{(\tilde{\phi}\phi_i)x}\right)\right)\right] + [\cdot]' - \gamma_x R_x \gamma_x'$$
$$+ \sum_i R_\chi(i)\hat{\phi}_i + \sum_{i,l} Q_{il} M(i,l) R_{xx\phi_i} M(i,l)'. \tag{2.27}$$

At a modal measurement:

$$\Delta P_{xx} = -\Delta\hat{x}\Delta\hat{x}' + \sum_k P_{xx\phi_k}\Delta\vartheta_k. \tag{2.28}$$

base-state, modal-state cross covariance

Between modal measurements:

$$\frac{d}{dt}P_{x\phi} = \sum_i \left(A_i P_{(x\phi_i)\phi} + B_i\left(u_t P_{\phi_i\phi} - \upsilon P_{(\tilde{\phi}\phi_i)\phi}\right)\right)$$
$$-\gamma_x H P_{\chi\phi} + P_{x\phi} Q + \sum_{i,l} Q_{il} M(l,l) R_{x\phi_i}. \tag{2.29}$$

At a modal measurement:

$$\Delta P_{x\phi} = -\Delta\hat{x}\Delta\hat{x}' + \sum_k P_{x\phi\phi_k}\Delta\vartheta_k, \tag{2.30}$$

where $\gamma_x = P_{x\chi}H'R_x^{-1}$. Equations (2.23)–(2.30) along with the equation for $\{P_{xx\phi_j};\ j \in \mathbf{S}\}$ delineate the PME for this particular case.

It is interesting to compare and contrast the PME with the $\mathcal{G}_t^\phi$-Kalman filter. The modal-state is known in the Kalman filter and does not require an estimate. The update in $\{\hat{\phi}_t\}$ has a form not unlike that found in the IMM (see (1.44) and (1.45)). Unanticipated observations are emphasized by $\{\Delta\vartheta_t\}$. The equations of evolution of the modal-state estimate are highly nonlinear.

The base-state estimate includes terms like those found in the $\mathcal{G}_t^\phi$-Kalman filter. The state matrix, $\boldsymbol{A}_i$, includes both the intramodal state matrix, A_i, and a contribution from the state discontinuity event, $\sum_l Q_{il}M(i,l)$. Formal inclusion of the discontinuity into a $\mathcal{G}_t^\phi$-Kalman filter would suggest a term $\sum_i \boldsymbol{A}_i x\phi_i\, dt$. Were we to neglect the correlation of zygostate errors, the base-state extrapolation might be expected to include $\sum_i \boldsymbol{A}_i\hat{x}\hat{\phi}_i\, dt$. However, the geometry of base-state $\times$ modal-state induces correlations, and the PME identifies the proper extrapolation to be $\sum_i \boldsymbol{A}_i R_{x\phi_i}\, dt$. When the modal estimate is a good one ($P_{x\phi} \approx 0$), $\sum_i \boldsymbol{A}_i R_{x\phi_i} \approx \sum_i \boldsymbol{A}_i\hat{x}\hat{\phi}_i$, and the PME mimics the Kalman filter.

The actuating signal makes a twofold contribution to $d\hat{x}_t$. By analogy with the Kalman filter, we would anticipate the contribution from the regulating signal would be a term $\sum_i B_i u_t\hat{\phi}_i\, dt = \hat{B}_t u_t\, dt$. This term does in fact appear in the PME . The influence of failure to apply the proper feedforward actuation is harder to intuit from the Kalman filter since this issue is ignored in the LGM model. The term $\sum_i B_i \boldsymbol{v} P_{\phi\phi_i}$ quantifies this effect. When the modal estimate is a good one (i.e., the sum is near zero), the feedforward contribution is negligible. Indeed, when the modal estimate is good, (2.25) is identical to (1.18).

The correction attributable to the base-state measurement is like that found in the Kalman filter with the not surprising difference that P_{xx} in the Kalman filter is replaced by $P_{x\chi}$ ($P_{x\chi} = P_{xx} + P_{x\phi}\boldsymbol{\chi}'$). The update attributable to the modal measurement, $\Delta\hat{x}_t = P_{x\phi}\Delta\vartheta_t$, has no analogue in the Kalman filter, though again if $P_{\phi\phi}$ (and hence $P_{x\phi}$) is small, this correction is negligible. In summary, the base-state estimator in the PME is of essentially the same complexity found in the Kalman filter and in fact reduces to the Kalman filter during extended modal sojourns when the regime identification has been accomplished.

The equation for the base-state error covariance has both anticipated and unforeseen terms. The base-state measurement reduces the error covariance in both the Kalman filter and the PME ($-\gamma_x R_x\gamma_x'\, dt$), though γ_x has a slightly different definition in the PME. Equation (1.19) contains a term related to the plant noise, $\sum_i R_\chi(i)\phi_i\, dt$, which is replaced in the PME by its mean $\sum_i R_\chi(i)\hat{\phi}_i\, dt$ (as expected). The Kalman filter contains another term, $\sum_i A_i P_{xx}\phi_i\, dt$. The PME replaces A_i with $\boldsymbol{A}_i$ (as expected). It is harder to intuit what moment would replace $P_{xx}\phi_i$. (It is $P_{(x\phi_i)x}$.) During long modal sojourns when the intercategory correlations are small, $\sum_i \boldsymbol{A}_i P_{(x\phi_i)x} \approx \sum_i \boldsymbol{A}_i P_{xx}\hat{\phi}_i$.

The error covariance in the Kalman filter has no control dependence because the actuating signal introduces no additional uncertainty. The PME has two such terms. The first is related to the uncertainty in the control matrix: $\sum_i B_i u_t P_{\phi_i x}\, dt$. If the control matrix is not mode dependent ($B_i \equiv B$), $\sum_i B_i u_t P_{\phi_i x} = 0$, or if the mode is known with confidence, $\sum_i B_i u_t P_{\phi_i x}$ is small. The second term is related to the feedforward signal: $-\sum_i B_i \boldsymbol{v} P_{(\tilde{\phi}\phi_i)x}\, dt$. This term becomes small under conditions of modal certainty as does the control aggregate.

Another term appearing in dP_{xx} is $\sum_{i,l} Q_{il}M(i,l)R_{xx\phi_i}M(i,l)'\, dt$. This positive contribution has no analogue in the Kalman filter but can be likened to pseudonoise. A similar replacement, called the white-noise equivalent of the discontinuity, has appeared in algorithms developed for applications of this type. However, the term used in the PME differs from a white-noise equivalent in that it is directional ($M(i,l)$), dependent on the likelihood of a modal transition (Q_{il}), and data dependent ($R_{xx\phi_i}$). When the mode is identified with confidence ($\hat{\phi}_t \approx \mathbf{e}_i$), this discontinuity term becomes

$$\sum_{i,l} Q_{il}M(i,l)R_{xx\phi_i}M(i,l)'\, dt \approx \sum_{l} Q_{il}M(i,l)R_{xx}M(i,l)'\, dt.$$

Despite the fact that $\{P_{xx}\}$ is a PD-process, $\{R_{xx\phi_i}\}$ (and hence $\{P_{xx}\}$) is not $\mathcal{Z}_t$-adapted.

The modal update does not appear in the Kalman filter. The increment in the base-state estimate is reflected in a correction in P_{xx}: $\Delta P_{xx} = -\Delta\hat{x}\Delta\hat{x}' + (\cdot)$. There is another term that relates the modal measurement to $\{P_{xx}\}$ though a third mixed moment: $\Delta P_{xx} = (\cdot) + \sum_k P_{xx\phi_k}\Delta\vartheta_k$. Again it should be noted that when the regime is known, these corrections become small. When $\hat{\phi}_t \approx \mathbf{e}_i$,

$$\frac{d}{dt}P_{xx} = \boldsymbol{A}_i P_{xx} + P_{xx}\boldsymbol{A}_i' - \gamma_x R_x \gamma_x' + R_\chi(i) + \sum_l Q_{il}M(i,l)R_{xx}M(i,l)'$$

(compare (1.19)).

There is no equation for $\{P_{x\phi}\}$ in the Kalman filter. The base-state, modal-state covariance appears in the base-state estimation algorithm as a gain for the modal measurement update. It appears in the base-state error covariance as a factor in the term related to the regulating signal. The base-state measurement reduces the cross covariance much as it does in the Kalman filter ($-\gamma_x H P_{x\phi}\, dt$). Equation (2.27) contains no term related to the plant noise, but this is replaced in the PME by a term related to the jump probabilities ($P_{x\phi}Q\,dt$). The collection, $\sum_i (A_i P_{(x\phi_i)x} + B_i(u_t P_{\phi_i\phi} - \upsilon P_{(\bar{\phi}\phi_i)\phi}))$, has the form of an analogous term in P_{xx}. The modal update in $\{P_{x\phi}\}$ is the analogue of that found in $\{P_{xx}\}$. When $\hat{\phi}_t \approx \mathbf{e}_i$,

$$\frac{d}{dt}P_{x\phi} = A_i P_{x\phi} - \gamma_x H P_{\chi\phi} + P_{x\phi}Q + \sum_l Q_{il}M(i,l)\hat{x}\mathbf{e}_l'.$$

The PME is not a *small-noise* approximation to the optimal nonlinear filter: In high SNR environments, ad hoc modifications of the Kalman filter suffice. The PME is unique in the way it integrates the nonlinear state dynamics into the estimates. For example, if the correlation between the state categories is ignored along with the influence of the control, (2.25) reduces to $\frac{d}{dt}\hat{x}_t = \hat{A}_t\hat{x}_t + \gamma_x\dot{\nu}_x$, with the equation for $\{P_{xx}\}$ matching the Kalman filter with "average" dynamics. However, the PME *does* utilize the geometry of the state path. In this way, the base-state error covariance is contingent upon both the estimated base-state path and the modal sequence. The PME neither averages the $\{\phi_t\}$ paths nor treats $\{\phi_t\}$ as additive. The correlations between errors in estimating the base- and modal-states are used directly to update the base-state estimate. This modal measurement to base-state update is possible only because the relevant second moment is computed as part of the PME.

In many applications, the base-state observations are discrete: ($y[k] = H\chi[k] + n[k]$; $E[n[k]n[k]'] = R_x > 0$). The PME can be adjusted to accommodate these measurements as is done in the Kalman filter. Between observations the $\mathcal{G}_t$-moments are extrapolated without dependence on the base-state observation ($R_x = \infty$). At a base-state measurement there is a correction. Observe that $E[(d\nu_x)d\nu_x' \,|\, \mathcal{G}_t^{\phi}]/\,dt = R_x$ for continuous observations and $E[\Delta\nu_k\Delta\nu_k' \,|\, \mathcal{G}_t^{\phi}] = HP_{\chi\chi}H' + R_x$ for discrete observations. In the PME with time-discrete base-state measurements, the gains are modified: $\gamma_x = P_{x\chi}H'(E[(d\nu_x)d\nu_x' \,|\, \mathcal{G}_t^{\phi}]/\,dt)^{-1}$ for continuous observations; $\gamma_x = P_{x\chi}H'E[\Delta\nu_k\Delta\nu_k' \,|\, \mathcal{G}_t^{\phi}]^{-1}$ for discrete observations. To implement the PME with discrete measurements, the same replacements will be made.

the PME: time-continuous plant; time-discrete measurement

Between observations:

$$\frac{d}{dt}\hat{\phi}_t = Q'\hat{\phi}_t,$$

$$\frac{d}{dt}\hat{x}_t = \sum_i M_1(i),$$

$$\frac{d}{dt}P_{x\phi} = \sum_i (M_2(i) + N_2(i)) + P_{x\phi}Q,$$

$$\frac{d}{dt}P_{xx} = \sum_i (M_3(i) + M_3(i)' + N_3(i) + R_\chi(i)\hat{\phi}_i).$$

At a modal observation:

$$\hat{\phi}^+ = \hat{\phi}^- * \Delta\vartheta,$$

$$\Delta\hat{x} = P_{x\phi}\Delta\vartheta,$$

$$\Delta P_{x\phi} = -\Delta\hat{x}\Delta\hat{\phi}' + \sum_k P_{x\phi\phi_k}\Delta\vartheta_k,$$

$$\Delta P_{xx} = -\Delta\hat{x}\Delta\hat{x}' + \sum_k P_{xx\phi_k}\Delta\vartheta_k.$$

At a base-state observation:

$$\Delta\hat{x} = \gamma_x \Delta\nu_x,$$

$$\Delta P_{x\phi} = -\gamma_x H P_{\chi\phi},$$

$$\Delta P_{xx} = -\gamma_x (H P_{\chi\chi} H' + R_x)\gamma_x'.$$

coefficient identities for the PME with time-discrete observations

$$\gamma_x = P_{x\chi} H'(H P_{\chi\chi} H' + R_x)^{-1},$$

$$M_1(i) = (A_i\hat{x}_t + B_i u_t)\hat{\phi}_i + A_i P_{x\phi_i} - B_i \upsilon P_{\phi\phi_i},$$

$$M_2(i) = (A_i\hat{x}_t + B_i u_t)P_{\phi_i\phi} + A_i\left(P_{x\phi\phi_i} + P_{x\phi}\hat{\phi}_i\right)$$

$$-B_i \upsilon P_{(\phi\phi_i)\phi},$$

$$N_2(i) = \sum_l Q_{il} M(i,l) R_{x\phi_i},$$

$$M_3(i) = (A_i\hat{x}_t + B_i u_t)P_{\phi_i x} + A_i\left(P_{xx\phi_i} + P_{xx}\hat{\phi}_i\right)$$

$$-B_i \upsilon P_{(\tilde{\phi}\phi_i)x},$$

$$N_3(l) = \sum_l Q_{il} M(i,l) R_{xx\phi_i} M(i,l)'.$$

3
Situation Assessment

3.1 Introduction to Situation Assessment

An abridged version of the polymorphic estimator was presented in Chapter 2. The PME is a finite-dimensional algorithm for generating the $\mathcal{G}_t$-conditional mean of the zygostate along with several moments that are interesting in their own right. The plant dynamics of the hybrid system are partitioned into a base-state, modal-state pair, and the observation has a compatible partition. In this chapter we will focus on the modal-state measurement subsystem in which the measurements are discrete.

The specific application to be considered is the study of situation assessment by human decision makers. Many geographically distributed systems include both human decision makers and a diverse collection of sophisticated hardware/software subsystems. With the unavoidable errors and distortions in data accumulation, transfer, and presentation, it is difficult to determine the appropriate human role, or even if the decision maker is properly filling the role given him. Athans referred to decision-making systems as "event driven" and observed that "the state variables . . . are both continuous and discrete" [Ath87]. In Athans's partitioning of the comprehensive state space, the discrete states represent global (or meta) states that modulate the local (or micro) aspects of the task environment. The decision maker's reaction to local phenomena tends to have a reflexive quality. It is in this reaction to macroevents that particularly human idiosyncrasies are manifest.

A primary task for a decision maker is to identify changing circumstances in an environment characterized by noise and clutter. This is called *situation assessment* and on its basis the decision maker takes actions or reports conclusions. The centrality of status identification is explicit in the recognition-primed decision-making (RPD) paradigm in which situation assessment is taken to be the primary cognitive task of the decision maker [Kle89, Kle91]. In RPD it is supposed that once the current status has been identified, the appropriate response is known by the decision maker from training and experience. RPD is based on the observation that human decision makers rely "on their abilities to recognize and appropriately classify a situation. Once they knew it was 'that' type of case, they usually also knew the typical way of reacting to it" [Kle89].

Creating a model flexible enough to represent the diverse population of human decision makers is difficult; decision makers have different styles, training, and temperament. The metastates are evolving variables. For example, if the metastate space is {friendly, hostile, neutral}, a situation classified as neutral for a period can suddenly become hostile. The metastates are also decision maker specific in a way that conventional system states are not. Position and velocity are objective quantities, but a situation that is hostile for one decision maker may be neutral to another. Hence, the metastate labels are descriptive, but they are neither absolute nor preestablished. Similarly, the decision maker perceives the tempo of an encounter both objectively and subjectively. The trained decision maker is expected to perform in diverse environments and over a wide range of scenarios. Realistic tests involving professionally trained decision makers are difficult to design and are costly to implement. The proper personnel are often not available when needed and are expensive when so utilized. This lack of authentic experimental capability precludes the multifaceted testing of different system configurations that are so common in electro-mechanical system design. Lacking empirical data, plausible, but ad hoc, system configurations are used all too frequently.

The PME can be used to create an analytical model of a decision maker performing a situation assessment task in a context requiring hierarchical processing. The modal-state estimation algorithm is a simple but formal description of the cognitive dynamics of a decision maker engaged in a task requiring situational identification in an environment containing considerable perceptual ambiguity, equivocal measurements, and sudden temporal change. In this modeling paradigm, the decision maker is viewed as comparing unprocessed observations with those predicted on the basis of a set of internal expectancies. He associates his current status with that model in a parametric family of models that most closely matches the anticipated with the actual observations.

The PME quantifies the dynamics of situation recognition within a natural taxonomy: complexity, tempo, and uncertainty. The *complexity* of a engagement is given by the number of alternative situational hypotheses (the metastates) acknowledged by the decision maker; the *tempo* is the pace of the metastate change, and *uncertainty* is related to the distinctiveness of the observation–metastate coupling. For example, a ship may engage a target. The target may be classified as friendly (metastate 1), hostile (metastate 2), or neutral (metastate 3). The prefix "meta" indicates that the state is classificatory. For each element in this trichotomy, it is the decision maker's *subjective* sense of the situational environment that is determinative rather than the objective environment. The confidence with which a decision maker responds to data depends upon the quality of the observation: both objective quality (determined by the available decision arts) and subjective (influenced by training, stress, and the credence the decision maker gives to his or her data source).

Experiments precluded by expense and scheduling must be replaced by simulation. Unfortunately, determining a tractable model of human decision maker response is difficult. This is due in large part to the fact that decision makers exhibit a wide range of behaviors as their tasks and operating environments change. In this chapter, we will explore the use of the PME in decision maker modeling.

3.2 Decision Maker Dynamics

In [SCK93], a simple mathematical model was proposed as a tool for quantifying the response of a human decision maker in a multimodal engagement. This work was based upon the aforementioned taxonomy in which the decision-making environment is quantified on the basis of tempo, uncertainty, and complexity [Vau90]. This partitioning is abstracted as follows:

complexity: A rational decision maker aggregates his environment in terms of a restricted taxonomy of metastates. These metastates form a set of alternative hypotheses concerning the status of the assigned task. Different decision makers partition their metaspace differently, and a given decision maker interprets his situation within his personal set of alternatives. Enumerate the hypotheses: $\{1, \ldots, S\} = \mathbf{S}$, and denote the current metastatus with a unit vector $\phi_t = \mathbf{e}_i$ if the ith metahypothesis is true at time t. An untrained decision maker will recognize few situations (S is small), whereas a trained individual can deal with a greater number. Hence, task complexity (the dimension of ϕ_t) has both objective and subjective components and is influenced by both experience and training.

tempo: The tempo of a decision-making task describes the pace at which metastate changes occur. Suppose that $\{\phi_t\}$ is a Markov process with $S \times S$ transition rate matrix Q. As with complexity, tempo is both objective and subjective. Even if the objective tempo is such that there are no changes in the metastate ($Q = 0$), the subjective sense of tempo may admit variation. For example, when the decision maker devalues old observations by comparison with current ones, he is acting as if the situational status could change even when he knows objectively that it cannot: When the metastate is known to be unchanging, the decision maker may still act as if it had a finite lifetime. Particularly under stressful conditions, this bias toward recent observations has been remarked upon.

uncertainty: The uncertainty regarding $\{\phi_t\}$ is resolved by the decision maker from spatiotemporal observations. It is accepted that "sensory data are decomposed into simpler elements concerning various attributes, which are analyzed and subsequently correlated by hierarchical processes of discrimination, recognition and classification" [Lig88]. To quantify the quality of an individual observation, suppose that each metahypothesis is associated with a perceptual signature by which it is identified; for instance, corresponding to $\phi_t = \mathbf{e}_i$ there is an associated measurement mark. The distinctiveness of the signature measures how easily the hypothesis can be identified from the observation. Note that this signature is as subjective and as idiosyncratic as the partitioning of the metastate space. A trained observer can detect and interpret nuances in a scene that completely escape the novice; the same objective stimulus will be weighted differently by different decision makers. Further, the strength of the signature may change as the encounter evolves. For example, an increase in decision maker stress may result in a coarser decomposition of the cognitive metaspace and a weaker response to stimuli, leading to a related change in the input–output behavior of the decision maker.

In the PME this is abstracted as follows:

metastate model

$$d\phi_t = Q'\phi_t\, dt + dm_t, \tag{3.1}$$

metastate measurement

$$dz_t = \lambda \mathbf{P}\phi_t\, dt + d\eta_t. \tag{3.2}$$

The process $\{\eta_t\}$ is an $\mathcal{F}_t$-martingale. The predictable quadratic variation is given in the Appendix 1 (see (A1.4)): $d\langle \eta, \eta; \mathcal{G}_t\rangle_t = R_\phi\, dt;\ R_\phi = \mathrm{diag}(\hat{\lambda}_t) > 0$. With a single source of measurement, the observation filtration is $\{\mathcal{Z}_t\}$.

The equation for the modal estimate is given in (2.23) and (2.24). The observation is first modified to create the piecewise constant process $\{\vartheta_t\}$ with increments

$$\Delta\vartheta_t = h(\hat{\lambda}_t^{-1} * \Delta z_t),$$

where $\hat{\lambda}_t^{-1}$ is interpreted componentwise. The equation for the metastate probabilities can then be written as shown:

metastate probability

Between metastate measurements:

$$d\hat{\phi}_t = Q'\hat{\phi}_t\,dt. \tag{3.3}$$

At a metastate measurement:

$$\hat{\phi}_t^+ = \hat{\phi}_t^- * \Delta\vartheta_t. \tag{3.4}$$

3.3 Order Bias in Human Decision Makers

An interesting empirical study illustrating a particular difficulty in decision maker modeling is presented in [AB91]. Adelman and Bresnick used a realistic experimental environment to test the way the order in which observations are received influences a decision maker's identification of his situational status. Trained air defense officers were placed in front of Patriot training simulators and were asked to classify a target as friend or foe. An officer was given five observations relating to the status of the target. An initial category measurement was made at target appearance. After an interval, the officer received an automatic friend-or-foe electronic (IFF) transmission. Further, the initial portion of the target path reflected on the nature of the target. These three data points were called the early-order data. Later, as the target approached the asset defended by the Patriot system, the observed path and the occurrence of jamming (or lack thereof) gave two more observations (late-order data) reflecting on the status of the target.

In the experimental protocol, the indicated observations could be received by the officers as a sequence of five individual data points, or it could happen that points two and three appeared simultaneously (early-order coincidence), and similarly with points four and five (late-order coincidence). Because of ambiguity in the observations, none of the elementary measurements was individually conclusive; each observation could correspond to either of the underlying hypotheses. Along with their other tasks, the subjects were asked to judge the likelihood that the target was friendly after each observation, and again at the conclusion of the exercise.

This experiment is interesting because of its naturalistic setting and the use of trained and motivated subjects – not student volunteers. The issues involved are concisely stated in [AB91]:

> There is substantial basic research ... demonstrating that people use heuristics ... to make many judgments and decisions .... For many tasks, these heuristics can result in systematic biases or errors in judgment.

Adelman and Bresnick proposed that

> when information is presented sequentially and a probability is obtained after each piece of information, people make new estimates by first anchoring on the current position, and then adjusting it by the degree to which the new information confirms or disconfirms this position. Moreover, the Hogarth–Einhorn model predicts that the greater the anchor, the greater the impact of the same piece of disconfirming information. For example, the higher ones probability that an unknown aircraft is a friend, the greater the negative impact of new information indicating it is a foe. Conversely, the smaller the anchor, the greater the impact of the same confirming information. Consequently, the order in which the same confirming and disconfirming information is sequentially presented is predicted to result in different final probability estimates.

Because the full data set is identical for any order of presentation, this effect (called *order bias*) is seen by many investigators as an egregious behavior, and one to be minimized if possible. After all, the final conclusion on whether to engage the target should be based upon the totality of measurements, and the order in which they are received should be superfluous.

Figure 3.1 portrays order bias in a decision maker. The probability that the target is friendly is labeled *sanguinity* in the figure. Suppose the target is initially thought to be more likely to be friendly. This is indicated by an initial sanguinity of 0.6. If the initial observation indicates friend (F shown in the figure with the icon ☺) the sanguinity indicator moves to 0.8, and a subsequent disconfirming observation (hostile H with icon ☹) moves sanguinity to 0.5. Alternatively, the (H,F) sequence

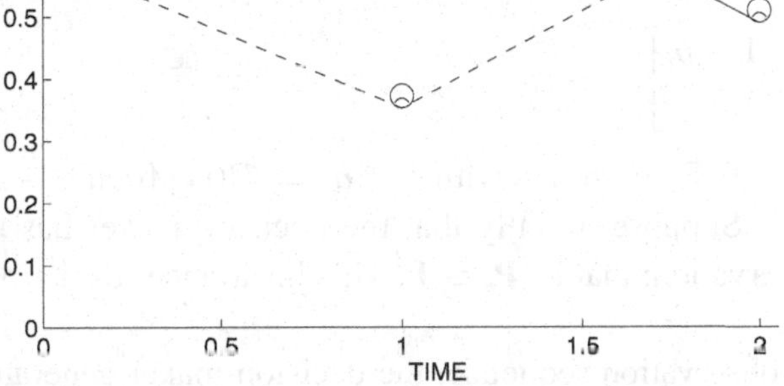

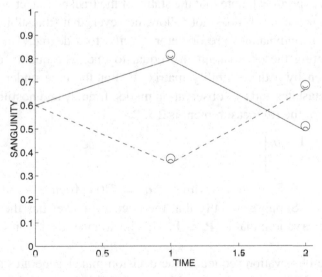

Figure 3.1. Sanguinity for observations F and H.

produces the lower curve in Figure 3.1, ending with a sanguinity of 0.7. Despite the fact that the status of the target does not change during the period, and the objective data are identical, the final appraisals differ with (H,F) being more strongly favorable to a nonthreat than is (F,H).

Figure 3.1 is a pictorial representation of the information order bias as described in the reference. Does order bias reveal a flaw in the way in which a trained decision maker integrates information, or is it a more basic behavioral characteristic of a decision maker in an ambiguous situation? Furthermore, if it is accepted as an authentic peculiarity of human response, can it be effectively delineated within existing modeling paradigms? Although order bias is an important issue and its quantification warrants considerable experimental effort, the expense of creating the environment described in [AB91] was considerable. The experiments themselves were difficult to control; for example, "despite all efforts of the participating personnel, two pieces of information (instead of one) appeared on the Patriot display at critical times prior to obtaining the participant's judgment" [AB91]. Because of the difficulties in carrying out the experiment (restricted though it was), Adelman and Bresnick could not explore the full variety of possible order effects that are of interest, and what is perhaps even more important, they were unable to examine the sensitivity of decision maker response to improved sensors and decision aids. Thus, it is of interest to see whether this (apparently irrational) order-bias behavior can be captured within the strict formalism of the PME.

The observations are generated in sequence and are so labeled. Because of errors in generation and interpretation, the observation marks are not a perfect indication of condition; the F mark may be assigned to either friend or foe, but the former is more likely. To abstract the experiment, note that the status of the tracked object will not change during the engagement. It does not follow, however, that (the subjective) $Q = 0$, because the metadynamics are operator specific to a degree. Time can be normalized by setting the observation rate equal to one. The quality of the measurements is given by a discernibility matrix, $\mathbf{P}$. For the case under study there are only two metastates and two observation marks: friendly and hostile. The discernibility matrix for the ith measurement is 2×2:

$$\mathbf{P}_i = \begin{bmatrix} a_i & 1 - b_i \\ 1 - a_i & b_i \end{bmatrix}, \tag{3.5}$$

where $a_i = \mathbf{P}_i(1, 1) = \mathcal{P}(\mathsf{F} \mid \text{friend})$ – with $1 - a_i = \mathcal{P}(\mathsf{H} \mid \text{friend})$ – and $b_i = \mathbf{P}_i(2, 2) = \mathcal{P}(\mathsf{H} \mid \text{foe})$. Suppose initially that the decision maker has the same confidence in each observation; that is, $\mathbf{P}_i \equiv \mathbf{P}$. This invariance restriction will be relaxed later.

On the basis of the observation sequence, the decision maker generates an assessment of the status of the engagement: $\mathcal{P}(\phi_t = \mathbf{e}_i \mid \mathcal{Z}_t)$. In particular, $(\hat{\phi}_t)_1$

is sanguinity in Figure 3.1. The PME has two free coefficients in Q (temporal freedom) and two free coefficients in $\mathbf{P}$ (observation freedom). Unfortunately, the sequential test of operator confidence performed in [AB91] is not sufficiently detailed to resolve the model uniquely. Because of the asymmetry noted in the experiment, an unbalanced confidence matrix will be used: $a = 0.68$ and $b = 0.77$; the probability of correctly interpreting a friendly target as such is 0.68 from a single observation.

This particularization of the PME requires only that Q be determined. If the decision maker expects the mean sojourn time in each metastate to be equal, this reduces to specifying the single parameter, Q_{11}. In [AB91], it was noted that the sequence (F, F, F, F, F) yields a sanguinity index of ($\hat{\phi}_1$) of 0.91, while (H, H, H, H, H) yields sanguinity index of 0.12. Selecting $Q_{11} = -0.2$ (mean sojourn in a metastate is 5), we show in Figure 3.2 the situational assessment graph for these two observation sequences.

Before considering order bias in this setting, it should be noted that the "update gain" in (3.4) uses the predictable version of $\{\hat{\phi}_t\}$: the value just before the observation. Predictability is a mathematical embodiment of decision maker *anchoring*. As the decision maker's confidence increases, the gain multiplying a confirming observation decreases. This effect is shown clearly in Figure 3.1, which was actually generated from Equations (3.3) and (3.4). Starting from a weakly favorable condition ($\hat{\phi}_1 = 0.6$), the upper curve shows the sequence confirm friend (F), disconfirm (H). Because the (F, H) sequence moves first into a high-$\hat{\phi}_1$ (low gain to a consequent F) region of the metaspace, the gain is higher for the subsequent

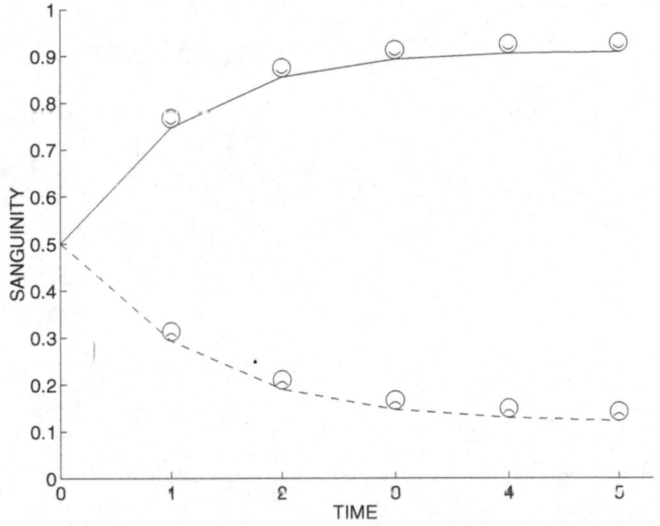

Figure 3.2. Sanguinity for two observation sequences: F, F, F, F, F and H, H, H, H, H.

disconfirm (high gain to the H actually received) than it would otherwise be. The reversed ordering of the same data, (H, F), moves the metastate down first, and the gain for the subsequent confirm is much higher. This results in the higher net increment in sanguinity as compared to (F, H).

Simultaneous receipt of data appears to violate the dynamic hypotheses that underlie the PME. It could be argued that multiple measurements are always separated by an nonzero interval, and they can therefore be treated sequentially. This is done in Kalman filtering applications in which a vector observation is reformed as a set of sequential scalar observations with a concomitant reduction in signal processing complexity. However, such an approach ignores hominal anchoring. The decision maker perceives propinquitous measurements as being simultaneous even when there is some minimal separation; the value of $\{\hat{\phi}_t\}$ is not updated between near coincident observations. The PME can be modified to account for coinstantaneous observations by enlarging the range space of $\{\Delta z_t\}$. For example, the observations could be ordered: $\mathbf{e}_1 = \mathsf{F}$, $\mathbf{e}_2 = \mathsf{FF}$, $\mathbf{e}_3 = \mathsf{FH}$, $\mathbf{e}_4 = \mathsf{HH}$, $\mathbf{e}_5 = \mathsf{H}$. A plausible $\mathbf{P}$ would be

$$\mathbf{P} = 0.5 \begin{bmatrix} a & 1-b \\ a^2 & (1-b)^2 \\ 2a(1-a) & 2b(1-b) \\ (1-a)^2 & b^2 \\ 1-a & b \end{bmatrix}.$$

In these terms it is possible to model some of the composite order-bias effects found by Adelman and Bresnick. Figure 3.3 shows the graphs of three observation

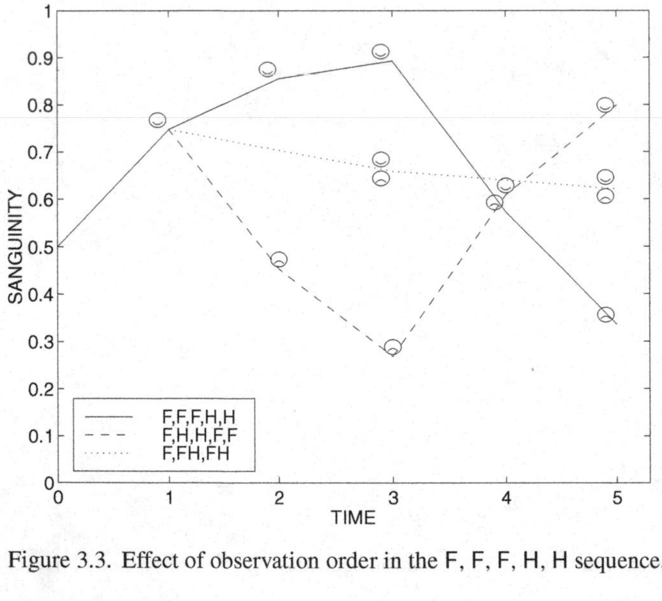

Figure 3.3. Effect of observation order in the F, F, F, H, H sequence.

sequences, each using the same objective measurements: (F, F, F, H, H), (F, H, H, F, F), and (F, FH, FH). Each sequence has an initial friend (F). This is followed with two confirming (F) and two disconfirming (H) observations. The recency effect on sanguinity is apparent. The (H, H) sequence has strong influence if received late in the interval, but the influence is much less when received earlier. Indeed, the two sequential patterns led to different final conclusions: (F, F, F, H, H) $\Rightarrow$ hostile target, (F, H, H, F, F) $\Rightarrow$ friendly target. Human anchoring is also seen when the observations are presented simultaneously. In this case the decision maker retains a vestige of his initial view that the target is friendly, but it is weakened by the contradictory data. Although operational difficulties prevented Adelman and Bresnick from examining the range of composite experiments discussed here, these results agree with the general thrust of those presented in the reference.

In the experiment, the target type does not change. The objective tempo is given by $Q = 0$; a friend remains a friend and so too a foe. The *true* Bayesian solution for this exercise is given by (3.3) with $Q = 0$. Figure 3.4 shows observation sequences as given in Figure 3.3: F, F, F, H, H and F, H, H, F, F along with the Bayes probabilities for each sequence. As expected, the Bayes solution does not exhibit order bias; after the five observations, the sanguinity is 0.8 for either observation sequence. Indeed, the Bayes solution path is close to the PME path for the sequence F, H, H, F, F. However, the PME and the Bayes solution paths differ significantly for the observation sequence F, F, F, H, H. The reason for this anomalous behavior lies in the observation dependence of the gains in (3.4). As $\hat{\phi}_1$ nears 1.0 (or 0.0) the

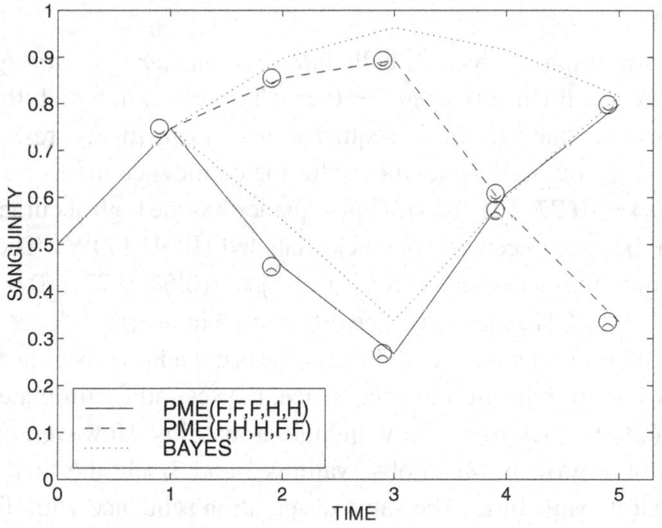

Figure 3.4. A comparison of the PME with the Bayesian estimate for the observations F, F, F, H, H.

gain of the Bayes update is small – an effect oft noted in Kalman filters with small process noise. This is apparent in Figure 3.4 near observation 3. With the small gain, the Bayesian assessment of target type is much slower to respond to the final F, F observations than is the PME. This leads to a large order bias after the final observation.

One measure of the utility of an observation is the degree to which it reduces decision maker confusion. A quantitative indication of situational uncertainty is obtained from $\{P_{\phi\phi}\}$. To illustrate the utility of the PME in ranking observation quality, consider again the Adelman and Bresnick experiment. The individual observations came from different sources and were not equally diagnostic. A more faithful model of this experiment would use a different $\mathbf{P}_i$ for each measurement. This modification is simply incorporated into the PME. Unfortunately, the data presented in [AB91] is not sufficiently detailed to particularize the $\{\mathbf{P}_i\}$ with conviction. However, the effect of the variability of the discernibility matrices can be illustrated by letting the sequence of confidence matrices change in concert with the type of observation; for example, the IFF measurement would yield one $\mathbf{P}_i$, whereas the early-order path measurement would yield another.

To illustrate the consequences of having measurements with different discernibility matrices, consider the early-order observation sequence (F, F, F). Suppose that $\mathbf{P}_1$ is as before, but observation two (respectively observation three) is more accurate than observation three (respectively observation two). The discernibility matrix for the ith observation is given by (3.5) where $i \in \{1, 2, 3\}$. To normalize the results, suppose that $a_1^2 = a_2 a_3$ and $b_1^2 = b_2 b_3$; improved acuity in one measurement is balanced by degraded acuity in the other. To avoid singularities with $a_1 = 0.68$ and $b_1 = 0.77$, a_2 and b_2 must be restricted in range: $a_2 \in [0.5, 0.92]$ and $b_2 \in [0.59, 1.0]$.

Figure 3.5 shows a graph of var(ϕ_1), called *decision maker uncertainty*, from its initial state of low confidence (var(ϕ_1) = 0.25). Two cases in which the decision maker receives the same (F, F, F) sequence are shown in Figure 3.5. The lower curve shows decision maker uncertainty for the confidence matrices: $\{a_i\} = (0.68, 0.92, 0.5)$, $\{b_i\} = (0.77, 1.0, 0.59)$. This sequence has the high accuracy measurement first with the lower accuracy one next (labeled HIGH-LOW). The upper curve reverses the measurements sequence: $\{(a_i, b_i)\} = \{(0.68, 0.77), (0.5, 0.59), (0.92, 1.0)\}$ (the LOW-HIGH sequence). The differences in operator doubt are apparent. The second measurement in the high-low case is not a flawless indication of friend, but it is flawless in indication of foe. Hence, the second F eliminates decision maker uncertainty by picking friend with high probability. However, operator *recency* causes doubt to grow between observations 2 and 3, and the final F is of such low quality that it avails little. The same observation sequence with different confidence matrices gains little from the second F, but it finds the third (high quality) F compelling. The decision maker ends the experiment sure that the object is

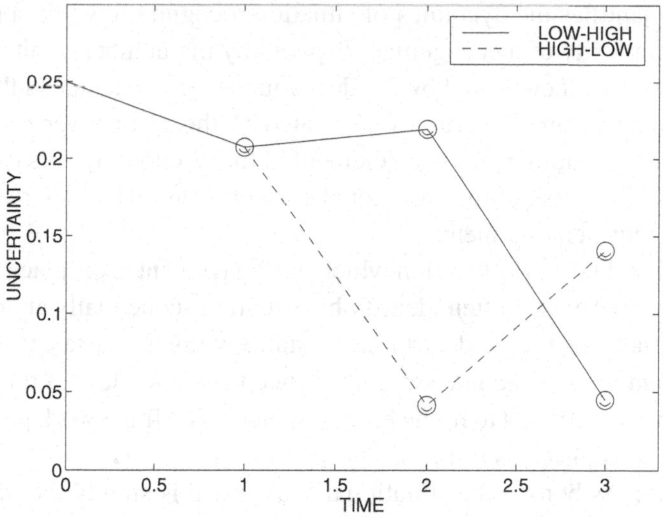

Figure 3.5. Decision maker doubt is related to accuracy order.

friendly. (The curve is not shown, but sanguinity for LOW-HIGH $\gg$ sanguinity for HIGH-LOW.)

The variation in operator uncertainty shown in Figure 3.5 mimics an anomalous behavior noted parenthetically by Adelman and Bresnick. They found that some operators engaged a target (decided it was hostile) before the full data set had been received. An "out-in" path was a strong indication of hostile intent, especially if it occurred in the late-order data. "Even though (the) participants later said the aircraft was a friend after it had returned to the safe-passage corridor, they would have long since shot it down" [AB91]. Precisely this kind of effect is indicated by the HIGH-LOW curve of Figure 3.5, in which the PME expresses complete confidence that the target is friendly at observation 2, only to waiver later. The supposition in the PME that the decision maker operates as if the time horizons of the metahypotheses were relatively short is confirmed in the decision maker's engagement policies. What some perceive as visceral behavior patterns are made quantitative in the PME and are made explicit in Figure 3.5.

3.4 Multilevel Situation Assessment

In the previous section, the modal-state estimation algorithm of the PME was generalized to include situations in which the discernibility matrix varies with the observation number and the dimension of the observation differs from the dimension of the modal state. These generalizations were sufficiently direct that there was no need to carefully review the development presented in the appendix. Let us consider a more subtle generalization of the modal estimator. As pointed out earlier, the PME quantifies the dynamics of situation recognition within a natural taxonomy. The complexity of a engagement is given by the number of alternative situational hypotheses acknowledged by the decision maker; the tempo is the pace of the metastate change; and uncertainty is related to the distinctiveness of the observation–metastate coupling. For each element in this trichotomy, it is the decision maker's *subjective* sense of the situational environment that is determinative rather than the objective environment.

When the situational framework within which an engagement takes place is multifaceted, a decision maker will often identify his situation sequentially at different levels of detail. Initially, he tries to determine his status within a coarse partitioning of his metaspace and ignores the fine structure. Once he is confident of the coarse level assessment, he will attempt to refine his assessment. The framework presented in Section 3.2 can be adjusted as follows:

complexity: As before, the situational state space is simply the union of the S canonical unit vectors in $\mathbb{R}^S$: $\phi_t \in \{\mathbf{e}_i; i \in \boldsymbol{S}\}$. However, when the number of metastates is large, a decision maker will simplify his metaspace by grouping collections of metastates into a smaller number of higher level groups called *metastate aggregates* (MSAs) (e.g., $A_i = \cup_j \mathbf{e}_j$ might be the ith MSA where j runs over an appropriate index set). For example, in a tracking context the MSA hostile-platform would include a variety of threats. Let α_t be an indicator of the MSA at time t; α_t is a canonical unit vector in $\mathbb{R}^k$ where k is the number of MSAs. The MSA can be written $\alpha_t = C\phi_t$, where $C_{ij} = 1$ if $\phi_j \in A_i$ and 0 otherwise. Note that the $\{A_i\}$ need not be disjoint at the phenomenological level: The same situational hypothesis can exist under more than one rubric. To accommodate this potential ambiguity, an elemental metastate will have multiple designations, one for each MSA in which it is classified. Although not a state variable in the strict sense, α_t will be called the *MSA-state*.

tempo: The metastates may change in the course of time. The pace of change is quantified by the frequency and sequence of metastate transitions. For the purposes of this section it will be supposed that the current MSA modulates the metastate dynamics and that changes in the MSA are independent of the realized metastate within an MSA. More specifically, suppose that in the collection of MSAs, $\{A_i; i \in \boldsymbol{k}\}$, the ith aggregate has $N(A_i)$ elements. By construction, the dimension of the metaspace satisfies $\sum_i N(A_i) = S$ although there may be some redundancy in the metastates. Let the individual unit vectors be ordered in accord with the $\{A_i\}$ (e.g., $\{\mathbf{e}_1, \ldots, \mathbf{e}_{N(A_1)}\}$ are associated with A_1 and so on). With this ordering, $\mathbf{C}$ will be block diagonal.

The decision maker perceives tempo subjectively and quantifies it in a manner compatible with the metastate–MSA decomposition. If the MSA is fixed (e.g., $\alpha_t = \mathbf{e}_i$), the metastate evolution will be represented with a Markov process whose state space is necessarily restricted to the metastates contained within the MSA (e.g., $\mathbf{e}_j \in A_i$). As each of the MSAs are considered in turn, there is generated a set of k transition rate matrices, $\{Q^i; i \in \boldsymbol{k}\}$, which describe the intra-MSA variation of the metastates. An individual Q^i-matrix describes the local tempo within a single MSA. The collection, $\{Q^i; i \in \boldsymbol{k}\}$, gives the dynamics of a set of isolated processes – ones having no mixing between different MSAs.

The MSA may also change during the course of an engagement: Changes in the MSA are produced by those transitions that take $\{\phi_t\}$ out of one local domain and place it in another. Suppose these higher level transitions are themselves represented by a Markov process. Since there are k distinct MSAs, this higher level process is parameterized by a $k \times k$ mixing matrix Q^m. The tempo of the composite encounter is formed from blending the local ($\{Q^i; i \in \boldsymbol{k}\}$) and global ($Q^m$) dynamics. This compound description of the metastate process delineates motion within an MSA and transitions across MSA boundaries. The metastate process is still Markovian and has an $S \times S$ generator Q', which can be found using conventional methods. In combining $\{Q^i; i \in \boldsymbol{k}\}$ and Q^m certain natural constraints must be observed. For example, if $\mathbf{e}_m \in A_i$ is phenomenologically identical to $\mathbf{e}_n \in A_j$, an $\mathbf{e}_m \mapsto \mathbf{e}_n$ transition is prohibited: If the intrinsic situational hypothesis does not change, the decision maker would not have cause to relabel the metastate in another MSA. For convenience, it will be supposed when an MSA transition occurs ($\mathbf{e}_i \mapsto \mathbf{e}_j$), the metastate is equally likely to occupy any permissible location in A_j.

To use the PME in this application, the structure of the discernibility matrix must be generalized. The decision maker perceives objective data in a subjective manner. The discernibility matrix is not fixed but itself depends on the decision maker's observations. If the decision maker is sure of his MSA, he will use the detail in his observation to distinguish alternatives within the MSA. Alternatively, if unsure of his MSA, he will use coarse processing to aid in MSA classification.

In an ambiguous environment, a decision maker first attempts to determine the proper A_i. Until α_t can be deduced with surety, he makes little attempt to isolate ϕ_t within its MSA. However, as the uncertainty regarding his high level status is reduced, he will begin a process of disaggregation; that is, he will attempt to determine the proper $\mathbf{e}_j \in A_i$. This hierarchical information processing is called *progressive deepening* in situation assessment. When a change in the metastates is such as to cause $\{\phi_t\}$ to leave the current MSA, the decision maker must return to his coarse processing mode to determine his MSA anew.

To describe the decision maker's cognitive process more precisely, note that an engagement takes the metastate process from unit vector to unit vector in an unpredictable manner. The decision maker's convictions regarding the situational status are given by $\{\hat{\alpha}_t\}$ and $\{\hat{\phi}_t\}$. To have a simple geometric description of this, let $\boldsymbol{\Psi}$ be the convex hull of the unit vectors in the metaspace: $\boldsymbol{\Psi}$ is an $(s-1)$-dimensional simplex generated by the S vertices $\{\mathbf{e}_i;\, i \in S\}$. From (3.3) and (3.4) it is evident that the set of all situational assessments (all $\hat{\phi}_t$) is precisely $\boldsymbol{\Psi}$. For this reason $\boldsymbol{\Psi}$ is called the *cognitive metaspace* of the decision maker: $\boldsymbol{\Psi}$ represents the totality of the judgments that the decision maker might have during the course of the engagement. The vectors in $\boldsymbol{\Psi}$ will be called *cognitive metastates* to contrast them with the metastates. The unit vectors that generate $\boldsymbol{\Psi}$ are the vertices of the cognitive metaspace, and when the cognitive metastate is near a vertex, the decision maker is sure of his situational status.

Multiplication of the decision maker's cognitive-metastate vector by C yields the *cognitive-MSA* state $\hat{\alpha}_t$. Define by $\boldsymbol{\Psi}_\alpha$ the cognitive MSA state space of the decision maker: It is the $(k-1)$-dimensional simplex generated by the k vertices $\{\mathbf{e}_i;\, i \in \boldsymbol{k}\}$. A vertex in $\boldsymbol{\Psi}_\alpha$ is the projection of a boundary set in $\boldsymbol{\Psi}$ formed as the convex hull of the unit vectors that make up the particular vertex. This set in $\boldsymbol{\Psi}$ will be called an A_i-face if the unit vectors forming it generate A_i. Thus at the MSA-level, status recognition could be described as being on the correct A_i-face ($\in \boldsymbol{\Psi}$) or at the ith vertex ($\in \boldsymbol{\Psi}_\alpha$).

To illustrate the geometric structure of the cognitive metaspace, consider the situational hypotheses: friendly ($\mathbf{e}_1$), hostile ($\mathbf{e}_2$), and neutral ($\mathbf{e}_3$). At time t, the decision maker's cognitive metastate is delineated by the three numbers: $\{\hat{\phi}_i = \mathcal{P}(\phi_t = \mathbf{e}_i);\, i = 1, 2, 3\}$. Suppose the decision maker aggregates friendly and neutral into nonthreatening. His MSA state space could be represented as nonthreatening ($\mathbf{e}_1$) and threatening ($\mathbf{e}_2$). The decision maker's cognitive-MSA state is: $\hat{\alpha}_1 = \hat{\phi}_1 + \hat{\phi}_3$; $\hat{\alpha}_1 = \hat{\phi}_2$. In this case $\boldsymbol{\Psi} \subset \mathbb{R}^3$ and $\boldsymbol{\Psi}_\alpha \subset \mathbb{R}^2$. Figure 3.6 shows the two cognitive spaces for this simple example. The cognitive metaspace is the shaded triangle shown in $\mathbb{R}^3$. It is the interior of the region bounded by connecting the three metastate vertices. At any point in time, the cognitive metastate vector is three dimensional, and it lies in the shaded region. If the decision maker is confident of his metastate identification,

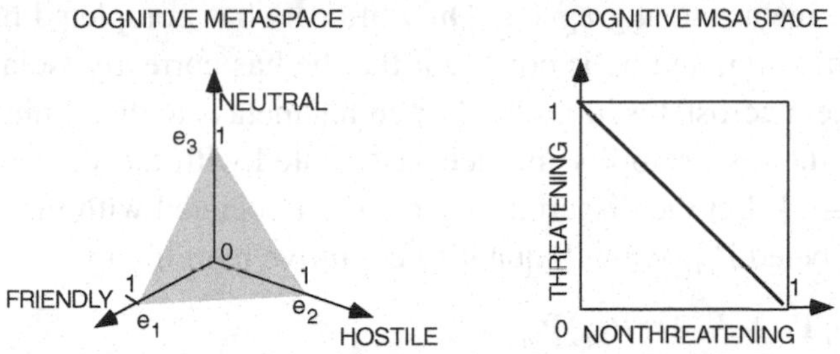

Figure 3.6. The cognitive metaspace and the cognitive MSA space for the example.

the cognitive metastate vector will be near one of thevertices. If in doubt, the cognitive metastate will be at a distance from all vertices. The decision maker could have reservations regarding whether the situation is friendly or neutral, but he may be sure that it is nonthreatening. In this case, the cognitive metastate is near the line joining $\mathbf{e}_1$ and $\mathbf{e}_3$: the nonthreatening face (A_1-face) of $\boldsymbol{\Psi}$. Though not near a vertex in $\boldsymbol{\Psi}$, the decision maker's cognitive MSA state is near the A_1-vertex in $\boldsymbol{\Psi}_\alpha$ shown in Figure 3.6.

The cognitive metaspace $\boldsymbol{\Psi}$ is the set of all allowable (that is, decision maker–specific) cognitive conditions in an engagement. The set of all vertices is the set of all surety conditions at the metastate-level, and the set of all A_i-faces is the set of all surety conditions at the aggregated level.† In most realistic engagements, the decision maker will remain in the interior of $\boldsymbol{\Psi}$; that is, he is never able to exclude absolutely any of the situational hypotheses. Still, when relatively near a vertex or an A_i-face, the decision maker will be described as being "at the vertex" or "on the face." When the decision maker is said to be in the central region of $\boldsymbol{\Psi}$, it is meant that he is truly puzzled.

uncertainty: A decision maker determines his current status by comparing his actual observations to those he expects. He interprets an observation in terms of an MSA (coarse processing) and in terms of a metastate (fine processing). The aspect that distinguishes this application from those that preceded it is the hierarchical variability of the discernibility matrix **P**. A specific form for **P** will be used. This choice is motivated by the fact that when a decision maker is uncertain as to his MSA status, he looks at the data at a coarse level, and he makes distinctions between

† There are faces of $\boldsymbol{\Psi}$ that are not A_i-faces. There are, for example, boundary surfaces formed by convex combinations of unit vectors, not all of which come from a single MSA. While such cognitive states are permissible, the likelihood of finding a decision maker in such a set is slight, and they are not of concern here. Indeed, because of the labeling convention, the named $\boldsymbol{\Psi}$-faces are disjoint, and every vertex is in one named face.

alternative metastate aggregates. Only after he has completed his high level partitioning, and he is confident that he has correctly isolated the appropriate macrostatus, does he deepen his metastate discrimination.

Denote the discernibility matrices associated with the vertices of Ψ by $\{\mathbf{P}_i; i \in S\}$. Let the discernibility matrix associated with the interior of Ψ be labeled $\mathbf{P}_m$. We will quantify cognitive merging by

$$\mathbf{P} = \frac{\sum_i \hat{\alpha}_i^2 \mathbf{P}_i + K_4 \mathrm{Tr}(P_{\phi\phi})\mathbf{P}_m}{\sum_i \hat{\alpha}_i^2 + K_4 \mathrm{Tr}(P_{\phi\phi})}, \qquad (3.6)$$

where

$$P_{\phi\phi} = \mathrm{cov}(\phi_t) = \mathrm{diag}(\hat{\phi}_t) - \hat{\phi}_t \hat{\phi}_t'$$

and $K_4 \geq 0$ is a distinguishability parameter. The rationale behind (3.6) is apparent. The nonnegative function $K_4 \mathrm{Tr}(P_{\phi\phi}) \to 0$ whenever $\hat{\phi}_t \approx \mathbf{e}_i$ for some $i \in S$. If the decision maker is confident that $\phi_t = \mathbf{e}_j \in A_i$ (i.e., $\hat{\alpha}_i \approx 1$ and $\mathrm{Tr}(P_{\phi\phi}) \approx 0$), he will interpret his observations according to $\mathbf{P}_i$. When the status changes, the decision maker first finds that the data do not confirm the ostensible MSA. Failing to match his expectations at a local level should change the weightings in $\mathbf{P}$ so as to enlarge the set of metahypotheses under consideration. This first leads the decision maker to explore other metastates within the current MSA and to consider new MSAs if need be.

The equation for the modal estimate is still given by (3.4) and (3.5). The discernibility matrix is now a random process but it is $\mathcal{Z}_t$-adapted. From $\{\hat{\phi}_t\}$, the MSA estimate can be deduced as well:

$$\hat{\alpha}_t = C\hat{\phi}_t.$$

3.5 Decision-Making Phenotypes: An Example

Consider the situation having five local metastates (labeled $\mathbf{e}_1$ through $\mathbf{e}_5$) and two higher order aggregates (labeled A_1 and A_2) as shown in Figure 3.7 [SC94].

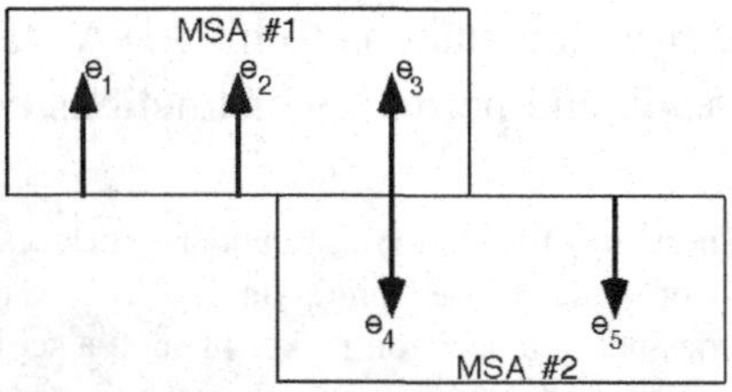

Figure 3.7. A situation involving five metastates and two MSAs.

To illustrate the variability of situation recognition on the cognitive state of the decision maker, suppose that when the decision maker's cognitive condition is such that $\hat{\alpha}_t \approx \mathbf{e}_1$, his discernibility matrix is approximately $\mathbf{P}_1$:

$$\mathbf{P}_1 = \begin{bmatrix} .85 & .025 & .025 & .025 & .2 \\ .025 & .85 & .025 & .025 & .2 \\ .025 & .025 & .85 & .85 & .2 \\ 0 & 0 & 0 & 0 & 0 \\ .1 & .1 & .1. & .1 & .4 \end{bmatrix}.$$

When the decision maker is convinced that $\phi_t \in A_1$, he uses his observations primarily to better isolate his metastate within A_1. This cognitive processing is illustrated in the first column of $\mathbf{P}_1$ (corresponding to the event $\phi_t = \mathbf{e}_1$):

- the observation will indicate $\phi_t = \mathbf{e}_1$ 85% of the time,
- the observation will be misclassified within A_1 5% of the time,
- the observation will be misclassified outside A_1 (within A_2) 10% of the time.

The MSA error rate is 10%, and the error is necessarily an $\mathbf{e}_5$ error since $\mathbf{e}_4$ is identical to $\mathbf{e}_3 \in A_1$. Indeed, if the decision maker thinks his metastate is in A_1, he will never classify an observation as being from the metastate $\mathbf{e}_4$ ($\mathbf{P}_{4.} \equiv 0$). Columns 3 and 4 are necessarily identical to indicate the fact that $\mathbf{e}_3$ and $\mathbf{e}_4$ create the same observation distribution, but the decision maker interprets the observation within the A_1 framework. Column 5 indicates the perspective of the decision maker when $\phi_t \in A_2$ ($\alpha_t = \mathbf{e}_2$), but he or she wrongly thinks that $\phi_t \in A_1$ ($\hat{\alpha}_t \approx \mathbf{e}_1$). Such observations are seen by the decision maker as both failing to confirm his notion of the current metastate (the classification across $\phi_t \in A_1$ is uniform) and as an increased probability of a classification in A_2 (40%).

Alternatively, suppose the decision maker's cognitive condition is such that $\hat{\alpha}_t \approx \mathbf{e}_2$, and the discernibility matrix is $\mathbf{P}_2$:

$$\mathbf{P}_2 = \begin{bmatrix} .2 & .2 & .05 & .05 & .05 \\ .2 & .2 & .05 & .05 & .05 \\ 0 & 0 & 0 & 0 & 0 \\ .3 & .3 & .8 & .8 & .1 \\ .3 & .3 & .1 & .1 & .8 \end{bmatrix}.$$

Again, look first at the first column of $\mathbf{P}_2$ (corresponding as before to $\phi_t = \mathbf{e}_1 \in A_1$). Despite the fact that the decision maker's situation is objectively the same as that described at the beginning of the preceding paragraph, the conclusions he draws from the observations (as quantified by $\mathbf{P}_2(., 1)$) are much different. If $\phi_t = \mathbf{e}_1$, the observations are classified as

- correctly indicating $\mathbf{e}_1$ 20% of the time (instead of 85% of the time),
- in A_1 40% of the time (instead of 90% and now making no distinction among the admissible states in A_1),
- in A_2 60% of the time (instead of 10%).

The observation is never seen as indicating $\mathbf{e}_3$ (instead of never seen as indicating $\mathbf{e}_4$). In this case, $\mathbf{e}_3$ receives no weight because $\mathbf{e}_3$ is not in A_2. Observations not confirming the MSA are not interpreted at the metastate level except for points in $A_1 \cap A_2$ (i.e., $\mathbf{P}_2(1, 1) = \mathbf{P}_2(2, 1)$ and $\mathbf{P}_2(4, 1) = \mathbf{P}_2(5, 1)$). Columns 3 and 4 are identical because they correspond to the same objective event, but in contrast to $\mathbf{P}_1$, Δz_t is interpreted within A_2. Columns 4 and 5 quantify the ability of the decision maker to distinguish the $\phi_t \in A_2$ (correctly 90% of the time).

When the decision maker is uncertain about his MSA status, he first will classify his observations at a coarse level (i.e., a measurement Δz_t is mapped to A_1 or A_2) but the decision maker makes little distinction at the metastate level. The matrix is

$$\mathbf{P}_m = \begin{bmatrix} .25 & .25 & .2 & .2 & .1 \\ .25 & .25 & .2 & .2 & .1 \\ .25 & .25 & .2 & .2 & .1 \\ .125 & .125 & .2 & .2 & .35 \\ .125 & .125 & .2 & .2 & .35 \end{bmatrix}.$$

Suppose again that the decision maker is not sure of his MSA but is actually in state $\phi_t = \mathbf{e}_1$. He will classify his next observation:

- correctly, as coming from A_1 75% of the time, with equal distribution across $\phi_t \in A_1$,
- incorrectly 25% of the time (placing the observation in A_2).

If $\phi_t = \mathbf{e}_3$ or $\mathbf{e}_4$, the metastate could be classified in either MSA. That $\alpha_t = \mathbf{e}_1$ is favored is a result of the fact that there are more metastates in A_1 than there are in A_2. The rest of $\mathbf{P}_m$ has a similar interpretation. Note that the first columns of $\mathbf{P}_1$, $\mathbf{P}_2$, and $\mathbf{P}_m$ all correspond to the same circumstance, but $\mathbf{P}_1(1, 1) = 0.85$, $\mathbf{P}_2(1, 1) = 0.2$, and $\mathbf{P}_m(1, 1) = 0.25$: The same objective event ($\phi = \mathbf{e}_1$ and $\Delta z_t = \mathbf{e}_1$) is interpreted differently at different points in the decision maker's cognitive metaspace.

To illustrate the situational evaluation problem in a dynamic environment, consider the framework described in Figure 3.7. The tempo of the engagement is measured by the decision maker's sense of metastate evolution; its parametric representation specifies how fast a particular decision maker *thinks* the engagement will develop. For the case portrayed, there are two MSAs ($\mathbf{\Psi}_\alpha$ is of dimension two) and five (four distinct) metastates ($\mathbf{\Psi}$ is of dimension five). The pace of local change is given by

- a 3×3 matrix Q^1 that describes local behavior within A_1,
- a 2×2 matrix Q^2 that describes local behavior within A_2.

The higher level pace is given by a 2×2 matrix Q^m that describes transition behavior between the MSA state spaces:

$$Q^1 = K_1 \begin{bmatrix} -1 & .6 & .4 \\ .6 & -1 & .4 \\ .5 & .5 & -1 \end{bmatrix},$$

$$Q^2 = K_2 \begin{bmatrix} -1 & 1 \\ 1 & -1 \end{bmatrix},$$

$$Q^m = K_m \begin{bmatrix} -1 & 1 \\ 1 & -1 \end{bmatrix}.$$

The transition matrices have a common factor (the bracketed elements), and a decision maker–specific factor (the coefficients $\{K_i\}$). An individual decision maker sees the engagement as follows. If the MSA is A_1, necessarily $\phi_t \in \{\mathbf{e}_1, \mathbf{e}_2, \mathbf{e}_3\}$ and the mean lifetime in each metastate is the same, $1/K_1$. Suppose $\phi_t = \mathbf{e}_1$. If there is a local transition, with probability 0.6 the transition will be $\mathbf{e}_1 \mapsto \mathbf{e}_2$, and with probability 0.4 the transition will be $\mathbf{e}_1 \mapsto \mathbf{e}_3$ (i.e., $\mathbf{e}_2$ is favored over $\mathbf{e}_3$ from $\mathbf{e}_1$). Similarly, $\mathbf{e}_1$ is favored over $\mathbf{e}_3$ from $\mathbf{e}_2$. Since $\mathbf{e}_3$ is equally likely to go to either of the metastates in A_1, the metastate is a chain in which the first two states are favored over the third. The A_1-tempo is particularized by K_1. The larger K_1 is, the more frequent the decision maker expects the state transitions to be, and the more difficult he will find them to resolve. In what follows, K_1 is considered to be a style attribute. Within A_2 the metastate process is a symmetric chain with sojourn times particularized by the style parameter K_2. Similarly, the tempo of macrochange is particularized by K_m. The comprehensive transition rate matrix Q, the parametric representation of the situational dynamics *as seen by the decision maker*, can be determined directly. Three coefficients, K_1, K_2, and K_m, are sufficient to distinguish the sense of tempo of the decision maker.

To be more specific, consider the following discernibility matrix primitives within the context of the sample engagement. The data may come from either a high quality or a lower quality source. The decision maker may or may not correctly recognize the source; for example, the objective data source may be of high quality, but the decision maker's subjective sense of the quality – and this is what parameterizes the PME – may be wrong.

High Quality Observations When the decision maker's cognitive metastate is on an A_i-face, he will interpret the data according to an A_i-specific discernibility matrix $\{\mathbf{P}_i : i = \mathbf{1}, \mathbf{2}\}$. When he is undecided about the MSA, his cognition is better described by $\mathbf{P}_m$.

Low Quality Observations As an alternative to the high quality data suppose the observations are described as indicated below where discernibility matrices are labeled with the additional subscript p.

$$\mathbf{P}_{p1} = \begin{bmatrix} .5 & .15 & .1 & .1 & .233 \\ .15 & .5 & .1 & .1 & .233 \\ .15 & .15 & .5 & .5 & .233 \\ 0 & 0 & 0 & 0 & 0 \\ .2 & .2 & .3 & .3 & .3 \end{bmatrix}.$$

$$\mathbf{P}_{p2} = \begin{bmatrix} .15 & .15 & .1 & .1 & .1 \\ .15 & .15 & .1 & .1 & .1 \\ 0 & 0 & 0 & 0 & 0 \\ .35 & .35 & .6 & .6 & .2 \\ .35 & .35 & .2 & .2 & .6 \end{bmatrix}.$$

$$\mathbf{P}_{pm} = \begin{bmatrix} .2 & .2 & .2 & .2 & .133 \\ .2 & .2 & .2 & .2 & .133 \\ .2 & .2 & .2 & .2 & .133 \\ .2 & .2 & .2 & .2 & .3 \\ .2 & .2 & .2 & .2 & .3 \end{bmatrix}.$$

In [SC94], the response of several decision making phenotypes were investigated. Three will be discussed here. The first, the *normative* decision maker (NOR), has a good understanding of the character of the engagement. The *novice* (NOV) has the same sense of the situation dynamics as does the normative decision maker, but lacking training, he is unable to make sharp cognitive differentiations at a metastate level. The *obtuse* (OBT) decision maker has the same recognitional skills as the normative decision maker, but he is not alert to the possibility of metastate change. Table 3.1 gives the parameters that particularize their representations.

To gain insight into how data accuracy influences decision maker response, consider the following engagement. A decision maker recognizes that a hostile aircraft is approaching. The aircraft may seek to destroy: Asset 1 (metastate $\mathbf{e}_1$) or Asset 2

Table 3.1. *Decision maker phenotypes.*

	MS Tempo	MSA Tempo	Distinguishability
	$K_1 = K_2$	K_3	K_4
Normative	0.05	0.1	4
Novice	0.05	0.1	16
Obtuse	0.0	0.0	4

(metastate $\mathbf{e}_2$) for which the decision maker is responsible; Asset 5 (metastate $\mathbf{e}_5$) for which the decision maker is not responsible; or Asset 3 for which the decision maker is jointly and partially responsible (metastate $\mathbf{e}_3$ if he accepts responsibility and metastate $\mathbf{e}_4$ if he does not). The metastate space is of dimension 5. The decision maker aggregates $\mathbf{e}_1$ through $\mathbf{e}_3$ into a threatening category and $\mathbf{e}_4$ through $\mathbf{e}_5$ into a nonthreatening category. The metastate will change if the aircraft changes its objective during the engagement.

Two scenarios will be investigated. In both cases, the time interval is $[0, 20]$ s with observations every second beginning with $t = 1$ s. At the beginning, the decision maker thinks that all of the metastates are equally likely. In both scenarios a single transition occurs in the metastate process at $t = 9^+$ s, and the engagement ends with the aircraft attacking Asset 1. The decision maker must first reduce his a priori metastate uncertainty and move toward the appropriate cognitive vertex. The decision maker must recognize when the asset under attack has changed and identify the terminal objective.

Scenario 5 Scenario 5 (S5) has a single metastate transition across an MSA boundary. Initially the aircraft begins an approach to Asset 5 ($\phi_t = \mathbf{e}_5$; the engagement is nonthreatening). At $t = 9^+$ s, $\{\phi_t\}$ makes an $\mathbf{e}_5 \mapsto \mathbf{e}_1$ transition and commences an attack on Asset 1. The decision maker's first 9 observations are associated with $\mathbf{e}_5$ with the rest associated with $\mathbf{e}_1$. In this prototypical MSA transition, the cognitive metastate will traverse the interior of Ψ, and the full complement of dynamical and discernibility matrices will be exercised. The omniscient decision maker sees variation at both the global and the local level: He would respond to this scenario with

$$\hat{\alpha}_t = \mathbf{e}_2 I_{[0,9]} + \mathbf{e}_1 I_{(9,20]}$$

and

$$\hat{\phi}_t = \mathbf{e}_5 I_{[0,9]} + \mathbf{e}_1 I_{(9,20]}.$$

Scenario 2 Scenario 2 (S2) has a single metastate transition within the threatening face. The aircraft begins an approach to Asset 2 ($\phi_t = \mathbf{e}_2$). At $t = 9^+$ s $\{\phi_t\}$ makes a $2 \mapsto 1$ transition; the aircraft is still threatening, but the objective is different. The decision maker's first nine observations are associated with $\mathbf{e}_2$ with the rest associated with $\mathbf{e}_1$. In this vignette, once $\hat{\alpha}_t \approx \mathbf{e}_1$, the cognitive metastate will traverse a boundary region of Ψ (the A_1-face). The omniscient decision maker would respond to this scenario with

$$\hat{\alpha}_t = \mathbf{e}_1 I_{[0,20]}$$

and

$$\hat{\phi}_t = \mathbf{e}_2 I_{[0,9]} + \mathbf{e}_1 I_{(9,20]}.$$

To better understand the interplay of decision making styles and data quality, the three decision makers delineated parametrically in Table 3.1 were made experimental subjects. Of particular interest is how performance with good data quality differs from that with poor data quality. The quality issue will affect some decision makers more than others and be a factor in some situations more than others. Another concern is that of objective versus subjective data quality. There are situations in which the decision maker misapprehends the quality of the data and treats observations as if they were coming from a source other than the actual one. The data generation protocol is as follows: If the data quality is good, the first nine data points in S5 are generated randomly with a probability mass function given by fifth column of $\mathbf{P}_2$ (respectively in S2 they are generated randomly with a probability mass function given by second column of $\mathbf{P}_1$), and the next 11 data points are generated similarly using the first column of $\mathbf{P}_1$. Each decision maker is a subject in each experiment, and each experiment in a series is independent of the others. An experiment using poor quality data is done in the same way, using the $\{\mathbf{P}_{pi}\}$ instead of the $\{\mathbf{P}_i\}$.

The PME is parameterized by both the individual style of the decision maker and the decision maker's conviction on the data quality. To label the different experimental conditions, the following convention will be used: An experiment is labeled (scenario)(objective data quality)(subjective data quality); for example, S5GG. To display decision maker behavior, a sample average of fifty independent runs is used to create an "average" response. For each experiment, a plot of the 50-sample mean of $\{\hat{\alpha}_1\}$ (labeled *Pr. of threat*) and $\{\hat{\phi}_1\}$ (labeled *Pr. Asset 1*) is given on [2, 20] s for each of the decision makers. This is an incomplete description of the evolution within the cognitive metaspace, but the decision maker's response is displayed in this way to reduce the detail. These plots give a sense of the decision maker's distance from the $\mathbf{e}_1$-vertex in $\mathbf{\Psi}$ (and the A_1-face) during the engagement.

Experiments S5GG The Asset 5 $\mapsto$ Asset 1 transition is an event in which the metastates are identified with different A_i. Figure 3.8 shows $\hat{\alpha}_1$ and $\hat{\phi}_1$ for the S5 scenario with good data quality. The decision makers move expeditiously from the interior of $\mathbf{\Psi}$ toward the $\mathbf{e}_5$ vertex (not shown). When the aircraft shifts to attack Asset 1, NOR and NOV move essentially in lock step and identify the aircraft as a threat at the same time: There seems to be little advantage to NOR's ability to disaggregate the MSA. OBT is much slower to identify the MSA. This is for two reasons. First, OBT has a slower sense of pace than do the other decision makers.

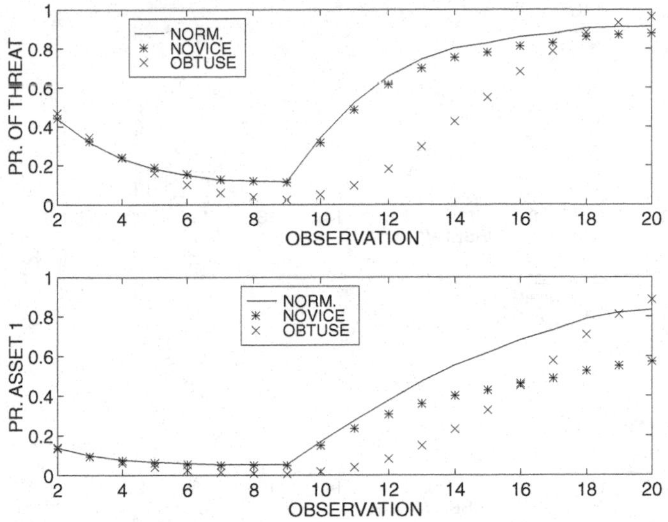

Figure 3.8. The response of three decision makers in the S5 engagement with the observations recognized as good.

Additionally, OBT starts the post transition interval in a worse initial state (i.e., the value of $\hat{\phi}_1(t = 0.9\,\mathrm{s})$ for OBT is smaller than that for either NOR or NOV).

At the metastate level, NOV and OBT have about the same delay with good data, significantly trailing NOR. Of course, OBT becomes more sure of the targeted asset as time passes, while NOV has limited ability to disaggregate. S5GG gives a good picture of NOV: reasonably fast in high level classification, but unable to detect the nuances in the observation.

Experiments S5GP In the preceding experiment, the objective data quality was known to the decision maker. This is the normative circumstance in which a well-trained decision maker is operating in familiar conditions. Stress, sensor degradation, improper training, etc. may result in the decision maker erring in his evaluation of the data quality. In this event, the performance curves shown earlier will not be representative; even with the same tempo, style, and objective data, the response of the decision maker is contingent on his *perception* of the data quality. In this second experiment, the objective data quality differs from that believed by the decision maker.

Suppose that the data are good, but the decision maker deems them to be poor (GP). Figure 3.9 shows the associated cognitive metapaths. When compared with S5GG, all of the decision makers are more conservative than they were – they don't believe the data. This conservatism serves them well when the transition occurs because they have a higher residual probability of threat (probability of Asset 1

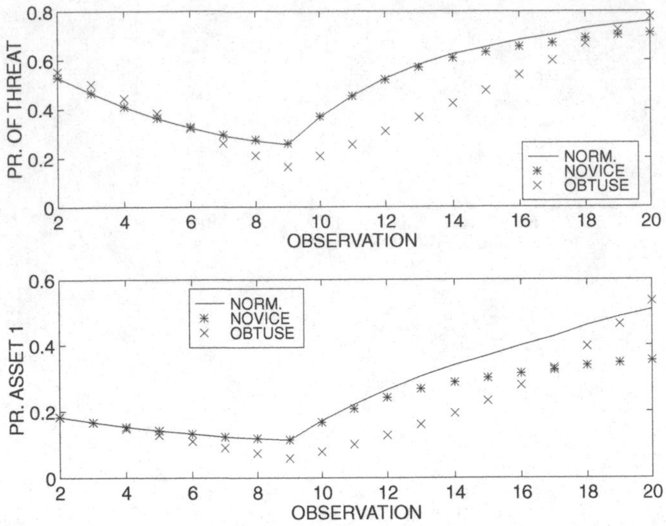

Figure 3.9. The response of three decision makers in the S5 engagement with the observations good but thought to be poor.

respectively) than was the case in S5GG. This misidentification of data quality preserves the decision maker order found in S5GG, but it acts to slow down all of the identifications (about a two second delay in threat status identification). Note, however, that this effect is not the same as would result from a tempo change in the PME. The decision makers are more data-driven than they are tempo-driven when the data quality is good. The performance at the metastate level is unacceptable: a 10 s delay for NOR and OBT and failure for NOV to identify at the 50% level.

Experiments S2PP The Asset 2 $\mapsto$ Asset 1 transition illustrates an engagement in which both the initial and the final objective are in the threatening MSA. Figure 3.10 shows $\hat{\alpha}_1$ and $\hat{\phi}_1$ for the S2 scenario and poor data quality. Since $\hat{\alpha}_1(t = 0) = 0.6$, the decision makers start in an advantageous position. OBT has an advantage in high level processing; the MSA does not change during the encounter and OBT does not expect it to. Both NOR and NOV improve their estimate of the MSA over time, but neither is the equal of OBT.

Identification of the metastate favors NOR and OBT. When the aircraft shifts its objective to Asset 1, all of the decision makers have determined the MSA with good accuracy. Both NOR and OBT are then able to disaggregate using $\mathbf{P}_1$. NOV finds disaggregation difficult because of the continued influence of $\mathbf{P}_m$ on $\mathbf{P}$. None of the decision makers are successful in identifying the metastate transition in a reasonable interval when the data are poor.

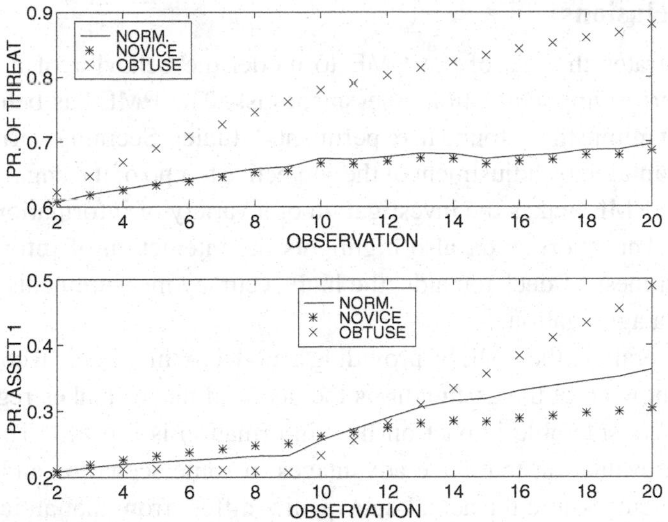

Figure 3.10. The response of three decision makers in the S2 engagement with the observations recognized as poor.

Experiments S2GP When the objective data quality is incorrectly evaluated, by the decision makers operate in a manner similar to that seen in the S5 engagement (see Figure 3.11). Because of (rather than despite) his stubbornness, OBT is better at situation assessment than are his fellows.

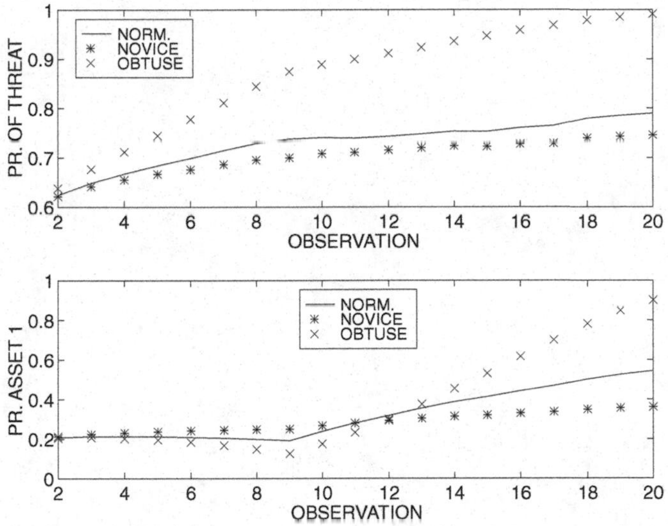

Figure 3.11. The response of three decision makers in the S2 engagement with the observations good but thought to be poor.

3.6 Conclusions

This chapter illustrates the use of the PME to model the behavior of a human decision maker performing a situation assessment task. The PME has behavioral characteristics that mimic those found in experimental studies. Seemingly irrational behaviors can be captured by adjustment of the subjective tempo of the engagement. Analysis using the PME makes the investigation of a variety of information-order effects very easy. This short study also highlights the interaction of information order and the usefulness of decision aids; the high accuracy measurements should be used late in data aggregation.

This chapter generalizes the PME by providing a model for hierarchical cognitive processing. It is shown that the assumptions the decision maker makes regarding data quality have considerable impact on how information is utilized. There are styles of situation evaluation that have advantages in some scenarios and not in others. In engagements where the actual data quality differs from that anticipated, some decision makers are better able than others to carry the recognition task to completion. The responses shown here quantify the influence of training (as indicated by the variability of the $\{K_i\}$) on speed of situation recognition.

4

Image-Enhanced Target Tracking

4.1 Tracking an Agile Target

One of the most thoroughly studied applications of hybrid estimation arises in the synthesis of tracking algorithms for agile targets. These targets, sometimes intentionally and sometimes inadvertently, have motion paths that make them difficult to follow. One example is that of a piloted vehicle whose location must be tracked from a fixed sensor.

> Whereas mobility describes the movement of the vehicle from one location to another in a given period of time, agility describes the vehicle's ability to alter its mean path during that time period.... A major component of agility arises from the driver's intention to maneuver. This is a product of training and the perception of threat. Any analytical approach to modeling a maneuvering vehicle will inevitably encounter a requirement to represent this intentional motion. [BPL82]

The operator of the vehicle exploits its maneuver capability to create a path that he hopes will cause the tracker to lose lock. Such paths have a familiar pattern, with nearly constant acceleration over intervals of unpredictable length followed by discrete maneuver mode changes. Even autonomous platforms can be designed to take such evasive maneuvers; for example, an antiship missile on approach to its objective may follow a preprogrammed jinking path. Such motions often have a constrained geometric structure. If the missile has limited speed control, it will focus its evasive efforts on turning motions, where the acceleration is perpendicular to the velocity. If the tracked object is an aircraft, its velocity is nearly along the longitudinal axis, and the dominant acceleration is nearly normal to the wings [HG90].

In such applications, the sample path of the maneuver acceleration is better described by a discontinuous process rather than a continuous one, and the path is clearly non-Gaussian (see [CEF89] and the references therein). A representative maneuver model is created by partitioning the possible turn rates into a finite number of levels and describing the maneuver process in terms of a Markov chain [GM77, MVM79]. The inclusion of discrete acceleration states complicates the filtering process because the sample characteristics of a Markov chain are unlike anything generated by an LGM shaping filter.

A problem of this type was introduced in Section 1.2. It was noted that the position–velocity model is intrinsically nonlinear. A simple model of an evasive target moving in the X–Y plane at essentially constant speed is given in (1.24):

$$d\begin{bmatrix} X \\ Y \\ V_x \\ V_y \end{bmatrix} = \begin{bmatrix} 0 & 0 & 1 & 0 \\ 0 & 0 & 0 & 1 \\ 0 & 0 & 0 & -\Phi \\ 0 & 0 & \Phi & 0 \end{bmatrix} \begin{bmatrix} X \\ Y \\ V_x \\ V_y \end{bmatrix} dt + \begin{bmatrix} 0 & 0 \\ 0 & 0 \\ 1 & 0 \\ 0 & 1 \end{bmatrix} d\begin{bmatrix} w_x \\ w_y \end{bmatrix}, \tag{4.1}$$

where $\{X, Y\}$ are position coordinates, and $\{V_x, V_y\}$ are associated velocities. In (4.1) there is neither endogenous control ($\upsilon_t \equiv 0$) nor plant state set point ($\chi = 0, \chi_t \equiv x_t$). The target is subject to two types of acceleration: a wide band omnidirectional acceleration represented by the $\mathcal{F}_t$-Brownian motion $\{(w_x, w_y)\}$ and a maneuver acceleration represented by the turn rate process $\{\Phi_t\}$.

In Chapter 1, location estimation was discussed in the context of a range-bearing sensor (often called a radar for convenience) located at the origin of the coordinate system. The outputs of the sensor are $\{r[k]\}$ and $\{\theta[k]\}$:

$$r[k] = \sqrt{X[k]^2 + Y[k]^2}, \tag{4.2}$$

$$\theta[k] = \tan^{-1}\frac{Y[k]}{X[k]}. \tag{4.3}$$

These range-bearing observations can be linearized about $\hat{x}_t$ in the usual way, and the equation for $y[k]$ follows ([HBS89] or [GA93]):

$$y[k] = Hx[k] + n[k]. \tag{4.4}$$

In what follows it will be assumed that this linearization has been done and it will be understood that H depends on the changing geometry of the measurement.

The simplest estimators use an EKF derived by neglecting the turn rate of the target (replace $A(\Phi_t)$ with $A = A(\Phi_t \equiv 0)$) but adding (white or colored) Gaussian noise to account for the energy and time correlation of the maneuver. The Brownian excitation in (4.1) fits well within the LGM framework, but the maneuver is troublesome. The discontinuities in the maneuver process do not harmonize well with a LGM state space model. The usual way to integrate the maneuver into an EKF would be to first determine the power spectral density (PSD) of $\{\Phi_t\}$, labeled $\mathbf{\Phi}(\omega)$. The PSD can be approximated with a shaping filter by adding states to the base-state equation (4.1) along with additional Brownian source for the maneuver, $\{w_a\}$. This maneuver noise is then added to $\{w_t\}$ in the nonmaneuver model $(A(\Phi_t \equiv 0))$. This leads to an EKF of higher dimension, which will be called $\mathsf{EKF}(\Phi(\omega))$ [Sin70, COD91].

Although this approach introduces the proper time correlations of the exogenous processes, the sample functions generated thereby are certainly not close analogues to pilot actions. An additive omnidirectional acceleration retains neither the sample path characteristics of the turn rate process nor the geometry of the acceleration–velocity vectors. In maneuver encounters, the lateral acceleration is the primary one, thus making the EKF formalism of doubtful applicability [SKT90]. If the maneuver is portrayed as additive, the motion model will fail to prompt the estimator to use the geometry of the encounter to best advantage.

More sophisticated estimators use multiple models to represent target motion with a family of equations, each tuned to a different turn rate hypothesis. Multiple-model techniques are motivated by the fact that if the maneuvers could be detected expeditiously, the plant model would be linear, and the associated EKF would simply be given by (1.20)–(1.23) with $\{A_t\}$ changing concurrently with the maneuver ([MS91] for example). Application of the IMM technique to tracking is explored in [BBS88], [HBS89], [AIK91], and [DM91].

Let us first look at the Kalman filter in this application:

Between observations:

$$\frac{d}{dt}\hat{x}_t = A\hat{x}_t, \tag{4.5}$$

$$\frac{d}{dt}P_{xx} = AP_{xx} + P_{xx}A' + R_\chi. \tag{4.6}$$

At an observation:

$$\Delta\hat{x}[k+1] = \gamma_x r[k+1], \tag{4.7}$$

$$\Delta P_{xx}[k+1] = -\gamma_x R_{yy}[k+1]\gamma_x', \tag{4.8}$$

where $\{x_t\}$ must include any states required to form the colored noise surrogate for $\{\Phi_t\}$. Equations (4.5)–(4.8) have been studied intensively and have been used in numerous tracking applications. Their application might be surprising since the A in (4.5)–(4.8) is not particularly close to $A(\Phi_t)$. However, one of the attractive propersties of the EKF is its generalized robustness; even if the actual environment differs somewhat from the model, by proper tuning, the Kalman filter gives a good estimate of the state [Gel84, Chapter 6]. Many examples of filter tuning are found in textbooks such as Maybeck's comprehensive three volume work [May79, May82a, May82b]. Maybeck shows clearly that sophisticated use of the degrees of freedom in the predictor–corrector framework in the Kalman filter creates serviceable solutions to numerous and seemingly unrelated estimation problems.

Sometimes the basic EKF must be adjusted by preprocessing the measurements. The primitive exogenous processes and the subordinate state and measurement processes in the LGM model are Gaussian, and a Gaussian density has a very thin tail. It may be that the samples of measurement noise are conspicuously different from those of the normative distribution and contain numerous outliers. The Kalman filter uses a linear weighting on the increments of the innovation process, and this has the effect of magnifying the influence of outliers; a single anomalous observation may overwhelm the effect of several more typical measurements. Although an isolated occurrence can be accommodated, if the filter time constants are long and the occurrences frequent, the estimate generated by the EKF will have significant error. Performance of the EKF estimator degrades significantly in this environment without a stage of preprocessing to reduce the influence of anomalous data points [HMZ87]. In the examples that follow, these outliers are created by target mismodeling, and the untoward influence of these anomalous observations will be apparent.

Conventional approaches have proven serviceable in those applications that only require a reasonably accurate computation of $\hat{x}_t$. If $\hat{\phi}_t$ is computed at all, it is done in a cursory manner, and the computation of $P_{x\phi}$ is not addressed at all. Particularly this last moment is useful in estimation and control, but the geometry is often so blurred in the model used to derive $\hat{x}_t$ that $P_{x\phi}$ cannot be recovered. With advances in sensor technology, the potential exists to address some of these deficiencies and to achieve significant improvement in estimation performance. For example, imaging sensors create measurements of quantities neglected in earlier algorithms. The raw image data is received as a sequence of matrices of gray levels, generated in packets or frames. These primitive data are neither in a form nor at a pace suitable for direct incorporation in a conventional estimation algorithm. In even a single frame, the fundamental target variables (e.g., target orientation) are buried in overwhelming spatial detail. An image processor recasts the data frame and extracts relevant features from it. The processor acts to both compress the data and focus attention on specific elements. Relevant geometric, topological, or spectral features of the image are computed, and categorization of the target is deduced upon their basis [Bha86].

When imaging sensors first made their appearance in tracking applications, the estimation architecture differed little from its lineal predecessors. For example, video and forward-looking infrared (FLIR) sensors create a sequence of pictures of a scene containing the target. No longer is the observation restricted to a point-equivalent object, but instead, features of spatial extent can be extracted from each image. In initial applications, however, the imager was viewed as a direct replacement for a nonimaging device. For example, in the air-to-surface *Maverick* missile, circa 1970, a video tracker was used to follow a mobile target. Scenes containing the target were generated at a fixed frame rate. From each data frame, the outline of the target was identified and its center located. In this way the diffuse image of the target and its surroundings was converted into a bearing angle of an equivalent point target. The temporal sequence of bearing angles formed the basis for estimation and guidance.

Imaging sensor-image processors have replaced conventional bearing sensors in an estimation architecture that differs in no essential way from that which would be used if the imager were replaced by a point sensor. Although the associated tracking algorithm is prosaic, experiment indicates that image-based estimators exhibit idiosyncrasies not common in other implementations. For example, if the rear portion of the target were suddenly obscured in the image frame, the center of the image would abruptly accelerate forward with obvious and detrimental effects for guidance. The source of this type of error resides in the rudimentary image processing used in the early applications. The processor locates the ostensible centroid of the target but ignores shape information because there is no means available to interpret it. Without compensation for changes in the spatial features of the target, the conventional estimator proves to be sensitive to structured obscuration. This peculiarity is not unique to image-based systems. Most orthodox filtering algorithms attenuate the noise by a generalized averaging. The motivation for this is clear in an LGM environment since the exogenous variables are unpredictable (Brownian motion). But the obscuration often found in image-based links has both a temporal and a strong spatial character; it is, consequently, less responsive to simple smoothing.

With more sophisticated processors, a pattern classifier can be employed to compute useful attributes of the image. A prespecified set of topical bins is specified and each image is placed in a bin. This reduces the output data rate to a manageable level (i.e., a bin number rather than the gray levels of a pixel array). The information obtained from the image is often complementary to that obtained from a point-location sensor. For example, the radar gives information on the motion of a point target, but the shape of the target gives information on target type and orientation. The errors inherent in the image link are, however, quite distinct from those found in the point-observation links. The relevant errors are misclassifications as occur when the target is placed in topical bin i despite the fact that the correct choice is bin j. This effect is not well portrayed by assuming that the exact (unquantized) variable is observed in additive Gaussian noise; that is, the image processor does not provide the true target type, say j, plus an $N(0, \sigma)$ measurement noise. Misclassifications depend upon the fidelity of the image and the sophistication of the processing, and these in turn depend upon range and geometry, the bin size, the sensitivity of the image elements, etc. The model of the image-observation is better

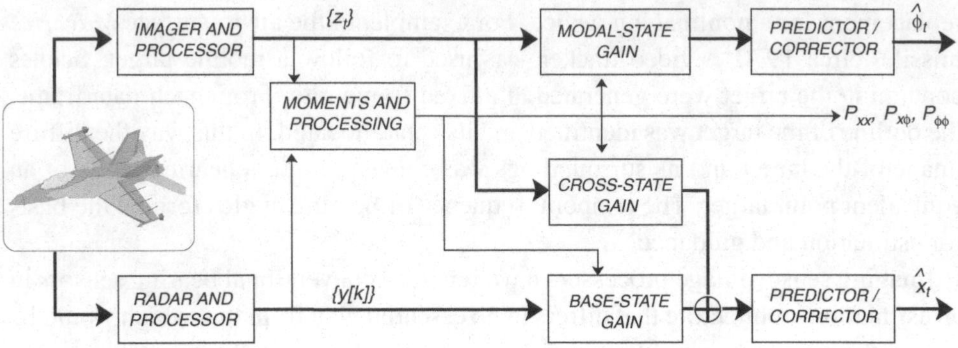

Figure 4.1. PME sensor fusion.

written as in (1.53):

$$dz_t = \lambda \mathbf{P} \phi_t \, dt + d\eta_t, \tag{4.9}$$

where the modal-state ϕ_t is suitable defined.

The PME provides an alternative form of sensor fusion, integrating a more comprehensive set of zygostate moments. It is akin to the single-model algorithms, but with the important distinction that the imager clearly provides a new and unique capability. The formal structure of the PME is shown in the block diagram of Figure 4.1 (see also [MS84], [KMR81]). The upper path is image-specific and uses observations of target shape to create an estimate of the modal-state $\{\hat{\phi}_t\}$. The imager is also used additively and multiplicatively to improve the utilization of the point-location data in the lower path. Both measurement processes are integrated in the moment generator block. Here, the canonical moments $\{P_{xx}\}$ and $\{P_{x\phi}\}$ are generated, along with $\{P_{xx\phi_i};\, i \in S\}$. These moments are important in their own right and are required to compute the base-state estimate. The lower path is similar to be the usual path from range-bearing to the estimate of the kinematic states, albeit with a gain that depends upon P_{xx} as computed in the moment generator. But the location estimate also depends explicitly upon the image measurement with a gain that depends upon the geometry through $P_{x\phi}$. In some instances, the measurement in the lower path is created, in part, by an imager operating in a *nonimaging* mode. For example, the imager may give a better bearing measurement than does the radar [CSB96b]. In this case, the former will be used to the exclusion of the latter without comment.

4.2 Image Modeling and Interpretation

In this chapter, we will use the PME to fuse image measurements with range-bearing (radar) measurements in a maneuvering target tracker. The target type is known and the regime variable is an indicator of the maneuver mode. The imaging sensor–processor generates a sequence $\{z_t\}$ used to determine the regime and to complement the radar measurement, $\{y[k]\}$, in location estimation. To illustrate, consider the problem of tracking a tank moving on an evasive path. The tracker contains a data base of tank images, stored and used as templates for classifying the received images. A *truth model* is created using scaled plastic replicas mounted on a gimbaled mechanism. The replica is rotated about its azimuth axis. Azimuth orientation is divided into L equal angular bins. In each bin, the replica is centered and a 2D-image template created. This set of projections forms the knowledge base upon which the image path in Figure 4.1 functions. The angular quantization is arbitrary, but the more image templates stored and the more detail preserved in the template, the harder the classification step becomes.

Figure 4.2 shows a pair of the images of a tank replica as it is rotated. Associate with each bin an orientation indicator vector ρ_t: $\rho_t = \mathbf{e}_i$; $i \in \boldsymbol{L}$, if the replica is in the ith angular bin. In the figures, the images are quite clear and it is evident that ρ_t could be deduced from the image.

Although the angular orientation of the target is a continuous variable, it is expedient to represent it with the discrete process also labeled $\{\rho_t\}$. This has the advantage of matching the angular state to the measurement, but it has the disadvantage that the turn rate cannot be written as the time derivative of $\{\rho_t\}$. Even so,

Figure 4.2. Images of a tank at two azimuth angles.

the angular state is not the modal-state. The rate of change in $\{\rho_t\}$ is closely related to the turn rate process: The faster the turn, the shorter the lifetime in an angular bin, with the direction determined by the sense of the turn. As $\{\rho_t\}$ moves from angular bin to angular bin, the turn rate can be inferred. There are intrinsic lags, however. As the target moves in azimuth, the image will be classified in a single bin until it crosses a bin boundary. Alternatively, if the image remains in a bin, has the target stopped rotating or is it merely in the process of crossing the bin?

These motion ambiguities are magnified by noise in the image. Disturbances in both the visible and the infrared bands produce changes in the apparent shape and internal structure of the target. For example, smoke and contrails in the visible band and plumes in the infrared band will cause the ostensible contour to differ from the true shape of an aircraft with a fixed geometric relation to the sensor. Shadows or internal reflections of sunlight can cause local edges in the interior of a visual image. Internal heating or external heat sources can likewise change the internal distribution of thermal signatures. Especially for ground vehicles, smoke and dust increase the likelihood of partial occlusion. Standard pattern recognition approaches that involve global features of the object (e.g., moment invariants or Fourier descriptors) degrade when confronted with even local changes in perceived shape.

Some of the potential image processing errors can be identified by studying Figure 4.2. A limited replica data base and the need to minimize the latency interval during which image classification is performed combine to restrict the precision with which the image models are stored. There are several types of error that arise in image interpretation, and these distinct error categories have demonstrably different effects on estimation accuracy. First, there are local ambiguities in the image that cause the processor to place the target in the orientation bin next to the correct bin. For example, slight distortions in the ostensible silhouette will lead the processing algorithm to misclassify an image by placing it in an adjacent bin. This is called a *nearest neighbor error* (NNE). Note that NNEs are particularly important in creating a false indication of a turn because a lateral acceleration is initially manifest in motion to the adjacent angular bin.

Another type of image interpretation error is illustrated by viewing the picture at a long range. When the image is of very poor quality (see, for example, image (4, 1) in Figure 1.7), the template matching procedure fits the image arbitrarily. To indicate the global nature of these errors, they will be referred to as *uniformly distributed errors* (UDE). Such errors are most common at ranges and geometries with few pixels-on-target or when there is significant image occlusion. The UDEs are most like the white noise disturbances often used in the radar models.

The image in Figure 4.2 is such that it is easy to see interior features of the tank. From these, it is possible to state that the tank is moving away from the observer. With lower image clarity, these features would become less distinct. Many classifiers emphasize the silhouette of the target. There is frequently more contrast at the target–background boundary, and the image processor uses this contrast to locate the silhouette. If the images templates are also reduced to silhouettes, the measured silhouette can be used to classify the orientation. Unfortunately, a silhouette is compatible with more than one angular bin. For example, the tank moving directly toward the sensor has the same silhouette as the tank moving directly away. When the image processor places an image in a bin associated with the same contour, but which is symmetrically placed about a plane perpendicular to the line of sight, a *projection error* (PE) is said to occur. Only the internal structure of the image can be used to reduce this aliasing.

The three error categories described above provide a basic error taxonomy useful in exploring the degree to which image enhancement can be used to improve estimation. In many encounters, both the magnitude and the relative distribution of the error types changes over time. For example, at long ranges the target subtends few pixels. Uniformly distributed error predominates, and the imager gives little useful orientation information. Fortunately, at these ranges, there is little advantage to maneuvering, and adequate estimation and prediction can be done without maneuver detection. At close ranges, the image is big enough that most of the image ambiguity is captured by the PE category. But these are precisely the ranges at which the internal structure can be used, if desired, to most effectively resolve this error. Furthermore, at close range maneuvers are of little consequence since the requisite prediction time is small. The NNEs are most prominent at intermediate ranges. Although this might seem to imply that NNEs are fairly benign, this is not the case. It is in this intermediate region that maneuvers are most effective and the NNEs have the most pronounced influence on false motion detection.

It is clear that an orientation measurement would be useful in determining changes in target motion. In a novel study of the utility of augmented sensor systems, Lefas noted that if the roll angle of an aircraft were transmitted to the tracker, improved performance would be attainable using a roll-angle adaptive filter [Lef84]. The problem that motivated his analysis involved estimating the trajectory of a cooperative airplane (air traffic control), and attention is focused on adaptive single-model filters in which the gain is adjusted whenever a maneuver is suspected. Although the estimation architecture would not be attainable in a hostile engagement, his results illustrate the efficacy of maneuver-adaptive estimation for tracking agile targets.

In an important series of papers, Andrisani and his coworkers studied orientation aiding within the EKF framework. The angular variable was added to the state space model in [AKKS91] and [AS87]. The measurement equation used in the references mimicked (4.4); that is, the center of reflection and the angular variables were both measured with an additive, white noise channel. The authors observed that "attitude information, obtained from an optical image processor, in a tracker's

Kalman filter gives valuable lead information when tracking a fixed wing aircraft by being able to determine the direction of the lift vector" [AKG86]. The *lead* in the image data is of fundamental importance in improved tracker performance. Position is two integrals removed from an acceleration. A maneuver manifests itself in a path change over time, but initially, this change is difficult to detect in the noisy location data available to the tracker. Orientation is a much faster indicator. After factoring in geometric effects, a change in orientation of a fixed-wing aircraft gives an immediate indication of a maneuver, and this can be used to adjust the gains in the EKF. For a target such as a tank, a jinking path is created by turning, and the turn rate can be deduced from changes in angular orientation. In both cases, direct measurement of target orientation gives a faster indication of a maneuver than could be obtained from noisy trajectory data alone.

4.3 Tracking Maneuvering Targets

In this section we will present algorithms that achieve fusion of unlike sensors. We wish to model the maneuver process with more fidelity than possible with an LGM shaping filter, and we wish to use a measurement model more representative of the idiosyncrasies of an image classifier. The previous section presents some issues associated with online image generation and processing. In the tracking application, the imager creates a measurement sequence that is discrete in both time and space. Time quantization is not unusual, but spatial quantization is unorthodox. Rather than generating a measurement whose natural range is an interval (e.g., bearing), the imager translates a data frame into a statement of category selected from a prespecified alphabet of topical symbols.

Equation (4.1) is the motion model of the target. If $\{\Phi_t\}$ is ignored, a radar-based EKF is simple to design but has obvious deficiencies. An increase in model complexity gives a more realistic motion equation that preserves some of the geometry of the engagement. Rather than neglecting the turn rate, replace it with a Gaussian surrogate. Select the initial probability distribution so that $\{\Phi_t\}$ is a wide sense stationary process with a power spectral density $\Phi(\omega)$. A shaping filter is used to represent the turn rate process. This increases the size of both the state vector and the Brownian motion [Ber83]. If the shaping filter is first order, the integrated motion model is:

$$d\begin{bmatrix} X \\ Y \\ V_x \\ V_y \\ \Phi \end{bmatrix} = \begin{bmatrix} 0 & 0 & 1 & 0 & 0 \\ 0 & 0 & 0 & 1 & 0 \\ 0 & 0 & 0 & -\Phi & 0 \\ 0 & 0 & \Phi & 0 & 0 \\ 0 & 0 & 0 & 0 & -\frac{1}{\tau} \end{bmatrix} \begin{bmatrix} X \\ Y \\ V_x \\ V_y \\ \Phi \end{bmatrix} dt + \begin{bmatrix} 0 & 0 & 0 \\ 0 & 0 & 0 \\ 1 & 0 & 0 \\ 0 & 1 & 0 \\ 0 & 0 & 1 \end{bmatrix} d\begin{bmatrix} w_x \\ w_y \\ w_a \end{bmatrix}, \tag{4.10}$$

where $\{w_a\}$ is a fictitious Brownian motion used to generate the ostensible maneuver path. The turn rate model is a two-parameter family, with τ the time constant of the shaping filter, and $R(0) = E[\Phi_t^2]$ the intensity of the turn rate process. The model (additive and multiplicative acceleration, nonlinear observation) leads naturally to a two-parameter family of filters, $\mathsf{EKF}(\mathbf{\Phi}(\omega);\ R(0), \tau)$, with the correct choice depending on the volatility of the encounter. Predictably, improved tracking can be obtained by tuning them to a pseudo-encounter (i.e., tuning for response rather than trying to mimic the encounter dynamics [CT84]). The attractive feature of (4.10) is that the acceleration dynamics are included explicitly, and thus the correlations in the acceleration direction are maintained during flight. The unattractive feature of $\mathsf{EKF}(\mathbf{\Phi}(\omega);\ R(0), \tau)$ is that it is based upon a localization of the motion model. The turn rate changes suddenly, and to replace the individuated equations with an average motion blurs the intrinsically partitioned nature of the equation of evolution. The estimates can be poor even if the first-order model of (4.10) is replaced with higher order models [SHK93].

To integrate the maneuver motion more precisely into the estimation algorithm, the encounter model must be completed. Because the turn rate tends to be nearly constant over intervals with sudden changes at unpredictable times, a useful model is created by partitioning the range of turn rates into K levels; $\Phi_t \in \{a_1, \ldots, a_K\}$ [RW78]. The indicated acceleration states are aggregates insofar as the actual acceleration is randomly placed within the associated bin. Replace the multivalued process $\{\Phi_t\}$ with a process taking on values in a set of unit vectors. Let $\{\alpha_t\}$ be the maneuver indicator process; $\alpha_t = \mathbf{e}_i$ if $\Phi_t = a_i$. The base-state is described by a family of K stochastic differential equations indexed by the maneuver acceleration and continuous at the modal transitions.

To complete the modal model, suppose the successive maneuver modes are represented by an $\mathcal{F}_t$-Markov chain with transition rate matrix Q^α. This subsumes the case in which the turn rate is an unknown constant, $Q^\alpha = 0$. The orientation process, $\{\rho_t\}$, will be modeled rather coarsely. Associated with every turn rate hypothesis (e.g., $\alpha_t = \mathbf{e}_i$), $\{\rho_t\}$ will be represented with an $\mathcal{F}_t$-Markov process with generator $(Q^i)'$. The rate matrices $\{Q^i;\ i \in \boldsymbol{K}\}$ can be selected to match the mean sojourns in each orientation bin and the bin-to-bin transitions. The latter are usually quite simple since in most cases the orientation process must transition to a contiguous bin. It will be supposed that bin transfers and maneuver mode transitions are not coincident. The modal state process is created by composing α_t and ρ_t: $\phi_t = \alpha_t \otimes \rho_t$. Unfortunately, this Markov model does not display relevant detail of the lifetime distributions in modal sojourns, either in bin or maneuver mode.

Although modeling the turn rate in this way makes analysis more difficult because it leads to non-Gaussian sample paths, Equation (4.1) is a linear equation with an unknown coefficient, α_t. This suggests that estimation could be divided into three distinct phases. "First the maneuver must be detected. Second the Kalman filter state is corrected to compensate for the previous maneuver. Third, after detection and correction, the Kalman filter parameters are correspondingly adjusted in anticipation of future maneuvers" [Bog87]. Although intuitively appealing, there are difficulties with this seemingly self-evident procedure. It was noted for example in the reference that "in one application . . . it was possible for the maneuver to be completed by the time it could be successfully detected" [Bog87]. The lags inherent in the maneuver detection and correction can so delay a response as to make the procedure unusable. Further, the need to adjust the parameters of the Kalman filter raises subtle issues. The filters are concatenated as the dynamic mode changes, and the gain of the EKF has the error covariance as a factor. This matrix is computed, forward in time, based upon the ostensible maneuver process; when a change in the maneuver mode is detected, the dynamic equation of $\{P_{xx}\}$ is changed in concert. But what should be done with $\{P_{xx}\}$ at the time of modal transition? This issue arose in the study of multiple-model algorithms. It has been proposed that "the covariance of the bias estimation filter (must be) reset to reflect the increased uncertainty in the bias estimation due to the transition" [WF88]. The uncertainty increases in the interval between a transition and its detection, but the degree to which it changes is difficult to quantify.

Various ways of resetting the covariance have been proposed. In [SH90], the turn rate was treated as additive disturbances, and its $\mathcal{Z}_t$-conditional mean was added to the conventional, nonmaneuvering, EKF. This proved to be a useful approach but exhibited poor performance subsequent to the maneuver. This deficiency was predictable and followed directly from the fact that a small value of R_χ in (4.6) will result in a slow decay of any errors created during the maneuver. To avoid this, covariance adaptation is required. As already noted, adding pseudonoise through W reduces the filter time constants and improves maneuver tracking but makes nominal operation more volatile. In one instance it was suggested that the elements of the W matrix be augmented proportionately "to the amount of acceleration along each axis" [Bek83]. However, W augmentation should be eschewed during a constant turn. Williams and Friedland [WF88] point out that it is actually the uncertainty in maneuver estimation that induces a need to change the EKF dynamics. Fortunately, the requisite information is readily available from $\{\hat{\alpha}_t\}$. The maneuver variance is given by

$$\mathrm{var}(\Phi_t) = \sum_i a_i^2 \hat{\alpha}_i - \left(\sum_i a_i \hat{\alpha}_i \right)^2 . \tag{4.11}$$

Equation (4.11) has an intuitive appeal. When the maneuver is resolvable from the image sequence ($\hat{\alpha}_t \approx \mathbf{e}_i$), acceleration uncertainty is small. Alternatively, as the a posteriori probabilities of the maneuver hypotheses become more diffuse, $\mathrm{var}(\Phi_t)$ grows. This is precisely the sort of situational adaptation required. The process noise intensity W can be augmented proportionally to acceleration uncertainty (rather than acceleration magnitude), transformed into the Cartesian coordinate system [SH92]. The corresponding value of $\{P_{xx}\}$ is the solution to (4.6) with the increased R_χ. Through this simple artifice, a maneuver adaptive EKF is created. The mean acceleration is added as a bias to the base-state drift, and the tracking time constants are modified as well.

The imager is a feature-matching block, and its errors are not well described by additive Gaussian noise. The imager collects data at a rate of λ frames/s and places each target image into one of L equally spaced orientation bins. The output of the image processor is written as an L-dimensional counting process $\{z_t\}$, the ith component of which is the number of times the target has been placed in bin i on the interval $[0, t]$. This sequence of symbols can be interpreted by a temporal processor to give the relative likelihoods of the various turn rate hypotheses. This development is carried out in [SH90] and can be summarized as follows. The composite modal-state of the target is given by $\{\phi_t\} = \{\alpha_t \otimes \rho_t\}$, the Kronecker product of maneuver and orientation. From ϕ_t both α_t and ρ_t can be deduced:

$$\alpha_t = (\mathbf{I}_K \otimes \mathbf{1}'_L)\phi_t, \tag{4.12}$$

$$\rho_t = (\mathbf{1}'_K \otimes \mathbf{I}_L)\phi_t. \tag{4.13}$$

The quality of the imager is determined by the frame rate λ and the $L \times L$ discernibility matrix $\mathbf{P} = [\mathbf{P}_{ij}]$, where $\mathbf{P}_{ij}$ is the probability that bin i will be selected by the processor if bin j contains the true target orientation at time of image creation: $\mathbf{P}_{ij} = \mathcal{P}(\Delta z_t = \mathbf{e}_i \mid \rho = \mathbf{e}_j)$ [SH89]. Note that the dimension of $\mathbf{P}$ is that of the orientation, and not that of the KL-dimensional modal-state, ϕ_t. The discernibility matrix can be expanded to accommodate ϕ_t in a direct manner:

$$\mathbf{P} \mapsto (\mathbf{1}_K \otimes \mathbf{I}_L)\mathbf{P}(\mathbf{1}'_K \otimes \mathbf{I}_L).$$

In what follows, we will talk about $\mathbf{P}$ as an $L \times L$ matrix while in the algorithms $\mathbf{P}$ will be $KL \times KL$. The modal-state is an $\mathcal{F}_t$-Markov process with generator Q' composed of the primitives Q^α and $\{Q^i; i \in \boldsymbol{K}\}$.

The fidelity of image interpretation is a function of the frame rate (λ), the ability of the image to correctly classify a single image ($\mathbf{P}$), and the tempo of the encounter (Q). These factors interact in subtle ways. For example, a rapid tempo requires a high frame rate if expeditious maneuver detection is to be accomplished. In the same way, to compensate for a low frame rate, the quantization must be fine (L large) and the processing accurate ($\mathbf{P} \approx \mathbf{I}$). In the tracker architecture shown in Figure 4.1, it is usually assumed that the image-based path gives a much more expeditious indication of a turn than does the conventional path. The motivation for this stems from the fact that the imager measures directly the target features that most clearly manifest changes when the target turns. In such cases, there is little error if turn rate estimation is based exclusively on image data. The $\mathcal{Z}_t$-conditional probabilities of the various turn rate hypotheses are given by the K-dimensional process $\hat{\alpha}_t$:

$$\begin{aligned}\hat{\alpha}_t &= [\mathcal{P}(\Phi_t = a_i \mid \mathcal{Z}_t)] \\ &= (\mathbf{I}_K \otimes \mathbf{1}'_L)\hat{\phi}_t. \end{aligned} \tag{4.14}$$

The L-dimensional, piecewise constant process $\{\vartheta_t\}$ is defined in Chapter 2 (see Section 2.3):

$$\Delta\vartheta_t = \lambda \mathbf{P}'(\hat{\boldsymbol{\lambda}}_t^{-1} * \Delta z_t).$$

The PME is made simpler because of the continuous base-state paths and the direct observation x_t. Recall that the covariance of the increment in the base-state innovations process is

$$R_{yy} = H P_{xx} H' + R_x = D_{yy}^{-1}.$$

Let $\gamma_x = P_{xx} H' D_{yy}$. We have the following set of equations:

the PME: continuous base-state, time-discrete measurements

Between observations:

$$\frac{d}{dt}\hat{\phi}_t = Q'\hat{\phi}_t,$$

$$\frac{d}{dt}\hat{x}_t = \sum_i A_i R_{x\phi_i},$$

$$\frac{d}{dt}P_{x\phi} = \sum_i A_i P_{(x\phi_i)\phi} + P_{x\phi} Q,$$

$$\frac{d}{dt}P_{xx} = \sum_i (A_i P_{(x\phi_i)x} + (\cdot)') + R_{\chi},$$

$$\frac{d}{dt}P_{xx\phi_m} = \sum_i (A_i P_{(x\phi_i)x\phi_m} + (\cdot)' + P_{xx\phi_i} Q_{im}).$$

At a modal observation:

$$\begin{aligned}
\hat{\phi}^+ &= \hat{\phi}^- * \Delta\vartheta, \\
\Delta\hat{x} &= P_{x\phi}\Delta\vartheta, \\
\Delta P_{x\phi} &= -\Delta\hat{x}\Delta\hat{\phi}' + \sum_k P_{x\phi\phi_k}\Delta\vartheta_k, \\
\Delta P_{xx} &= -\Delta\hat{x}\Delta\hat{x}' + \sum_k P_{xx\phi_k}\Delta\vartheta_k, \\
\Delta P_{xx\phi_m} &= -\Delta\hat{\phi}_m\Delta\hat{x}\Delta\hat{x}' - \Delta\hat{\phi}_m P^+_{xx} - \Delta\hat{x} P^+_{\phi_m x} - P^+_{x\phi_m}\Delta\hat{x}' \\
&\quad + \sum_k P_{xx\phi_m\phi_k}\Delta\vartheta_k.
\end{aligned}$$

At a base-state observation:

$$\begin{aligned}
\Delta\hat{x} &= \gamma_x\Delta\nu_x, \\
\Delta P_{x\phi} &= -\gamma_x H P_{x\phi}, \\
\Delta P_{xx} &= -\gamma_x R_{yy}\gamma_x', \\
\Delta P_{xx\phi_m} &= -\gamma_x H P_{xx\phi_m} - P_{xx\phi_m} H'\gamma_x.
\end{aligned}$$

4.4 An Example: An Antiship Missile

To illustrate the utility of image enhancement in tracking a maneuvering target, suppose an antiship missile is launched at a range of 80 km, an altitude of 1 km, and speed of 300 m/s. After a free fall to 780 m, the missile approaches a ship at a speed of 335 m/s. Nearing the ship at constant altitude, the missile performs a series of 7 g jinks, coasting for 10 s and then making a final 3 g turn toward its intended destination. This trajectory, along with several other motion paths, was created by investigators at the Naval Surface Warfare Center (NSWC), Dahlgren Division to provide realistic benchmark tests for proposed tracking algorithms [BWC94]. The sensor suite consists of a radar and a collocated imager on the ship. The radar errors are Gaussian with standard deviation 40 m in range and 1.75 mr in bearing (at 50 km this translates into about a 90 m cross-range error). The nominal radar interdwell time is 1 s. Because the duration of a turn is only a few seconds, the radar has neither the update rate nor accuracy to resolve turns well. Radar-exclusive, input-identification algorithms are destined to fail because of the lack of timely motion data.

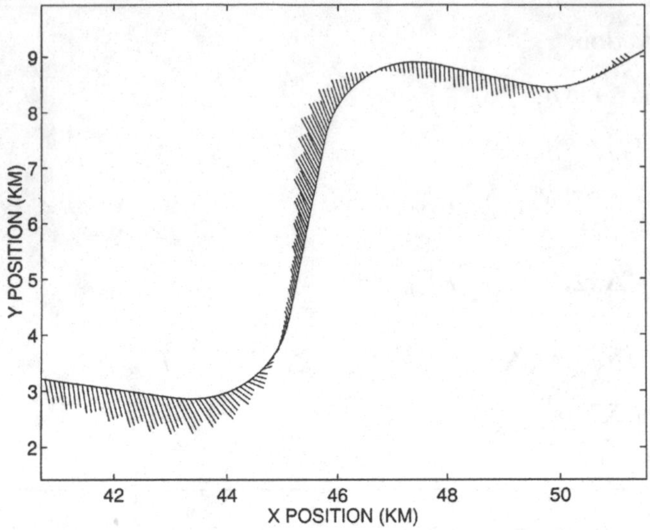

Figure 4.3. The target path along with estimates of position generated by EKF(1sR).

The radar-exclusive EKFs selected for comparison are found by neglecting the maneuvers (i.e., $\Phi_t \equiv 0$) and setting $W = 1$. In the algorithms that follow, the initial covariance is diagonal with 100 m standard deviation in position and 20 m/s in velocity. All of the estimators are initialized on the true base-state of the target at the time of detection, though the estimators don't know this.

A sample of a portion of the path is shown in Figure 4.3 along with a feather plot of the output of the nominal EKF (denoted by EKF(1sR)). The target is detected at $t = 75$ s, just before the first turn, which begins at $t = 77.3$ s. This gives the estimator 2.3 s to initialize itself on the initial coast segment.

The EKF is seen not to be adequate in this application. Because of the omniscient initialization, EKF(1sR) tracks into the first turn, but it begins to lag thereafter. By chance, the target later turns toward the estimate. This fortuitous event causes the error in the EKF to be reduced, but the error soon builds up again in the other direction. The intermediate coast is not long enough for the EKF to return to quiescent operation. Again the error is small only when the target happens to turn toward the estimate.

More intensive use of the radar will lead to improved performance, although this has the disadvantage that it may preclude the use of the radar for tracking other targets. With an interdwell interval of 0.1 s, an EKF using the same radar can be developed (labeled EKF(0.1sR)). The mean radial errors of EKF(1sR) and EKF(0.1sR) are shown in Figure 4.4. This figure (as were all plots of mean error) was generated as the sample average of ten independent runs. Certain unappealing aspects of an EKF are apparent from the graph. With perfect initialization, both

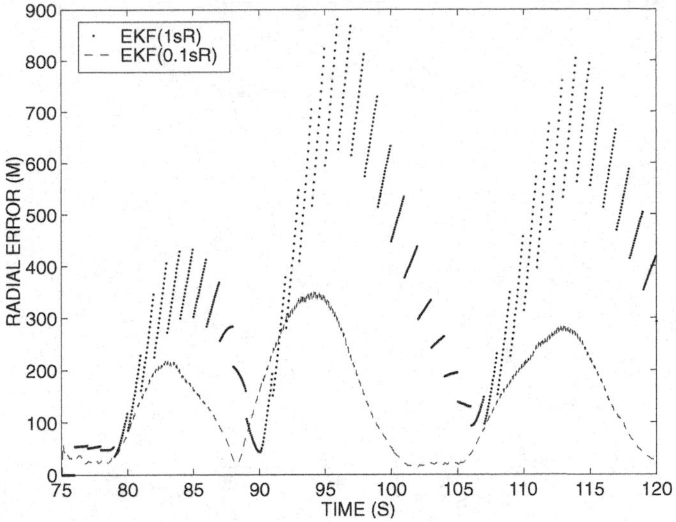

Figure 4.4. Mean radial error for EKF(1sR) and EKF(0.1sR).

EKFs begin with small error. The first turn, beginning at 77.3 s, causes the radial error to increase, initially at about the same rate for both EKFs, but the rates soon separate themselves. Indeed the error before an update builds to 900 m for EKF(1sR) and 350 m for EKF(0.1sR). This occurs despite the 100 m accuracy for the measurements. This excess error is caused by the failure of the EKFs to avoid a buildup of velocity error during the first turn. Velocity error confuses both of the EKFs so that when the target returns to coast motion (~90 s and ~105 s), neither tracker returns to quiescent conditions. This response is typical on paths with mode transitions. Mismodeling leads to errors early in the encounter that are reflected throughout the path, and not just in the portion of the path during which the mismodeling is most egregious.

It should be noted that the radar used in the benchmark studies loses lock at about 400 m radial error at this range. With the indicated errors, the trackers would likely lose lock on most runs. Certainly, better performance is achieved when the interdwell time on the radar is short. A shorter extrapolation time yields less error during the turn for both position and velocity, but interestingly this reduction is not nearly proportional to the increase in sensor utilization (by a factor of 10). The EKF at 10 dwells/s is simply not acceptable.

There are several contributors to the poor performance of the EKFs in this application, but a primary one is related to the imprecision of the covariance calculation, P_{xx}. Figure 4.5 shows the path of the target along with the one-σ error ellipses (formed from the error covariance P_{xx} and shown every 0.2 s). The small ellipses are associated with EKF(0.1sR) and the larger with EKF(1sR). The ellipses are centered on the associated estimate of the EKF. The improvement in performance

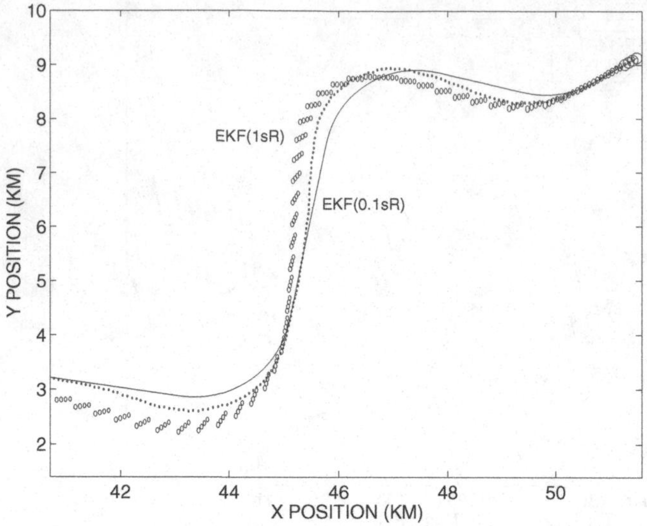

Figure 4.5. Computed error ellipses for EKF(1sR) and EKF(0.1sR).

associated with more frequent measurements is apparent in this figure too. However, both of the EKFs have an overly optimistic view of their performance as evidenced by the area of the ellipses. Neither of the EKFs use the innovations efficiently, leading therefore to tracking errors far in excess of the measurement errors. The failure of the error ellipses (or even two-σ ellipses) to enclose the path has important implications. Adaptive control of the sampling rate, of the window size, or of the radar SNR are often based upon the computed error covariance matrix. When P_{xx} is too small, the adaptation is unsuitable.

To illustrate the utility of image enhancement, suppose this same radar is augmented with an imager. The imager has the same sample rate as does the radar: $\lambda = 1$ f/s. Very coarse bins will be used: $L = 12$ (30° angular bins). Both the frame rate and the angular discretization are well within current practice [SVH93]. The **P** matrix characterizes the fidelity of the imager–image processor for a single frame. The error taxonomy separates the imager error into three categories and constructs **P** from these (exclusive) primitives. Local image degradation is caused by pixel noise and quantization. This will be represented by assuming that the image is misclassified symmetrically into a neighboring bin (spillover error, NNE). Other sources of image distortion act globally (e.g., severe occlusion of the image). This results in errors of large magnitude and will be represented by assuming that the image is classified with uniform probability across all possible angular bins (UDE). The most distinctive image error occurs when the ostensible orientation bin is symmetrically placed about a line perpendicular to the line of sight (PE). This error source represents the intrinsic ambiguity that arises when the orientation classifier relies heavily on the silhouette of the extended object.

Table 4.1. *Error parameters for good and poor imagers.*

Imager	UDE	NNE	PE
IG	0.05	0.05	0.1
IP	0.2	0.1	0.4

Two imagers will be studied. The first (IG) is of good quality, correctly classifying target orientation 80% of the time. The second (IP) is of poor quality, providing correct classification only 30% of the time. The specific parameters of the imagers are given in Table 4.1.

Although the standard deviation of the imager errors does not give a clear measure of imager accuracy in the same way it does for the radar, it is interesting to compare the size of the angular errors for the two sensors. If the true target orientation is uniform in the bin, and if the projection error, which can vary from 0° to 180°, is set at 90°, the standard deviation of IG is 37°, while that of IP is 74°. Both error figures are so big as to suggest the imager would be of little use in tracking (compare the radar bearing error of 0.1°). However, the tracker has a well-defined model of angular motion.

To particularize the PME, let

$$\Phi_t \in \{0.2 \text{ r/s}, \alpha_t = \mathbf{e}_1;\ 0 \text{ r/s}, \alpha_t = \mathbf{e}_2;\ -0.2 \text{ r/s}, \alpha_t = \mathbf{e}_3\}.$$

The mean sojourn time in each of the maneuver states will be assumed to be 5 s, with each turn being followed by a coast. From coast, the turns are equally likely. The initial maneuver modes will be assumed to be equally likely. From this the modal dynamics can be deduced. The PME using imager IG is labeled PME(IG), and the PME using imager IP is labeled PME(IP). A comparison of sensor utilization is given in Table 4.2.

Mean radial error using PME(IG) is shown against that for EKF(1sR) in Figure 4.6. The radar sample rate for PME(IG) is equal to that of the nominal radar:

Table 4.2. *Radar rate and imager quality for EKF and PME.*

Algorithm	Radar Rate	Imager
EKF(1sR)	1 dwl/s	none
EKF(0.1sR)	10 dwl/s	none
PME(IG)	1 dwl/s	good
PME(IP)	1 dwl/s	poor

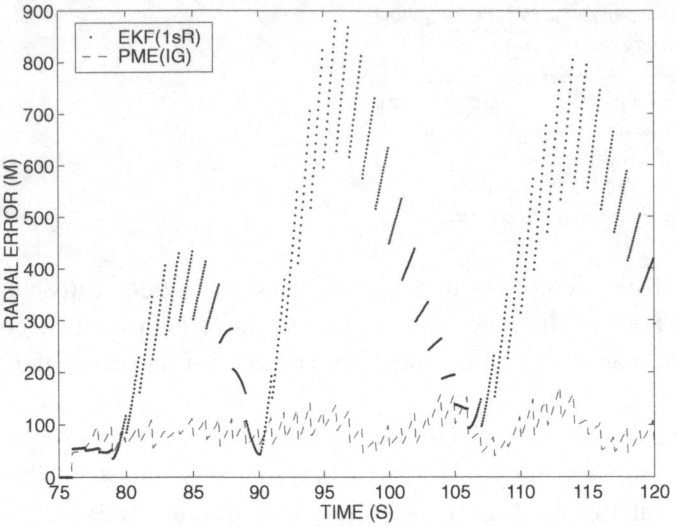

Figure 4.6. Mean radial error for EKF(1sR) and PME(IG).

1 dwell/s. With the perfect initialization, the EKF is superior to the PME into the first turn: EKF(1sR) errors of 50 m on [75, 80] s compare with 100 m errors for PME(IG). The EKF is superior to PME(IG) because it does not expect a turn and none occurs. However, as the turn proceeds, the EKF has errors that grow to 900 m before an update and to 600 m after an update. The PME keeps errors during the turn to less than 150 m, a significant improvement given that they both use the same radar. Again it should be noted that the points along the curve at which the EKF is superior to the PME(IG) are those points that the target chances to turn into the estimate. The relative performance of PME(IG) and PME(IP) varies during the turns (not shown), but not to the degree nor in the direction that one might expect. A good imager leads to improved performance for the most part, but not uniformly. Over the full path, the PME keeps its errors low: less than 100 m after a radar update and less than 60 m over most of the path for PME(IG) with slightly higher numbers for PME(IP) (see [SBCV99] for the comparison). Both display tracking performance that is superior to an EKF with ten times the radar sample rate.

As noted, the problems encountered by the EKFs are due in part to their inability to properly ascertain the error covariance along the path. Because their gains are lower than they should be, they are unable to correct the large errors spawned by the turns. Figure 4.7 displays a single sample of the centered error ellipses generated by PME(IG). These are shown along with the previously displayed ellipses of EKF(1sR). Note that the error ellipses of PME(IG) are larger than those associated with EKF(1sR) as befits the recognition of the uncertainty created by maneuvers. Further, the ellipses lie closer to the true path – the two-σ ellipses would enclose the

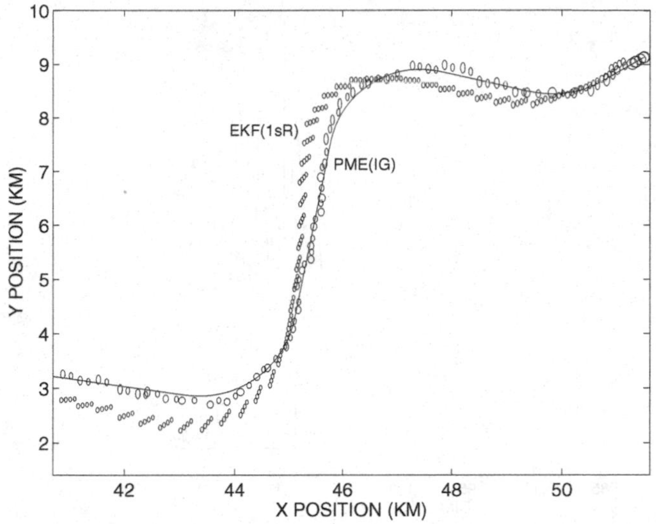

Figure 4.7. Computed error ellipses for EKF(1sR) and PME(IG).

true path in all but a few instances. The computed error variance is more responsive to changing conditions. The corresponding ellipses for PME(IP) look much the same as those of PME(IG).

The PME achieves improved performance with its explicit inclusion of the turn dynamics in the path model. The accuracy of the image-based maneuver identification is dependent upon the imager accuracy (i.e., on the discernibility matrix **P**). Figure 4.8 shows a sample of the $\mathcal{Z}_t$-probability of a turn: $\{1 - \alpha_2\}$. Both the poor

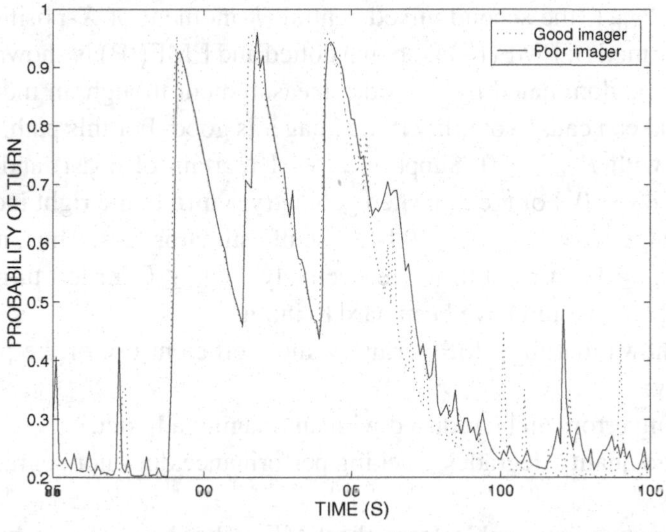

Figure 4.8. The computed probability of a turn PME(IG) and PME(IP).

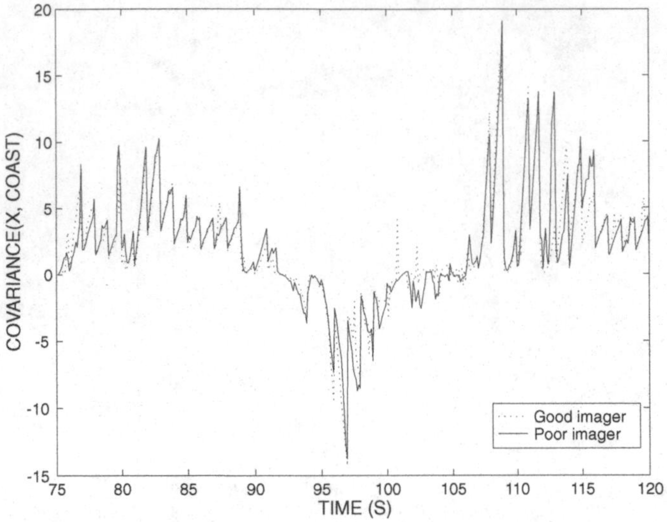

Figure 4.9. The cross error moment of X-position and coast maneuver mode.

imager (shown solid) and the good imager (shown dotted) are displayed. The good imager is both faster and more certain in its identification of a return to coast as compared to the poor imager. It is not, however, nearly as much better as the relative sizes of the imager errors would suggest. The good imager is more confident, but when it makes an error, it tends to be a bigger blunder.

The geometry of motion enters into the extrapolation and update through $\{P_{x\phi}\}$, the second mixed central $\mathcal{G}_t$-moment of the zygostate. Most algorithms do not compute this moment, and so $\{P_{x\phi}\}$ is not used to improve the estimate. Figure 4.9 shows a sample path of $\{P_{X\alpha_2}\}$ (the second mixed central $\mathcal{G}_t$-moment of X-position and the coast maneuver mode): PME(IG) is shown dotted and PME(IP) is shown solid. The local variation is dominated by the radar measurement though an individual image measurement can cause volatility if the imager is good. For this path, a right turn is associated with $P_{X\alpha_2} \geq 0$. Suppose $\hat{\alpha}_2 \approx 1$ (certain of coast) and a turn begins: $\tilde{\alpha}_2 = \alpha_2 - \hat{\alpha}_2 < 0$. For the indicated geometry, a turn to the right increases the X-velocity; that is, $\tilde{x}_t = x_t - \hat{x}_t \leq 0$ because of estimator lags. Thus, it is not surprising that $P_{X\alpha_2} \geq 0$ for right turns (respectively $P_{X\alpha_2} \leq 0$ for left turns), but the actual path of $\{P_{x\phi}\}$ would have been hard to intuit.

This example shows that the PME permits a more efficient use of the primary sensor. Specifically:

- The tracking error can be reduced with an imaging adjunct.
- If the image quality degrades, tracking performance does not degrade proportionally.
- The performance attainable from the PME is hard to achieve by more intensive use of the primary sensor.

4.5 Renewal Models for Maneuvering Targets

While relatively uncomplicated, Equation (1.12) has a major drawback as a pattern for deliberate maneuvers. It is well known that if $\{\alpha_t\}$ is a Markov process, the sojourn times in each maneuver mode are exponentially distributed, and as a consequence, short lifetimes predominate. A sample function of a Markov process provides convincing evidence of the occurrence of maneuvers too short to accomplish any evasive intent (see, for example, [LF91]).

With the inherent lags in the target dynamics and in the internal actuating loops, it is difficult to imagine that a pilot would be inclined to select maneuvers with duration below some limit determined by the agility of the vehicle and the responsiveness of the actuators. To do so would result in much pilot effort and little meaningful path deviation. In fact, pilots are trained to fly evasive jinking maneuvers in the form of concatenated periods of constant turn-rate maneuvers with the duration of each segment lasting seconds. The short sojourns predicted by the Markov model seldom appear in exercises. During quiescent motions, the modal lifetimes are nominally longer, and there is a low probability of the occurrence of overly brief sojourns even with the Markov model.

To correct this readily apparent deficiency, a modal model should be used that preserves the unpredictability of the turns but reduces the incidence of excessively short sojourns. Figure 4.10 shows three different probability density functions with the same mean: 1 s. The first ($R = 1$) is exponential and clearly shows a preponderance of short lifetimes along with a rather fat tail. The second ($R = 2$) shows a dramatic reduction in the frequency of very short lifetimes, along with some

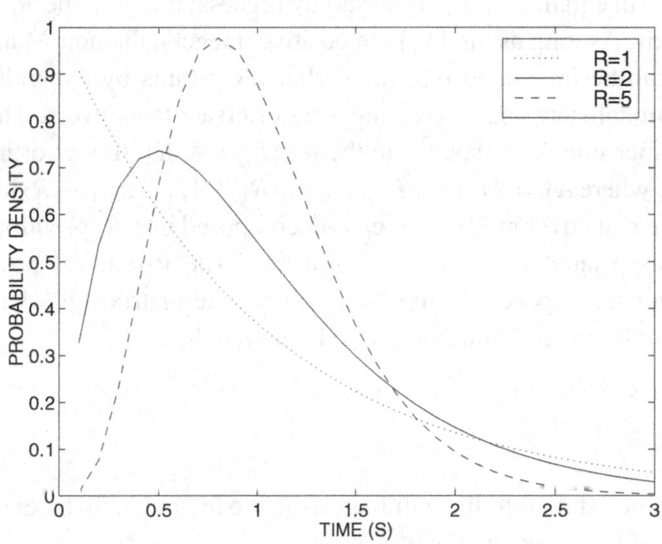

Figure 4.10. Gamma densities for three values of R.

decrease in protracted ones. The third curve ($R = 5$) is sharply peaked, indicating more quasi-predictable intervals.

The three curves shown in Figure 4.10 are gamma density functions. The gamma density is a two-parameter family, $\gamma(t;\, R, \lambda)$; $R, \lambda \geq 0$, in which R controls shape of the density and λ controls time scale:

$$\gamma(t;\, R, \lambda) = \begin{cases} \lambda(\Gamma(R))^{-1}(\lambda t)^{R-1}e^{-\lambda t}, & t \geq 0, \\ 0, & \text{otherwise.} \end{cases} \tag{4.15}$$

The mean of a gamma distributed random variable is $\nu = R/\lambda$, and because it has a more intuitive connotation, ν will be used instead of λ to indicate the time scale. The plots in Figure 4.10 are for $\gamma(t;\, R, \nu = 1)$.

The modal lifetime distribution using a gamma density is more reasonable in applications than is that using an exponential density. Suppose therefore that, instead of using a Markov process, we model $\{\Phi_t\}$ with a finite state *renewal process* having modal sojourns distributed according to a γ density. A renewal process is distinguished by two things: the sequencing of modes, for example, $\mathbf{e}_i \mapsto \mathbf{e}_j$ at time t ($\Delta\alpha_t = \mathbf{e}_j - \mathbf{e}_i$), and the lifetime in each mode. Both are unpredictable (random). In a renewal model, the modal-state sequence is Markovian with transition matrix P: P_{ij} is the probability that $\mathbf{e}_i \mapsto \mathbf{e}_j$ given that $\alpha_{t-} = \mathbf{e}_i$ and the maneuver mode changes at time t, with $\mathsf{P}_{ii} = 0$ for all $i \in \boldsymbol{K}$ to complete the matrix. The sojourn times are independent given the past sequence of modal transitions. Though not identically distributed across modes, it will be assumed that the sojourn times in a maneuver state are given by the gamma density (4.15). Instead of the single transition rate matrix Q in (1.12), the maneuver dynamics are now characterized by the pair $\{\mathsf{P}, \gamma(t;\, R_i, \nu_i);\, i \in \boldsymbol{K}\}$.

If $\{R_i;\, i \in \boldsymbol{K}\}$ are all equal to 1, $\{\alpha_t\}$ is a Markov process; if any of the R_i are not 1, $\{\alpha_t\}$ is not Markov. As long as the $\{R_i\}$ are positive integers, the non-Markovian maneuver model can be integrated into the earlier algorithms by expanding the dimension of the modal-state space; creating a set of ersatz maneuvers. The state space of α_t is of dimension K. Associate to the set $\{R_i;\, i \in \boldsymbol{K}\}$, the set of integers $\boldsymbol{A} = \boldsymbol{A}_1 \cup \ldots \cup \boldsymbol{A}_K$, where $\boldsymbol{A}_1 = \{1, \ldots, R_1\}$, $\boldsymbol{A}_2 = \{R_1 + 1, \ldots, R_1 + R_2\}$, and so on; each substantive maneuver mode, $\alpha = \mathbf{e}_i$, is decomposed into R_i pseudomodes, each of which is associated with the same turn rate. The γ-renewal process is Markov in the larger state space. To find its transition rate matrix, Q^γ, define for every $p \in \boldsymbol{K}$ the maximum and minimum pseudomode indices:

$$r_p = \min_k (k \in \boldsymbol{A}_p),$$

$$s_p = \max_k (k \in \boldsymbol{A}_p).$$

Then r_p is the entrance state into the pth maneuver mode, and s_p is its exit. Note that the dimension of the γ-renewal model is s_K.

To maintain both the mean sojourn times and the sequencing of substantive maneuver modes displayed in Q^α, Q^γ should be chosen as follows:

$$\begin{aligned} Q^\gamma_{ij} &= R_p Q^\alpha_{pp}, && \text{if } i, j \in \mathbf{A}_p \text{ and } j = i; \\ &= -R_p Q^\alpha_{pp}, && \text{if } i, j \in \mathbf{A}_p \text{ and } j = i + 1; \\ &= R_p Q^\alpha_{pk}, && \text{if } i = s_p,\ j = r_k, \text{ and } p \neq k; \\ &= 0, && \text{otherwise.} \end{aligned}$$

With these changes, the maneuver state dynamics can be found by replacing the $K \times K$ transition rate matrix for the maneuver process, Q^α, by the $(s_K \times s_K)$-matrix Q^γ. The dimension of $\{\hat{\alpha}_t\}$ goes up as s_K/K.

4.6 Performance Contrasts with Different Lifetime Modeling

To illustrate the how tracking performance changes as the modal model is varied, consider a low-speed vignette. After an extended interval of coasting, a target makes a single turn of nearly 180°. The target is detected at $(X_0, Y_0) = (1.0, 6.4)$ km, with initial velocity $(V_{x0}, V_{y0}) = (5.0, -13.3)$ m/s. The target moves at essentially constant velocity (coast) on $t \in [0, 10)$ s. A 0.5 g turn is executed on $t \in [10, 20)$, after which the target returns to constant velocity motion. The omnidirectional accelerations are slight: $W = 0.1\ (\text{m/s}^2)^2$.

To determine the processor dynamics in the upper path in Figure 4.1, the maneuver tempo must be quantified. The lateral accelerations in the target will be given by

$$\begin{aligned} \Phi_t \in \{a_1 &= 17.3°/\text{s}, \alpha_t = \mathbf{e}_1;\ a_2 = 0°/\text{s}, \alpha_t = \mathbf{e}_2; \\ a_3 &= -17.3°/\text{s}, \alpha_t = \mathbf{e}_3\} \end{aligned}$$

with the chain $\{\alpha_t\}$ symmetric about the coast mode. The elements of the Q^α-matrix are determined jointly by the mean sojourn time in each acceleration mode, $\{\nu_i, i = 1, 2, 3;\ \nu_1 = \nu_3\}$, and the transition probabilities from a maneuver to the nonmaneuvering mode. Let $q = \mathcal{P}(\alpha_t = \mathbf{e}_2 \mid \alpha_{t-} = \mathbf{e}_1 \text{ and } \Delta\alpha_t \neq 0)$. Then q measures the fraction of times that a maneuver ends in a coasting motion (e.g., $q = 0$ implies pure jinking motion, and $q = 1$ always interjects coasting between turns). The sojourns in the coast mode have exponential lifetimes ($R_2 = 1$) and the lifetime distributions in the turn mode are the same ($R_1 = R_3 = R$).

The sensors are located at the origin (0, 0) of the coordinate system. A radar provides range-bearing measurements, $\{y[k]\}$, at a 10 samples/s rate. The measurement errors are Gaussian with standard deviation 5 m and 0.25° (about 4 mr) respectively.

An imaging sensor is collocated and makes orientation measurements of the target at the same rate (i.e., $\lambda = 10$ frames/s). The bin width is 30° ($L = 12$). The imager error taxonomy includes errors of three types as before: UDE, in which the output symbol is uniformly distributed without regard to the true orientation; NNE, in which the ostensible orientation is placed in the neighboring angular bin; and PE, in which the image processor places the target orientation in the bin situated symmetrically with respect to a line perpendicular to the line of sight. The indicated frame rate and bin size are well within current standards. The error rates depend upon the sensor used.

There are various ways to approach this tracking problem. The simplest would be to ignore the turns and the imager and select an EKF based upon the geometry and the radar. This would lead to a four-dimensional tracker. A more sophisticated implementation of the EKF would use a shaping filter to generate a surrogate for the maneuver process and add this contribution to the Brownian disturbance. This increases the dimension of the base-state model because of the state augmentation in the shaping filter. The augmented model is also highly nonlinear.

Denote the augmented EKF by EKF($\boldsymbol{\Phi}(\omega)$; q, R). To implement this filter, we must find the power spectral density of $\{\Phi_t\}$, use $\boldsymbol{\Phi}(\omega)$ to create a shaping filter, and design an EKF in the augmented state space. This requires some computation because of the convoluted form of the maneuver process. The derivation of $\boldsymbol{\Phi}(\omega)$ for a stationary version of the maneuver process is presented in [SKVH95]. Here, we will sensitize the EKF to a turn; set $q = 0$. The coast mode ($\phi_t = \mathbf{e}_1$) is transient and does not influence $\boldsymbol{\Phi}(\omega)$. The PSDs are shown in Figure 4.11 for three values

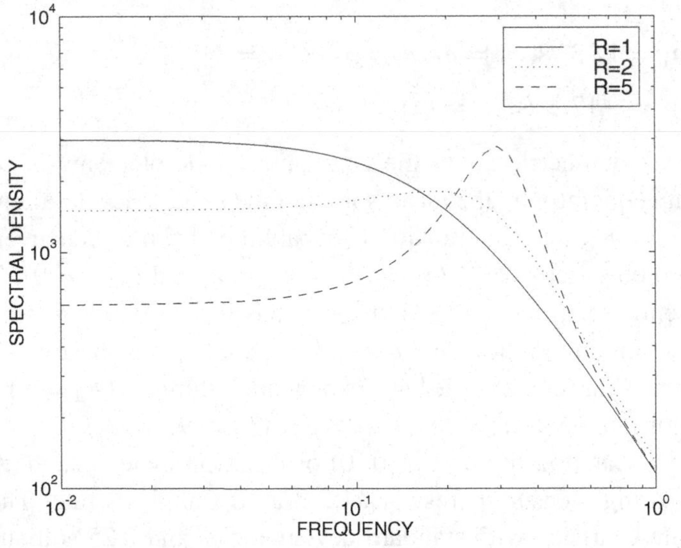

Figure 4.11. Power spectral density for $q = 0$; $R = 1, 2,$ and 5.

Table 4.3. *Sensor quality for two imagers.*

Imager	UDE	NNE	PE
I1	0.01	0.07	0.3
I2	0.2	0.1	0.4

Table 4.4. *Tempo models.*

Tempo	$\nu_1 = \nu_3$	ν_2	q	R
T1	10	40	0.5	5
T2	4	20	1	1
T3	10	20	1	$1 \rightarrow 4$

of R. The figure shows the considerable flexibility that exists in shaping the PSD with R. All of the maneuver mode chains have the same mean sojourn times and the same stationary distribution. As we move from $R = 1$ (an exponential lifetime) to $R = 5$ (a quasiperiodic process), the resonant peak of $\Phi(\omega)$ increases significantly. The power spectral densities shown in Figure 4.11 are not simply represented using a low-order shaping filter. The curves can, however, be approximated using first- ($R = 1$) or second- ($R = 2$ or $R = 5$) order LGM models, and this is done in what follows.

The first two EKFs are radar exclusive, and the maneuver is treated as an additive disturbance: white or colored. If the imager is used, the EKF framework can be modified to utilize the new data. First the image data is translated into an estimate of $\{\alpha_t\}$ using the PME modal filter. The mean maneuver acceleration is added to the increments in the velocity estimates. To achieve higher gains during and soon after a turn, covariance modification is required. This is accomplished by augmenting the base-state noise proportionally to the acceleration uncertainty as measured by the variance of Φ_t: An image-based pseudonoise is added to the target dynamics proportional to $\text{var}(\Phi_t)$ (see (4.11)) [HS92]. Through this simple artifice, the filter gain is increased during transient intervals while remaining small at other times. This estimator will be labeled $\text{EKF}(\gamma; q, R)$.

A more complete integration of the radar and the imager measurements is achieved using the PME. The γ-renewal model for the maneuver process leads to the algorithm $\text{PME}(\gamma; q, R)$. The PME algorithm depends upon the specific imager and a maneuver model used in the encounter. The characteristics of two imagers are given in Table 4.3. Three different scenario models are presented in Table 4.4.

Let us look first at the maneuver mode estimation of Imager I1 and Tempo T1. Figure 4.12 shows a sample average of $\{\hat{\Phi}_t\}$ as computed by $\text{PME}(\gamma; q =$

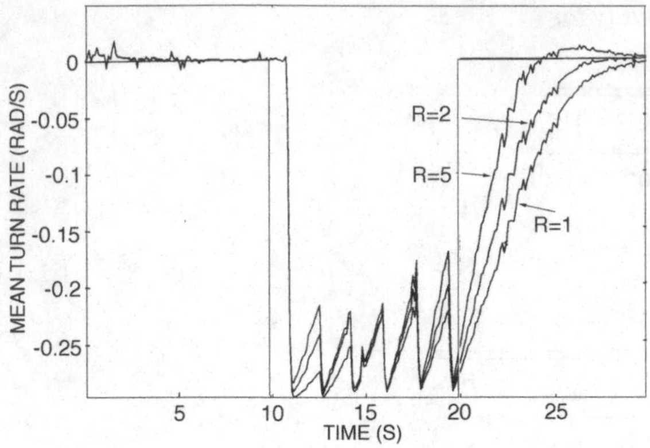

Figure 4.12. Turn rate estimates for $\mathsf{PME}(\gamma; 0.5, R)$.

0.5, R) as R varies from 1 to 5. All maneuver models have the same intrinsic state space and modal lifetimes. They differ only in the shape of the sojourn density selected to describe the turn. During the initial coast phase, all PMEs are good and $\hat{\Phi}_t \approx 0$. The turn begins at $t = 10$, but it cannot be recognized until $t = 11$ because the target heading must first traverse the initial angular bin. Crossing a bin boundary is recognized by all the PMEs as connoting a turn and all so indicate. In the interval [10, 20) the target has a constant turn rate. The modal-state processor uses bin crossings to identify turns. Consequently, as $\{\rho_t\}$ moves across a bin, all the estimators produce a decay in $\{\hat{\Phi}_t\}$ with a decay rate that is determined by Q. At first the decay rate is far bigger in $\mathsf{PME}(\gamma; 0.5, 1)$ than it is in $\mathsf{PME}(\gamma; 0.5, 5)$: A Markov model generates more short sojourns than does the $R = 5$, γ-renewal process. A quick return to coast is not unexpected if $R = 1$. The roles are reversed when $t = 20$ s: $\hat{\Phi}_{20}$ is bigger for $\mathsf{PME}(\gamma; 0.5, 5)$ than it is for $\mathsf{PME}(\gamma; 0.5, 1)$ because the ($R = 5$)-renewal process is expecting a return to coast relatively soon while the Markov process does not keep track of elapsed time. After return to coast ($\sim$20 s), $\mathsf{PME}(\gamma; 0.5, 5)$ is much faster than $\mathsf{PME}(\gamma; 0.5, 1)$ in identifying it.

There are several choices for algorithms that actually track this target. Consider first some of the variants on the EKF:

- $\mathsf{EKF}(W{=}0.1)$: This algorithm neglects the maneuver. It is the simplest with base-state dimension four.
- $\mathsf{EKF}(\Phi(\omega); q = 0, R = 5)$: This uses the tempo model to determine the correct shaping filter. The dimension of this filter is six. To sensitize the algorithm to turns, the value of q has been reduced to $q = 0$.
- $\mathsf{EKF}(\gamma; q = 0.5, R = 5)$: This uses the image path to adjust the gain and the drift in the EKF.

Rather than looking at the sample paths of the algorithms, consider their biases. In LGM estimation, we expect $\{\hat{x}_t\}$ to be centered about $\{x_t\}$ with a sample path

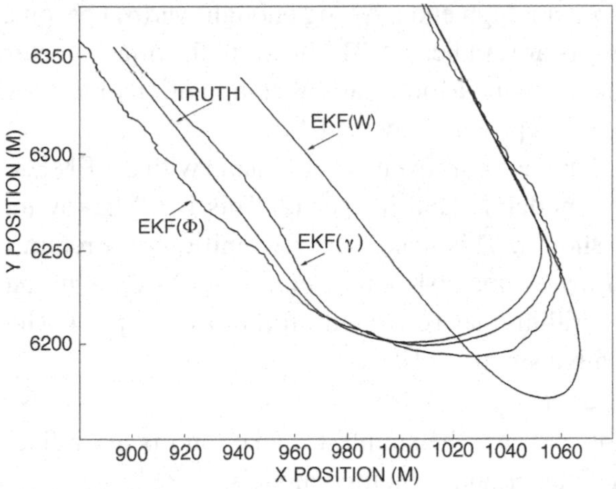

Figure 4.13. Path following bias from EKF(W=0.1), EKF($\Phi(\omega)$; 0, 5), and EKF(γ; 0.5, 5).

deviation of size related to $\{P_{xx}\}$. Earlier examples showed this not always to be true: An unmodeled feature of the path (a turn) causes a bias in the estimate. Figure 4.13 shows this bias clearly. Twenty-sample averages of the paths of three different EKFs are displayed with the true path as a reference.

EKF($W = 0.1$) seems nearly oblivious to the turn. It is slow to detect a turn when it occurs. It loops out with a 90 m error. When the target returns to coast, and the correct motion mode underlies the EKF, the small gain in EKF(W=0.1) prolongs the post maneuver return to the coast path. Indeed, it is so slow to correct for the turn that there is considerable excess error at the end of the simulation.

Adding colored noise to the model improves performance at the cost of increased filter dimension. EKF($\Phi(\omega)$; 0, 5) is an improvement over EKF(W=0.1). The maximum error during a turn is reduced by a factor of three. Interestingly, the post maneuver return to the coast is still unsatisfactory. This is due to the fact that an LGM shaping filter does not generate sample paths that mimic $\{\Phi_t\}$ and the modal estimate in EKF($\Phi(\omega)$; 0, 5) is not good. Failure to resolve the turn in an expeditious manner leads to the protracted meander at the end of the turn.

Of the three estimators, the image-enhanced algorithm, EKF(γ; 0.5, 5), is best able to balance the conflicting demands imposed by this multimodal tracking environment. The errors are much less in the turn. A peculiar bias arises after return to coast. The estimate generated by EKF(γ; 0.5, 5) continues to follow a turn for so long that it overshoots when the target returns to coast. The overshoot is due to the delay in determining the end of the turn from the image information (see Figure 4.12). Covariance augmentation raises the gain after the turn ends and this causes the tracking error to decay. But the decay is slow because of the failure of the gain augmentation to remain high enough long enough; var(α_t) becomes small before the turn-induced errors are eradicated. This leads to the dilatory return to the coast path. To refine this adjustment, a more careful computation of the covariance process is necessary, and this is provided by the PME.

To demonstrate the performance improvement attainable with the PME, consider the same path with tempo model T2 and imager I2. This is a Markov maneuver model. The lifetimes are shorter. This makes the recognition of a return to coast comparable with the Markov estimate shown in Figure 4.12. Recognition is also improved by telling the algorithms that a coast must follow a turn: $q = 1$. The imager is also somewhat better with a smaller NNE.

Consider the following trackers:

- EKF(W=0.1): This is again the simplest tracker and is not influenced by the changes in the imager and the tempo models.
- EKF($\gamma; q = 1, 1$): This uses the faster tempo model and the imager is improved.
- PME($\gamma; q = 1, 1$): This uses the image path to adjust the gain and the drift in the EKF.

The effect of these changes in the maneuver model are shown in Figure 4.14. A twenty-sample average of the path estimates is shown along with the true path. The response of EKF(W=0.1) is as it was in Figure 4.13: EKF(W=0.1) is not influenced by the maneuver model. EKF(γ; 1, 1) responds more quickly to the motion, but this is due primarily to the improved imager. EKF(γ; 1, 1) retains the slow decay to the true path at the end of the experiment.

The PME has much smaller bias even though it utilizes the same $\{\hat{\alpha}_t\}$ as does EKF(γ; 1, 1). The overshoot and slow return to coast observed in EKF(γ; 1, 1) is avoided by the PME. The average radial errors are displayed in Figure 4.15.

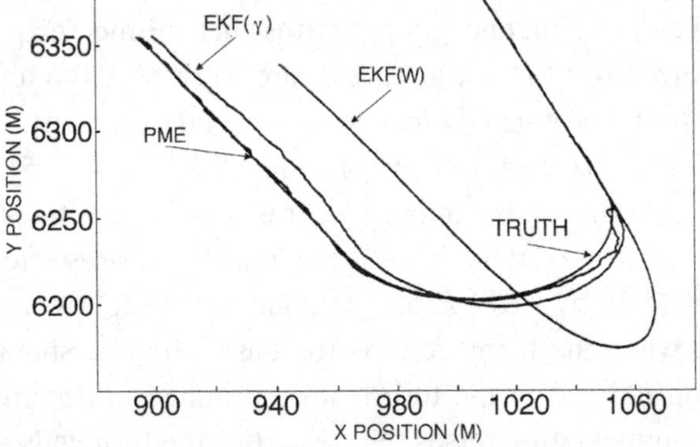

Figure 4.14. Path following bias from EKF(W=0.1), EKF(γ; 1, 1), and PME (γ; 1, 1).

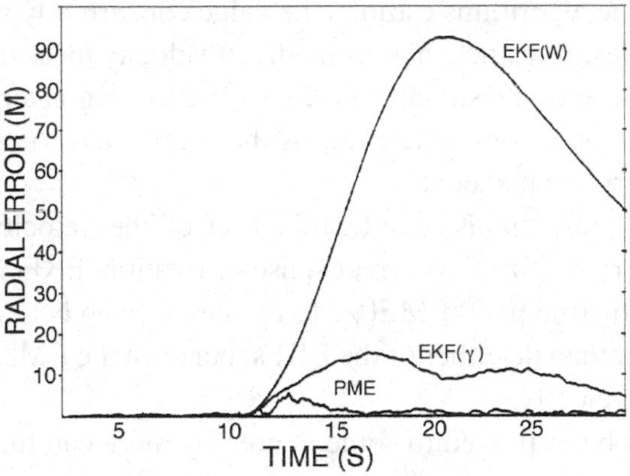

Figure 4.15. Average radial error for $\mathsf{EKF}(W{=}0.1)$, $\mathsf{EKF}(\gamma;\,1,1)$, and $\mathsf{PME}(\gamma;\,1,1)$.

$\mathsf{EKF}(W{=}0.1)$ is unacceptable. The terminal response of $\mathsf{PME}(\gamma;\,1,1)$ is seen to be far better than that of $\mathsf{EKF}(\gamma;\,1,1)$.

High quality estimates of velocity are important in applications that require accurate prediction of future target position (e.g., fire control or guidance). The simplest predictors extrapolate from the current location estimate in the direction of the current velocity estimate. Velocity errors therefore produce sizable errors in predicted position if the prediction interval is large.

Figure 4.16 shows sample mean velocity profiles of the three trackers in the velocity plane. Velocity is a slack variable in the estimation algorithms. It is included

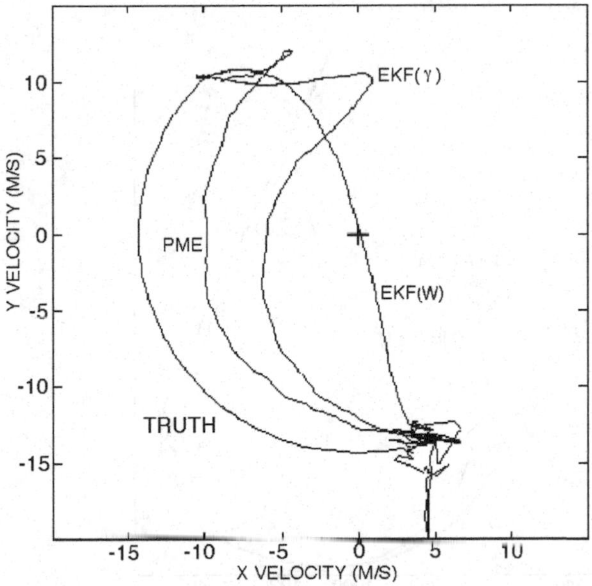

Figure 4.16. Average velocity profiles for $\mathsf{EKF}(W{=}0.1)$, $\mathsf{EKF}(\gamma;\,1,1)$, and $\mathsf{PME}(\gamma;\,1,1)$.

in the base-state model, and the algorithms estimate its value concurrently with estimates of the location variables. But since there is no direct velocity measurement, the trackers tend to assign observation residuals to velocity to a much greater degree than one might expect. This is true to some extent during the nonmaneuvering phase, but the effect is magnified during a maneuver.

In this scenario, there is conspicuous misidentification of the velocity, with EKF(W=0.1) failing even to recognize the correct sense of rotation. EKF(γ; 1, 1) is better but is still far from the true path. PME(γ; 1, 1) does a much better job of following the velocity contour than do either of the EKFs, but even the PME makes significant errors in tracking velocity.

The velocity estimates can be improved to some degree by more careful tempo modeling. Let us return to imager I1 and use tempo T3 to delineate the maneuver process.

Consider now the following trackers:

- EKF(γ; $q = 1$, $R = 1$): This uses the Markov model to produce the image-based corrections.
- EKF(γ; $q = 1$, $R = 4$): This uses the γ-renewal model to produce the image-based corrections.
- PME(γ; $q = 1$, $R = 1$): This uses the Markov model for sensor fusion.
- PME(γ; $q = 1$, $R = 4$): This uses the γ-renewal model for sensor fusion.

Figure 4.17 shows the average track bias for the last three algorithms. The PME with the renewal model most descriptive of the encounter is the best performer.

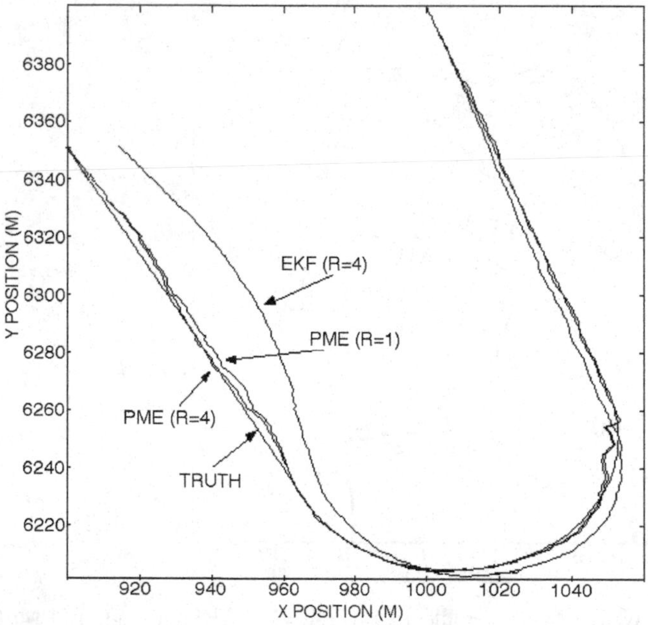

Figure 4.17. Path following bias from EKF(γ; 1, 4), PME(γ; 1, 1), and PME(γ; 1, 1).

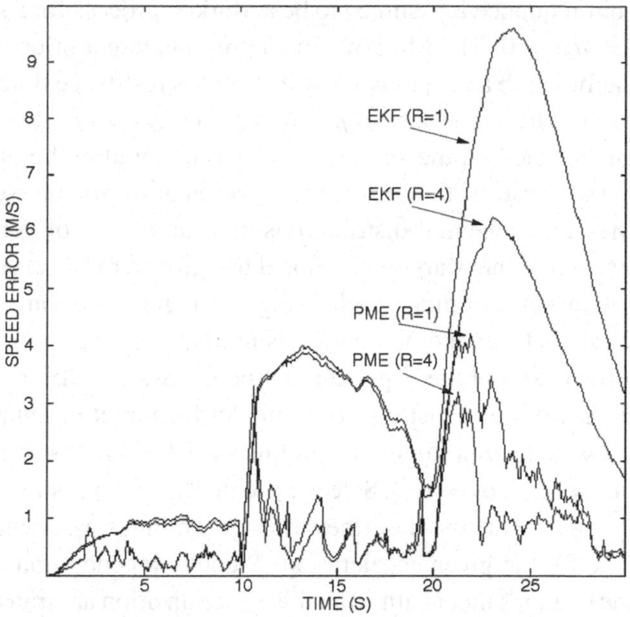

Figure 4.18. Average speed errors for EKF(γ; 1, 1), EKF(γ; 1, 4), PME(γ; 1, 1), and PME(γ; 1, 1).

The improvement over the Markov PME is not great however. For the EKFs, the performance improvement associated with a γ-renewal model is considerable. EKF(γ; 1, 1) (not shown) has twice the overshoot of EKF(γ; 1, 4). Still, EKF(γ; 1, 4) is inferior to the simplest PME.

The radial error has a double peak, with maximum errors subsequent to the initiation of the turn and to the cessation of the turn. Figure 4.18 shows corresponding speed error profiles, and the curves for the EKFs are not nearly as good as those for the PMEs. Both of the PMEs have about the same maximum speed errors, but PME(γ; 1, 4) reduces the error much faster than does PME(γ; 1, 1).

Performance improvement of PME(γ; 1, 4) must be balanced against its complexity (i.e., R). Though not shown, the relative response of $\{\hat{\alpha}_t\}$ for longer or shorter sojourns is easily found. For off-nominal conditions, PME(γ; 1, 4) is surprisingly robust. The path of $\{P_{xx}\}$ computed for PME(γ; 1, 4) differs little from that computed for PME(γ; 1, 1) [SV94]. The improvement in PME(γ; 1, R) with R derives more from the more accurate computation of the mixed moment, $P_{x\phi}$.

This example illustrates the use of a γ-renewal model for the maneuver modes. Parasitic effects also make the sojourn times in an angular bin random even when the nominal turn rate is known. Further, the finite state space for acceleration is an abstraction used to describe what is actually a continuum of possible actualizations. For example, a_1 is but a single element in an interval (bin) of rotation rates; a_1 is close to, but likely not equal to, the actual value of Φ_t when Φ_t is in the a_1 bin. In the basic

PME, the orientation bin sequence is assumed to be a Markov process for a specific turn rate. Suppose $\Phi_t = a_m > 0$. The Markov model presents target orientation as remaining in an angular bin with an exponentially distributed residence time (mean $\frac{2\pi}{La_m}$), after which the bin indicator process, $\{\rho_t\}$, transitions $\mathbf{e}_k \mapsto \mathbf{e}_{k+1}$; the faster the turn, and/or the smaller the bin, the shorter the mean time within the bin, with the direction fixed by the sense of the turn. This representation for bin sojourns has an obvious weakness: Exponential distributions have an excess of very short and very long lifetimes. While the Markov bin model has proven to be serviceable, it does not manifest the quasi-periodicity underlying the angular motion well; for example, if the frame rate is high, the orientation is unlikely to enter a bin at one frame time and exit it the next, but the exponential model masks this trait.

It is possible to create a more realistic sojourn model for target orientation by using a γ-renewal process for the bin sojourns. To adjust the PME for this orientation model, first fix the turn rate (e.g., $\alpha_t = \mathbf{e}_m$). Select a γ-density (with parameters (λ_m, R), the latter integer valued) that best describes the sojourn times associated with an angular bin residence for the given acceleration. R could depend upon the turn rate (i.e., m) too, but the changes that result from this generalization are transparent. Partition the sequence of integers RL into L, equally spaced, contiguous blocks: $\boldsymbol{A} = \boldsymbol{A}_1 \cup \ldots \cup \boldsymbol{A}_L$; $\boldsymbol{A}_1 = \{1, \ldots, R\}$, $\boldsymbol{A}_2 = \{R+1, \ldots, 2R\}$, and so on. The orientation model replaces each substantive angular bin with R sub-bins; that is, as the orientation moves across a single *true* bin (the kth say), the model has it traverse R *ersatz* bins (labeled $\boldsymbol{A}_k$). Let $\{r_t\}$ be an R-dimensional indicator. In this new framework, the orientation is given by the unit vector $\rho_t \otimes r_t$, the former denoting the true angular bin, and the latter, the sub-bin. To preserve the mean angular change of the target, the mean time in each of the sub-bins must be reduced by the factor R ($\lambda_m = \frac{2\pi}{RLa_m}$). The joint (acceleration $\times$ orientation) process is now Markov even for the γ-renewal residence times, albeit in a state space of higher dimension: The maneuver state of the target is $\phi_t = \alpha_t \otimes \rho_t \otimes r_t$. The transition rate matrix for the orientation model associated with the mth maneuver mode can be deduced in a direct manner, and the PME follows [SBCV99].

5

Hybrid Plants with Base-State Discontinuities

5.1 Plant State Discontinuities

Analytical design of complex dynamic systems is based upon a formal mathematical description of the plant and the observations. A flexible representation of plant evolution phrases the plant state in terms of a set of nonlinear stochastic differential equations:

$$d\chi_t = \mathbf{f}(\chi_t, \upsilon_t, \Phi_t)\,dt + \mathbf{g}(\chi_t, \upsilon_t, \Phi_t)\,dw_t, \tag{5.1}$$

where $\{\upsilon_t\}$ is the plant input, and $\{\chi_t\}$ is the plant state.

Unfortunately, Equation (5.1) does not provide a particularly hospitable framework for the synthesis of algorithms for estimation and control. For this reason, in hybrid synthesis the single plant model is replaced with the pair of equations:

base-state model

$$dx_t = \sum_i (A_i x_t + B_i(u_t - \upsilon\phi_t)\,dt + C_i\,dw_t)\phi_i$$
$$+ \sum_{i,l} (M(i,l)x_t + (\chi_i - \chi_l) + M(i,l)\chi_i + \rho(i,l))\phi_i \mathbf{e}_l' \Delta\phi_t, \tag{5.2}$$

modal-state model

$$d\phi_t = Q'\phi_t\,dt + dm_t. \tag{5.3}$$

It is in the interplay of these dissimilar processes that hybrid estimation takes on its peculiar flavor.

The observations are matched to the zygostate processes:

base-state measurement

$$y[k] = H\chi[k] + n[k], \tag{5.4}$$

modal-state measurement

$$dz_t = \lambda \mathbf{P}\phi_t\, dt + d\eta_t. \tag{5.5}$$

In the previous chapter, we considered one version of the hybrid estimation problem:

- The endogenous actuating signal was absent: $\upsilon_t \equiv 0$.
- The base-state sample paths were continuous: $M(\cdot,\cdot)$, $\rho(\cdot,\cdot)$, and χ were all zero.

These restrictions had the effect of reducing the base-state equation to

$$dx_t = \sum_i (A_i x_t\, dt + C_i\, dw_t)\phi_i. \tag{5.6}$$

The plant state set points were equal to zero and this avoided the modal-state contamination of the base-state observation

$$y[k] = Hx[k] + n[k].$$

In the problem of tracking a maneuvering target, the PME proved to be a good fusion algorithm. The sample paths of $\{\hat{x}_t\}$ and $\{\hat{\phi}_t\}$ followed the underlying state paths rather well.

The advantageous performance of the PME stems from the fact that the PME computes a relevant set of zygostate cross moments. For example, $\{P_{x\phi}\}$ assimilates the motion geometry and allows the image measurements to be integrated into the base-state update. This image information is particularly important near the time of regime change events. The Kalman filters failed to follow even the base-state path because their gains did not adapt to changing regime conditions. The Kalman estimates tended to drift from the true path over time, and the base-state error became so large that it exceeded the tracking window width.

In this chapter, we will study an estimation problem that is in many ways more challenging: The plant experiences a base-state discontinuity at a change in regime. For example, a hostile target may use motion discontinuities to enhance the deleterious effect of changes in maneuver mode and to bring about degraded estimation accuracy. When the base-state discontinuities are hidden in the plant state measurements by the conflated modal estimation error, the plant state path becomes even more difficult to resolve.

To be specific, return to the base-state model and again assume that there is no endogenous control:

$$\begin{aligned} dx_t = {} & \sum_i (A_i x_t\, dt + C_i\, dw_t)\phi_i + \sum_{i,l} (M(i,l)x_t + (\chi_i - \chi_l) \\ & + M(i,l)\chi_i + \rho(i,l))\phi_i \mathbf{e}_l' \Delta\phi_t. \end{aligned} \tag{5.7}$$

As before, we wish to fuse the measurement sequences so as to generate a good estimate of the plant state, $\{\hat{\chi}_t\}$, along with relevant higher moments.

The PME provides a fusion algorithm, but the comprehensive implementation is quite complex. In this chapter we will study three specializations of the plant-observation framework. In each, a different kind of discontinuity is isolated. These individuated cases will provide insight into the influence of discontinuities and will suggest approximations useful when the PME cannot be fully implemented.

Suppose $\{\phi_t\}$ makes the transition $\mathbf{e}_i \mapsto \mathbf{e}_j; i \neq j$. The three specializations are:

- plant state rotation: $\Delta x_t = M(i,l)x_t$,
- plant state translation: $\Delta x_t = (\rho_j - \rho_i)'$,
- Variable set point: $\Delta x_t = \chi_i - \chi_j$.

5.2 Plant State Rotation

Although a base-state model with continuous paths is representative of certain applications, there are situations in which the regime transition creates plant state discontinuities (e.g., an aircraft may slow when it executes a turn). Suppose the plant state reference points are zero ($\chi = 0$). The base-state model, (5.7), becomes

$$dx_t = \sum_i (A_i x_t\, dt + C_i\, dw_t)\phi_i + \sum_{i,l} M(i,l)x_t(Q_{il}\, dt + dm_l)\phi_i. \tag{5.8}$$

Equation (5.8) is nonlinear both within and between modes with a discontinuity: If $\{\phi_t\}$ transitions $\mathbf{e}_i \mapsto \mathbf{e}_l$, then $\Delta x_t = M(i,l)x_t$. The PME algorithm can be simplified in this application [SB97a]. Again let

$$\mathbf{A}_i = A_i + \sum_l Q_{il} M(i,l).$$

We then have:

the PME : base-state rotation; $\chi = 0$

Between observations:

$$\frac{d}{dt}\hat{\phi}_t = Q'\,\hat{\phi}_t,$$

$$\frac{d}{dt}\hat{x}_t = \sum_i \mathbf{A}_i R_{x\phi_i},$$

$$\frac{d}{dt}P_{x\phi} = \sum_i \left(\mathbf{A}_i P_{(x\phi_i)\phi} + \sum_l Q_{il} M(i,l) R_{x\phi_i} \mathbf{e}_l' \right) + P_{x\phi} Q,$$

$$\frac{d}{dt}P_{xx} = \sum_i \Big(\boldsymbol{A}_i P_{(x\phi_i)x} + (\cdot)' + R_\chi(i)\,\hat{\phi}_i + \sum_l Q_{il} M(i,l) R_{xx\phi_i} M(i,l)' \Big),$$

$$\frac{d}{dt}P_{xx\phi_m} = \sum_i \big((\boldsymbol{A}_i P_{(x\phi_i)x\phi_m} + Q_{im} P_{(x\phi_i)x} M(i,m)) + (\cdot)' + R_\chi(i) P_{\phi_i\phi_m} + Q_{im}(P_{xx\phi_i} + M(i,m)(P_{(xx\phi_i)\phi_m} + R_{xx\phi_i}) M(i,m)) \big). \tag{5.9}$$

At a modal observation:

$$\begin{aligned}
\hat{\phi}^+ &= \hat{\phi}^- * \Delta\vartheta, \\
\Delta\hat{x} &= P_{x\phi}\Delta\vartheta, \\
\Delta P_{x\phi} &= -\Delta\hat{x}\Delta\hat{\phi}' + \sum_k P_{x\phi\phi_k}\Delta\vartheta_k, \\
\Delta P_{xx} &= -\Delta\hat{x}\Delta\hat{x}' + \sum_k P_{xx\phi_k}\Delta\vartheta_k, \\
\Delta P_{xx\phi_m} &= -\Delta\hat{\phi}_m\Delta\hat{x}\Delta\hat{x}' - \Delta\hat{\phi}_m P^+_{xx} - \Delta\hat{x}P^+_{\phi_m x} - P^+_{x\phi_m}\Delta\hat{x}' \\
&\quad + \sum_k P_{xx\phi_m\phi_k}\Delta\vartheta_k.
\end{aligned} \tag{5.10}$$

At a base-state observation:

$$\begin{aligned}
\Delta\hat{x} &= \gamma_x\Delta\nu_x, \\
\Delta P_{x\phi} &= -\gamma_x H P_{x\phi}, \\
\Delta P_{xx} &= -\gamma_x R_{yy}\gamma_x', \\
\Delta P_{xx\phi_m} &= -\gamma_x H P_{xx\phi_m} - P_{xx\phi_m} H'\gamma_x.
\end{aligned} \tag{5.11}$$

Equations (5.9)–(5.11) give the zygostate estimator for a system with rotational discontinuities. The structure of the algorithm is identical to that presented in Section 4.3, but (5.9) has an internal modification: $A_i \mapsto \boldsymbol{A}_i$. This change is not unexpected. The drift in $\{\hat{x}_t\}$ is adjusted by adding the mean discontinuity $E[\sum_{il} M(i,l) Q_{il} x_t \phi_i \,|\, \mathcal{G}_t]$. The canonical moments all have the same replacement for $\{A_i\}$, but they are influenced by the jump in $\{x_t\}$ in additional ways. Because each of the higher moments satisfies a nonlinear stochastic equation, it is difficult to give a natural rationale for the size of the discontinuity-induced changes. However, some insight into the impact of the base-state jump can be gained by looking at individual terms in the moment equations that arise from a base-state discontinuity.

P_{xx}: The base-state error covariance satisfies an equation with a structure like that found in Section 4.3 with an adjustment in one term and the addition of another:

$$\frac{d}{dt}P_{xx} = \cdot + \sum_i \left(R_\chi(i)\,\hat{\phi}_i + \sum_l Q_{il} M(i,l) R_{xx\phi_i} M(i,l)' \right).$$

The base-state is exposed to two types of exogenous disturbance. The Brownian motion leads to the term $\sum_i R_\chi(i)\,\hat{\phi}_i$. This is a routine substitution in which the (conditional) mean intensity of the Brownian motion replaces the fixed intensity. The term $\sum_{il} Q_{il} M(i,l) R_{xx\phi_i} M(i,l)'$ is more subtle. It provides an additive, positive increment to the noise intensity. This contribution is related to both the rates of different jumps (quantified by Q_{il}) and the expected product of the base-state discontinuities (quantified by $M(i,l) R_{xx\phi_i} M(i,l)'$).

Suppose that there is a good estimate of the modal state: $\hat{\phi}_t \approx \mathbf{e_r}$. In this case, the contribution due to the discontinuity reduces to

$$\frac{d}{dt}P_{xx} = \cdot + R_\chi(r) + \sum_l Q_{rl} M(r,l) R_{xx} M(r,l)'.$$

The influence of Brownian motion is augmented by another positive term: an outer product of the jumps expected in $\{x_t\}$ weighted by their size (R_{xx}). This term can be likened to the well-known pseudonoise augmentation, though it should be noted that the increase in P_{xx} is observation dependent. When the modal estimate is not good, the same argument holds, but there is a blending of the modal- and base-state in $P_{(x\phi_i)x}$ and $R_{xx\phi_i}$.

$P_{x\phi}$: The second mixed moment is particularly important in the PME because it acts as the gain for the modal measurements in (5.10). Comparing (5.9) with the analogous equation in Section 4.3, we see that the immediate contribution of the discontinuity to $\{P_{x\phi}\}$ is

$$\frac{d}{dt}P_{x\phi} = \cdot + \sum_{i,l} Q'_{il} M(i,l) R_{x\phi_i}\,\hat{\phi}_i \mathbf{e}'_l.$$

Again suppose that there is a good estimate of the modal state: $\hat{\phi}_t \approx \mathbf{e_r}$. The contribution to $\{P_{x\phi}\}$ due to the discontinuity reduces to

$$\frac{d}{dt}P_{x\phi} = \cdot + \sum_l Q_{rl} M(r,l)\hat{x}_t \mathbf{e}_l'.$$

The discontinuity leads to a term that is essentially a product of independent factors in the spatial variable $(M(r,l)\hat{x}_t)$ and in the modal variable $(Q_{rl}\mathbf{e}_l')$. As in $\{P_{xx}\}$ there is a blending of these effects when ϕ_t is not precisely known.

$P_{xx\phi_m}$: This third mixed central moment appears in numerous places in the PME . It quantifies the relation between modal errors and the base-state covariance (e.g., it is the modal-measurement gain in P_{xx}). Again looking only at the terms that depend on the state discontinuity, we have

$$\frac{d}{dt}P_{xx\phi_m} = \cdot + \sum_i \big((Q_{im} P_{(x\phi_i)x} M(i,m)) + (\cdot)' \\ + Q_{im} M(i,m)\big(P_{(xx\phi_i)\phi_m} + R_{xx\phi_i}\big)M(i,m)\big).$$

This combination of terms is hard to motivate in general. Again suppose that there is a good estimate of the modal state: $\hat{\phi}_t \approx \mathbf{e_r}$. The dominant terms in the relevant moments are $P_{(x\phi_i)x} \approx P_{xx}$, $R_{xx\phi_i} \approx R_{xx}$, and $P_{(xx\phi_i)\phi_m} \approx 0$. In this case, the contribution due to the discontinuity reduces to

$$\frac{d}{dt}P_{xx\phi_m} = \cdot + (Q_{rm} P_{xx} M(r,m)) + (\cdot)' + Q_{rm} M(r,m) R_{xx} M(r,m)'.$$

This term becomes even simpler in the most common situation in which $R_{xx} \gg P_{xx}$:

$$\frac{d}{dt}P_{xx\phi_m} = \cdot + Q_{rm} M(r,m) R_{xx} M(r,m)'.$$

5.3 A Maneuvering Aircraft with Variable Drag

To illustrate the behavior of the PME with plant state rotation, return to the example presented in Section 1.2. A target is detected at a range of 3.6 km ($t = 0$) traveling in the X–Y plane at a speed of 300 m/s. The target coasts for three seconds, makes a 7 g turn for six seconds, coasts for two seconds and makes a 7 g turn in the other direction for five seconds, and then returns to coast. A turn causes the target to slow by 40% of the speed that it has entering the turn with a 40% increase in speed when

Table 5.1. *Error parameters for imager IG.*

Imager	UDE	NNE	PE
IG	0.05	0.05	0.1

a turn transitions to coast. A motion model is given in (1.24):

$$d\begin{bmatrix} X \\ Y \\ V_x \\ V_y \end{bmatrix} = \begin{bmatrix} 0 & 0 & 1 & 0 \\ 0 & 0 & 0 & 1 \\ 0 & 0 & 0 & -\Phi \\ 0 & 0 & \Phi & 0 \end{bmatrix} \begin{bmatrix} X \\ Y \\ V_x \\ V_y \end{bmatrix} dt + \begin{bmatrix} 0 & 0 \\ 0 & 0 \\ 1 & 0 \\ 0 & 1 \end{bmatrix} d\begin{bmatrix} w_x \\ w_y \end{bmatrix},$$

where (X, Y) are position coordinates, and (V_x, V_y) are the associated velocities. The target acceleration is a wide band, omnidirectional acceleration described by the Brownian motion $\{w_x, w_y\}$ summed with the maneuver acceleration represented by the turn rate process $\{\Phi_t\}$. The speed is slowly varying when the turn rate is constant: $C_i = \mathbf{e}_2 \otimes \mathbf{I}_2$ for all i, and $W = \mathbf{I}_2$. The jinking behavior is captured by

$$\Phi_t \in \{0.2 \text{ r/s}, \alpha_t = \mathbf{e}_1; 0 \text{ r/s}, \alpha_t = \mathbf{e}_2; -0.2 \text{ r/s}, \alpha_t = \mathbf{e}_3\}.$$

With this ordering, $\alpha_t \in \{\mathbf{e}_1, \mathbf{e}_2, \mathbf{e}_3\}$ and the turn rate is given by $\Phi_t = a'\alpha_t$.

At the origin of the coordinate system there is a radar and collocated imager. The radar measures the position of the target every second with Gaussian errors of 40 m in range and 1.75 mr in bearing. The imager model is that given in IG in Table 5.1. The sample rate is 10 frames/s; the bin width is 30°.

The most straightforward approach to this tracking problem would be to neglect the turn process and design a radar-exclusive EKF based upon the specification of radar quality given above. This was done in Section 1.2 and the filter was labeled EKF(W=1). The initial covariance is taken to be diagonal with standard deviation in position (100 m) and velocity (20 m/s). Figure 5.1 shows the response of a sample path of the target along with the one-σ error ellipses centered on the estimate as generated by EKF(W=1). (This is a composition of Figures 1.1 and 1.2, decimated for clarity.) All of the algorithms are initialized on the true state.

To design the PME, the maneuver dynamics must be specified. Suppose the mean sojourn in any maneuver mode is 5 s; that every turn ends in a coast ($q = 1$ in the notation of Chapter 4); and the chain is symmetric about coast. First look at the performance of the PME developed in Chapter 4 (labeled PME(M=0) since it ignores the speed changes). Figure 5.2 shows a sample of the tracking performance. PME(M=0) is superior to EKF(W=1). The one-σ circles are closer

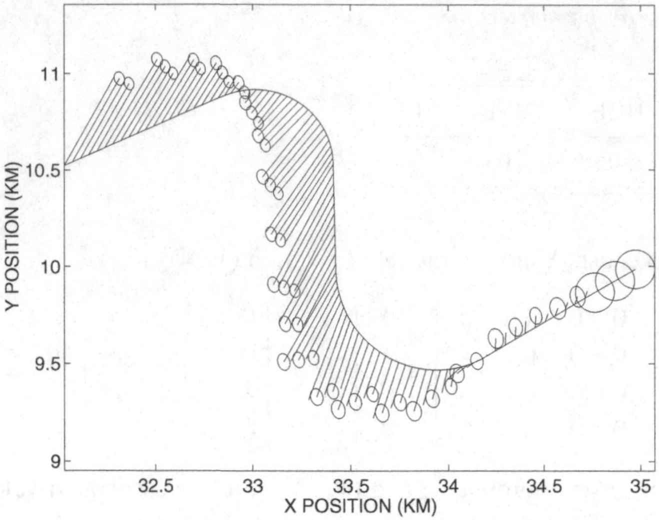

Figure 5.1. The EKF(W=1) for a path with variable drag.

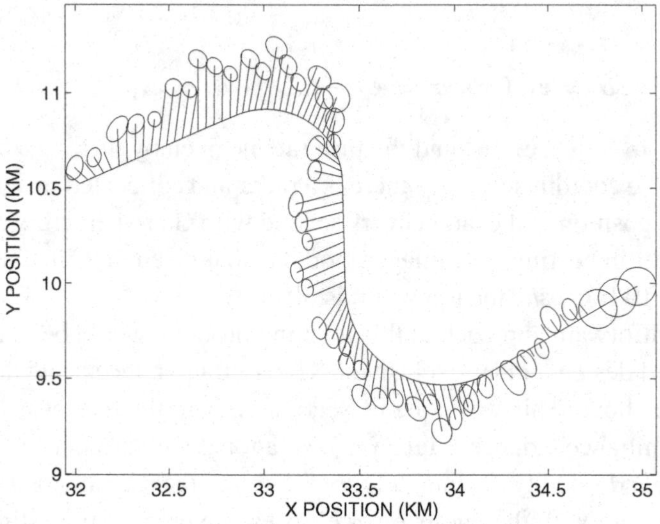

Figure 5.2. The PME(M=0) for a path with variable drag.

to the path than those of EKF(W=1) and adapt better to changing conditions, for example, by increasing their radius when a turn is suspected and decreasing their radius during quiescent operation. The response of the PME lags the path after the drag increases (compare with Figure 4.7), and the error circles still fail to cover the path. PME(M=0) finds the deceleration confusing, but it is faster than EKF(W=1) to correct the velocity estimate during the intermediate coast.

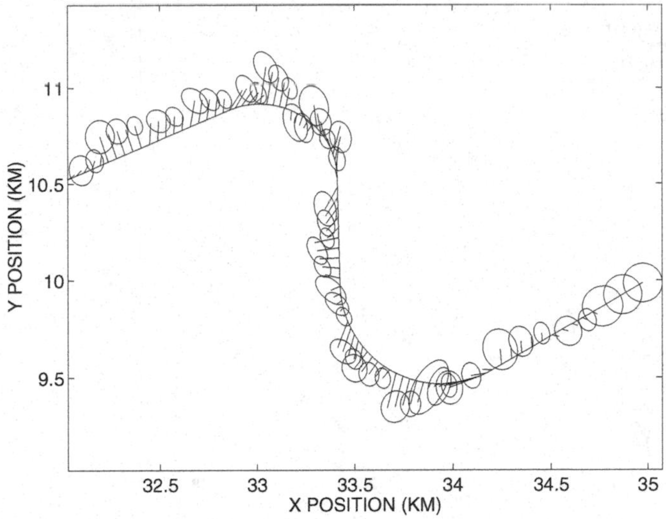

Figure 5.3. The PME(M) for a path with variable drag.

On the final coast, PME(M=0) reduces the error more expeditiously than does EKF(W=1).

The error ellipses of PME(M=0) fail to cover the true path both because of irreducible lags in the modal-state-estimator (on the order of 2 s) and because it has no reason to suspect that the speed will change at modal transitions. No image-based estimator can do anything about the former because the target must rotate across a bin boundary in order for a turn to be recognized. To alert the PME to the latter, set

- $M(i,l) = 0.6\mathbf{E}_2 \otimes \mathbf{I}_2$ if $i = 2, l \in \{1, 3\}$,
- $M(i,l) = 1.4\mathbf{E}_2 \otimes \mathbf{I}_2$ if $i \in \{1, 3\}, l = 2$,
- $M(i,l) = 0$ otherwise

Figure 5.3 (see Figure 1.6) shows the sample performance of this PME. There is an increase in the area of the error ellipses following every modal transition, but the ellipses are corrected reasonably well after the transition is identified. The centers of the error ellipses are closer to the true path than either PME(M=0) or EKF(W=1), and their radii are such as to consistently approach the true path (the two-σ ellipses would encircle the path quite well). The radii of the ellipses display considerably more adaptivity than those associated with PME(M=0).

Although mean radial error (MRE) is a reasonable index of estimator performance, MRE for this multimodal path depends upon the phase of the path. Because of the approximations involved, the computed error covariance is not a guaranteed predictor of MRE. Figure 5.4 (see Figure 1.5) shows the mean radial error as calculated from a sample of size ten. Three filters are shown: EKF(W=1), PME(M=0),

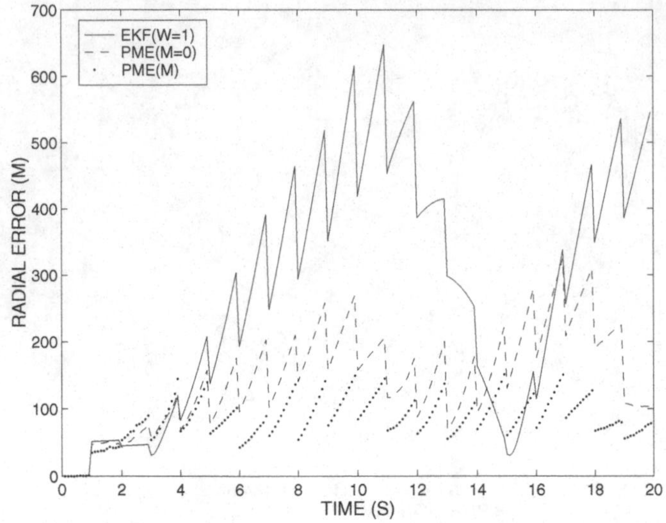

Figure 5.4. The mean radial error for EKF(W=1), PME(M=0), and PME(M) for a path with variable drag.

and PME(M). Although the location sensor has a one-σ error radius of 75 m, the estimation error grows to nearly 700 m when EKF(W=1) is used, and the estimator is oblivious of this fact: Its computed $\{P_{xx}\}$ is far too small. The superior performance of the EKF(W=1) ($\sim$15 s) is due to the fact that the path happens to pass through the estimated path and not because EKF(W=1) is accurate in this interval. PME(M=0) is superior to the EKF with maximum errors of 300 m for the most part. PME(M=0) is unable to effectively extrapolate between radar measurements because the speed changes confuse it. With a 400 m radar window at this range, both estimators would lose lock.

PME(M) is superior to either of the others: It is somewhat worse on the initial coast because of the omniscient initialization for the estimators and it is worse than EKF(W=1) because of the conjunction of the paths in the second turn. The MRE for PME(M) builds up during interradar intervals but not to the degree found in PME(M=0). Since PME(M) requires little additional computation (over PME(M=0)), it would be the algorithm of choice in this application.

5.4 Plant State Translation

The previous example shows how rotational discontinuities increase the difficulty of estimation. In this section we will look at another class of discontinuities: plant state translations. Again suppose the plant state set points are zero ($\chi = 0$) and that there is no endogenous control. When there is a modal transition, the base-state translates. Let ρ be a matrix with rows $\{\rho_i\}$ ($\rho = [\rho_i]$ such that if $\{\phi_t\}$ makes the transition $\mathbf{e}_i \mapsto \mathbf{e}_j$, $\Delta x_t = \rho_j' - \rho_i'$). This is a specialization of the translation array $\{\rho(i, j); i, j \in S\}$ introduced in Section 1.1. This type of behavior arises when there is a null level and all translations are referenced to it. This distinctive form permits the formulas for the PME to be simplified [SB97b].

The base-state model becomes

$$dx_t = \sum_i (A_i x_t\, dt + C_i\, dw_t)\phi_i + \rho' d\phi_t. \tag{5.12}$$

The PME is then written:

Between observations:

$$\frac{d}{dt}\hat{\phi}_t = Q'\hat{\phi}_t,$$

$$\frac{d}{dt}\hat{x}_t = \sum_i A_i R_{x\phi_i} + \rho' Q'\hat{\phi}_t,$$

$$\frac{d}{dt}P_{x\phi} = \sum_i (A_i P_{(x\phi_i)\phi} + V(\mathbf{e}_i)\hat{\phi}_i) + P_{x\phi}Q + \rho' Q' P_{\phi\phi},$$

$$\frac{d}{dt}P_{xx} = \left[\sum_i A_i P_{(x\phi_i)x} + \rho' Q' P_{\phi x})\right] + (\cdot)' + \sum_i (R_\chi(i)\hat{\phi}_i + \rho' V(\mathbf{e}_i)\hat{\phi}_i \rho),$$

$$\frac{d}{dt}P_{xx\phi_m} = \left[\sum_i A_i P_{(x\phi_i)x\phi_m} + \rho'\left(Q' P_{\phi x\phi_m} + \sum_i V(\mathbf{e}_i)_{\cdot m} P_{\phi_i x}\right)\right] + (\cdot)' + \sum_i (R_\chi(i) P_{\phi_i\phi_m} + Q_{im}(P_{xx\phi_i} + \rho'(V(\mathbf{e}_i)P_{\phi_i\phi_m} + U_m(\mathbf{e}_i)\hat{\phi}_i)\rho)). \tag{5.13}$$

At a modal observation:

$$\hat{\phi}^+ = \hat{\phi}^- * \Delta\vartheta,$$

$$\Delta\hat{x} = P_{x\phi}\Delta\vartheta,$$

$$\Delta P_{x\phi} = -\Lambda\hat{x}\Lambda\hat{\phi}' + \sum_k P_{x\phi\phi_k}\Delta\vartheta_k,$$

$$\Delta P_{xx} = -\Delta\hat{x}\Delta\hat{x}' + \sum_k P_{xx\phi_k}\Delta\vartheta_k,$$

$$\Delta P_{xx\phi_m} = -\Delta\hat{\phi}_m\Delta\hat{x}\Delta\hat{x}' - \Delta\hat{\phi}_m P^+_{xx} - \Delta\hat{x} P^+_{\phi_m x} - P^+_{x\phi_m}\Delta\hat{x}' + \sum_k P_{xx\phi_m\phi_k}\Delta\vartheta_k. \tag{5.14}$$

At a base-state observation:

$$\Delta\hat{x} = \gamma_x \Delta\nu_x,$$

$$\Delta P_{x\phi} = -\gamma_x H P_{x\phi},$$

$$\Delta P_{xx} = -\gamma_x R_{yy}\gamma_x',$$

$$\Delta P_{xx\phi m} = -\gamma_x H P_{xx\phi_m} - P_{xx\phi m} H'\gamma_x. \tag{5.15}$$

Equations (5.13)–(5.15) give the PME for a plant with translation discontinuities, which we shall denote by PME(ρ). It is interesting to compare PME(ρ) with the Kalman filter and also with the PME that does not acknowledge discontinuities (PME($\rho = 0$)). The term $\sum_i A_i \hat{x}_t \phi_i$ in the Kalman filter is replaced by $\sum_i A_i R_{x\phi_i}$ in both PMEs, and the term R_χ in the Kalman filter is replaced with $\sum_i R_\chi(i)\hat{\phi}_i$ in the PMEs. Additionally, in PME(ρ), the drift in $\{\hat{x}_t\}$ contains a term $\rho' Q'\hat{\phi}_t$ to reflect the expected discontinuity (an analogous term appears in PME(M=0)). This term does not appear in PME($\rho = 0$) but is plausible from the form of (5.12).

Comparing (5.13)–(5.15) with similar equations in PME($\rho = 0$), the influence of the base-state discontinuity can be isolated. First, the update equations (5.14) and (5.15) in PME($\rho = 0$) are identical to those in PME(ρ). The discontinuity enters into PME(ρ) only in the equations that propagate the moments between measurements.

P_{xx}: The inclusion of a discontinuity introduces the following terms into the equation for the base-state error covariance:

$$\frac{d}{dt}P_{xx} = \cdot + \rho' Q' P_{\phi x} + P_{x\phi} Q\rho + \rho' \sum_i V(\mathbf{e}_i)\,\hat{\phi}_i \rho.$$

One effect of the discontinuity is to increase the intensity of the exogenous disturbance. This is reflected in $\rho' \sum_i V(\mathbf{e}_i)\,\hat{\phi}_i\rho$. The paired terms, $\rho' Q' P_{\phi x} + P_{x\phi} Q\rho$, relate the base-state covariance to the correlation of the modal- and base-state error. Note that if the modal estimator has a high quality ($\hat{\phi}_t \approx \mathbf{e_r}$), (5.14) reduces to

$$\frac{d}{dt}P_{xx} = \cdot + \rho' V(\mathbf{e}_r)\,\hat{\phi}_r\rho,$$

a near analogue to the *white noise equivalent* of the jump discontinuity. This term reflects the obvious need to account for the sudden change in the base-state, and within an LGM framework, the exogenous disturbance is represented as additional white noise. Of course the terms and updates involving the modal measurement never appear in the Kalman filter since the modal-state is ignored in the LGM model.

$P_{x\phi}$: The mixed moments do not appear in the Kalman filter. Comparing PME(ρ) with PME($\rho = 0$), we see that the discontinuity adds the terms

$$\frac{d}{dt}P_{x\phi} = \cdot + \rho' Q' P_{\phi\phi} + \sum_i V(\mathbf{e}_i)\,\hat{\phi}_i.$$

The change in $\frac{d}{dt}P_{x\phi}$ is related to both the frequency of jumps (Q, $V(\mathbf{e}_i)$) and their size (ρ). Again suppose that there is a good estimate of the modal state ($\hat{\phi}_t \approx \mathbf{e_r}$). In this event, the contribution due to the discontinuity reduces to

$$\frac{d}{dt}P_{x\phi} = \cdot + \rho' V(\mathbf{e}_r)\,\hat{\phi}_r.$$

$P_{xx\phi_m}$: This moment quantifies the relation between modal errors and the base-state covariance. Again looking at the terms that depend on the state discontinuity, we get

$$\frac{d}{dt}P_{xx\phi_m} = \cdot + \left(\rho'\left(Q' P_{\phi x \phi_m} + \sum_i \left(V(\mathbf{e}_i)_{.m} P_{\phi_i x}\right)\right)\right) + (\cdot)'$$
$$+ \sum_i Q_{im}\rho'\left(V(\mathbf{e}_i) P_{\phi_i\phi_m} + U_m(\mathbf{e}_i)\,\hat{\phi}_i\right)\rho.$$

The rationale behind this combination of terms is hard to fathom, but again suppose that there is a good estimate of the modal-state. The dominant term due to the discontinuity is simply

$$\frac{d}{dt}P_{xx\phi_m} = \cdot + Q_{rm}\rho' U_m(\mathbf{e_r})\rho,$$

the square of the base-state jump ($\rho' \cdot \rho$) times a factor related to its likelihood ($Q_{rm}U_m(\mathbf{e_r})$).

5.5 A Maneuvering Aircraft with Sudden Translations

The examples in Chapter 4 have shown that when the base-state path is continuous, an oblique measurement of the regime may suffice. For example, the intrinsic regime variable for the maneuvering target in Chapter 4 was turn rate (i.e., $\{\alpha_t\}$). The image measurement was not sensitive to $\{\alpha_t\}$ but rather to its integral: The change in angular orientation is used to infer turn rate. The error in $\{\hat{\alpha}_t\}$ produced a drift in $\{\tilde{x}_t\}$ slow enough that satisfactory tracking could be accomplished. The modal$\times$ base-state cross gain contained sufficient information on motion geometry to avoid losing the target. However, when the path is discontinuous, the lag inherent in an orientation measurement may be unacceptable and a direct regime measurement may be required.

To illustrate the advantage of including the translation dynamics in the motion model, return to the maneuvering aircraft studied earlier but now give it the capability to translate suddenly. As before, a missile approaches at nearly constant altitude and a speed of 300 m/s with detection at a range of 35 km. Let us look only at the first 7 g turn to the right with return to coast. When the missile turns, it simultaneously translates 200 m in the Y direction. The position discontinuity accentuates the turn as seen from the sensor. After turning for 6 s, the missile returns to the coast mode and reverses the discontinuity.

The proposed motion is, of course, only an approximation to an actual path. A missile can neither change its angular rate nor its position discontinuously. A 200-m translation would require a 40 g acceleration even if achieved in 1 s. Nevertheless, this motion provides an approximation to an irregular path. Also, with the extended final coast, this path provides the opportunity to determine how well the trackers readjust to quiescent operation.

The sensor suite contains the radar we have used before, but its sample rate is slower; the standard errors are 40 m in range and 1.75 mr in bearing, and the interdwell time is 2 s. Again, the EKF selected for comparison, EKF(W=1), is found by neglecting the maneuvers. A sample of the path is shown in Figure 5.5 along with a performance plot of EKF(W=1). As has been the case, the EKF is not adequate in this application. EKF(W=1) tracks well into the first turn. But the discontinuity at the turn is reflected immediately in the tracking error. The radar measurement at $t = 4$ s improves things, but the errors grow because the velocity has not been properly determined: The EKF tacks away from the turn and the translation compounds the problem. After return to coast, EKF(W=1) gradually identifies the missile velocity. But even at the end of the part of the path shown, the error is quite large.

Now add a modal sensor to the suite. It will be assumed that this latter measures the azimuth angle of the missile (an imager) with frame rate $\lambda = 10$ frames/s. The

Table 5.2. *Modal sensor parameters.*

Error Type	Probability
UDE	0.1
NNE	0.1
PE	0.3
ME	0.2

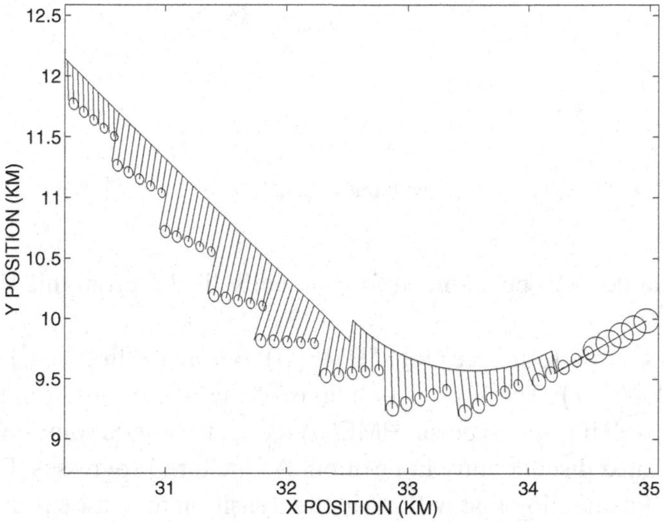

Figure 5.5. The EKF(W=1) for a path with translation.

image classifier again uses 30° bins and its errors are intermediate between imager IG and imager IP in Table 4.1. The error parameters are given in Table 5.2. (With the usual assumptions, the standard deviation of orientation classifier is 76°.)

Additionally, the PME avails itself of a direct maneuver measurement. The translational acceleration might be distinguishable from some peculiarity in measured image or the thruster might have a distinctive spectral resonance. Errors in maneuver classification (labeled ME) will be assumed to be uniform across the appropriate bins (see Table 5.2).

To synthesize PME(ρ) the maneuver modes are described as before:

$$\Phi_t \in \{0.2 \text{ r/s}, \alpha_t = \mathbf{e}_1; 0 \text{ r/s}, \alpha_t = \mathbf{e}_2; -0.2 \text{ r/s}, \alpha_t = \mathbf{e}_3\}.$$

The mean sojourn time in each of the maneuver states will be assumed to be 5 s, with each turn being followed by a coast. From coast, the turns are equally likely. The

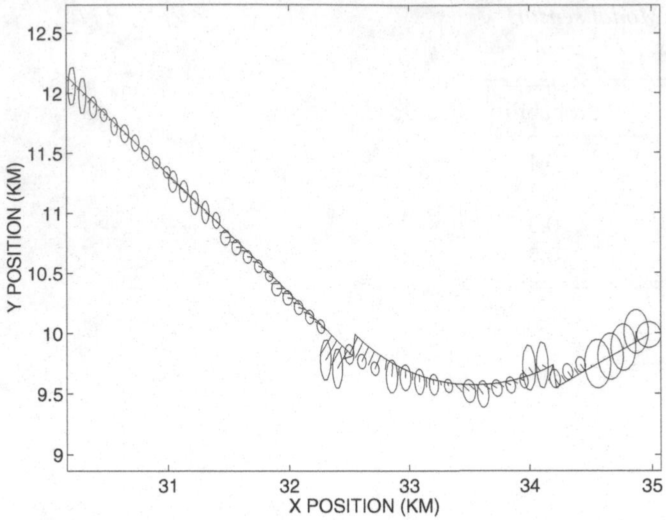

Figure 5.6. The PME(ρ) for a path with translation.

initial maneuver modes will be assumed to be equally likely. From this PME(ρ) can be deduced.

Figure 5.6 shows the performance plot of PME(ρ). Before the first turn, PME(ρ) is worse than EKF(W=1): The EKF is oblivious of the possibility of a turn and will perform quite well if no turns occur. PME(ρ) recognizes (with some delay) the turn and the associated discontinuity in position. As the turn progresses, PME(ρ) biases its error in the direction that will make the transition to coast easier. So the return to coast creates less error.

PME(ρ) uses a better algorithm for computing $\{P_{xx}\}$ than does the Kalman filter, EKF(W=1). The error ellipses of EKF(W=1) are smaller than those of PME(ρ). This indicates that the EKF imagines itself to be doing a good job. But the EKF is deluding itself. The true path is not within the 1-σ ellipses after the turn begins (or the 2-σ or even the 5-σ in some cases). PME(ρ), as befits its recognition of the uncertainty created by maneuvers, has larger ellipses and in the main they encircle the true path. It is interesting to note that after the maneuver transition events, the PME(ρ) error ellipses grow significantly. The computed error variance of PME(ρ) is thus responsive to changing conditions.

Neglecting the path translations leads to a slightly simpler tracker. Figure 5.7 shows the performance plot of PME ($\rho = 0$). PME ($\rho = 0$) tracks well through the turn, but it has difficulty recovering on the final coast. PME(ρ) biases its estimates in preparation for a modal transition, while PME ($\rho = 0$) does not. This bias is not an unmixed blessing, but the error ellipses generated by PME(ρ) are more representative of the uncertainty than are those generated by PME ($\rho = 0$).

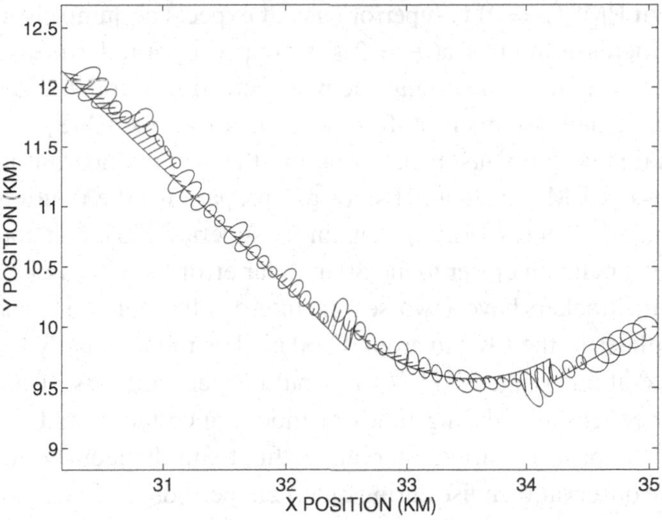

Figure 5.7. The PME($\rho = 0$) for a path with translation.

PME(ρ) is somewhat more complicated than the basic PME($\rho = 0$). The latter acknowledges the possibility of a turn but is blind to the path discontinuities. Figure 5.8 shows the mean radial tracking errors for the two PMEs. The figure was generated as the sample average of ten independent runs. Certain aspects of PME(ρ) are apparent from the graph. With perfect initialization, both PMEs begin

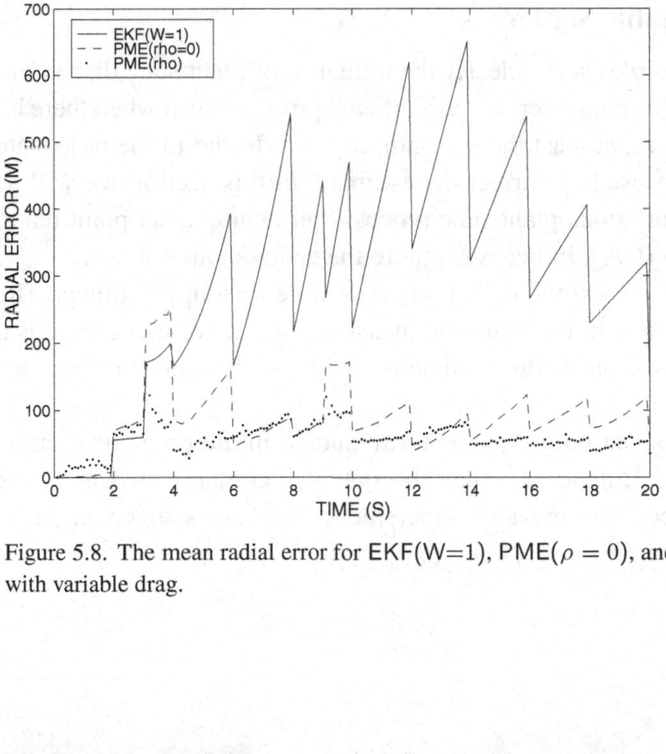

Figure 5.8. The mean radial error for EKF(W=1), PME($\rho = 0$), and PME(ρ) for a path with variable drag.

with small error, but PME($\rho = 0$) is superior since it expects no jump changes and none occur. (The increase in error at $t = 2$ s is created by a radar measurement moving both PMEs away from their omniscient initialization.) Both trackers experience an unavoidable increase in error after the turn, at $t \approx 3$ s. PME(ρ), with its sophisticated modal measurements, is better during the turn but not uniformly so. As the turn progresses, PME(ρ) biases its error to prepare for the return to coast. After the transition, at $t \approx 9$ s, PME(ρ) is again far superior. Note that both of the PMEs keep their errors much closer to the 80 m radar error than does EKF(W=1) even though all of the trackers have a two-second interdwell time: The 80 m/sample radar error accumulates in the EKF to around 800 m. The radar actually loses lock at about 400 m, and at this range EKF(W=1) would experience loss-of-lock.

PME(ρ) increases its gains during times of modal uncertainty and is quicker to respond during the post-transition intervals without simultaneously increasing amplification of the observation noise during quiescent periods. Even severe motion anomalies can be countered by improved modeling and sensor integration.

We conclude that:

- Proper integration of modal measurements significantly improves tracking performance as compared to the EKF.
- The maximum tracking error using PME(ρ) is much smaller than that associated with a tracker that ignores the discontinuities, PME($\rho = 0$).
- The PME(ρ) has a better idea of its own errors and hence is more suitable for use in a measurement-adaptive implementation.

5.6 Variable Set Points

The previous examples show clearly the influence of plant state discontinuity. Although abrupt state changes create a significant problem even when there is a direct modal measurement, at least the discontinuity is reflected in the base-state observation and can be used to correct the estimate. In this section we will look at a system with a continuous plant state process, but having a set point that changes with regime: $\chi \neq 0$. Again, we will ignore the endogenous control.

The plant state is continuous, but this does little to simplify things. The modal change is now hidden in the base-sate measurement. Also, when there is a modal transition, the base-state is discontinuous: If $\{\phi_t\}$ makes the transition $\mathbf{e}_i \mapsto \mathbf{e}_j$, $\Delta x_t = \chi_i - \chi_j$.

This type of discontinuity has the form studied in the previous section (it is a simple translation). But the base-state observation is contaminated, and it is difficult to determine the correct innovations increment. The base-state equation is

$$dx_t = \sum_i (A_i x_t \, dt + C_i \, dw_t)\phi_i - \chi \, d\phi_t. \tag{5.16}$$

The base-state observation is

$$y[k] = H\chi[k] + n[k],$$

where $\chi[k] \neq x[k]$. The PME (PME(χ)) becomes:

Between observations:

$$\frac{d}{dt}\hat{\phi}_t = Q'\hat{\phi}_t,$$

$$\frac{d}{dt}\hat{x}_t = \sum_i A_i R_{x\phi_i} - \chi Q'\hat{\phi}_t,$$

$$\frac{d}{dt}P_{x\phi} = \sum_i \left(A_i P_{(x\phi_i)\phi} + V(\mathbf{e}_i)\,\hat{\phi}_i\right) + P_{x\phi}Q - \chi Q' P_{\phi\phi},$$

$$\frac{d}{dt}P_{xx} = \left(\sum_i A_i P_{(x\phi_i)x} - \chi Q' P_{x\phi}\right) + (\cdot)' + \sum_i \left(R_\chi(i) + \chi \sum_i V(\mathbf{e}_i)\chi'\right)\hat{\phi}_i,$$

$$\frac{d}{dt}P_{xx\phi_m} = \left(\sum_i A_i P_{(x\phi_i)x\phi_m} - \chi\left(Q' P_{\phi x\phi_m} - \sum_i \chi V(\mathbf{e}_i)_{.m} P_{\phi_i x}\right)\right) + (\cdot)' + \sum_i \Big(R_\chi(i) P_{\phi_i\phi_m} + Q_{im}\big(P_{xx\phi_i} + \chi\big(V(\mathbf{e}_i) P_{\phi i\phi_m} + U_m(\mathbf{e}_i)\,\hat{\phi}_i\big)\chi'\big)\Big). \tag{5.17}$$

At a modal observation:

$$\hat{\phi}^+ = \hat{\phi}^- * \Delta\vartheta,$$

$$\Delta\hat{x} = P_{x\phi}\Delta\vartheta,$$

$$\Delta P_{x\phi} = -\Delta\hat{x}\Delta\hat{\phi}' + \sum_k P_{x\phi\phi_k}\Delta\vartheta_k,$$

$$\Delta P_{xx} = -\Delta\hat{x}\Delta\hat{x}' + \sum_k P_{xx\phi_k}\Delta\vartheta_k,$$

$$\Delta P_{xx\phi_m} = -\Delta\hat{\phi}_m\Delta\hat{x}\Delta\hat{x}' - \Delta\hat{\phi}_m P^+_{xx} - \Delta\hat{x} P^+_{\phi_m x} - P^+_{x\phi_m}\Delta\hat{x}' + \sum_k P_{xx\phi_m\phi_k}\Delta\vartheta_k. \tag{5.18}$$

At a base-state observation:

$$\begin{aligned}\Delta \hat{x} &= \gamma_x \Delta \nu_x, \\ \Delta P_{x\phi} &= -\gamma_x H P_{\chi\phi}, \\ \Delta P_{xx} &= -\gamma_x R_{\chi\chi} \gamma_x', \\ \Delta P_{xx\phi_m} &= -\gamma_x H P_{\chi\chi\phi_m} - P_{\chi\chi\phi_m} H' \gamma_x.\end{aligned} \tag{5.19}$$

The base-state estimator in PME(χ) differs from the Kalman filter in the usual ways: The expected discontinuity ($-\chi Q'\hat{\phi}_t$) is added to a term utilizing the correlation of ϕ_t and x_t; the gain depends on the covariance $P_{\chi\chi}$ (rather than on P_{xx}); the filter gain is $P_{x\chi} H' D_{\chi\chi}$ rather than $P_{xx} H' D_{xx}$. The update equations are identical in form to previous algorithms with the base-state innovations replaced by

$$\nu[k] = y[k] - H(\hat{x}[k] + \chi \hat{\phi}[k]).$$

These changes appear innocuous but they mask the significant increase in difficulty in estimation.

5.7 Estimating the Temperature of a Solar Panel

To illustrate the performance of PME(χ), consider a specific example. On the California desert stands a 10 MWe solar electric generating system. Movable mirrors (heliostats) are used to focus the sun's energy on a group of boiler panels on a central tower. A steam temperature regulator controls the feedwater flow rate to the panels so as to maintain the proper temperature of both the outlet steam and receiver panel itself.

On a partly cloudy day the insolation changes suddenly and unpredictably. The panel dynamics change as well: At low insolation, the panel dynamics are slow; at high insolation, the panel dynamics are fast. Let τ_m be the panel metal temperature perturbation variable. (The nominal metal temperature of the panel must be added to τ_m to obtain the actual metal temperature, T_m.) A first-order model of the panel is given by

$$d\tau_m = \alpha_\phi \tau_m \, dt + \beta_\phi \, dw_t, \tag{5.20}$$

where the coefficients depend upon insolation (i.e., ϕ_t) and $\{w_t\}$ is phrased in terms of an equivalent flow rate (lb/min of feedwater).

In a study of the control of the panel subsystem, T_m was assumed to be measured without error [SR83]. A more accurate description of the panel would include the

Table 5.3. *Panel coefficients.*

Mode	Insolation	υ	χ	α	β	a
$\mathbf{e}_1$	1.8×10^5 BTU/min	186 lb/min	1,017°F	−1.8	1.58	.9
$\mathbf{e}_2$	2.7×10^4 BTU/min	30 lb/min	1,000°F	−0.36	1.9	.25

dynamics of a metal-temperature sensor with output T_c. We can represent this sensor with a first-order lag,

$$\frac{d}{dt}T_c = -a_\phi(T_c - T_m), \tag{5.21}$$

where a_ϕ is a coefficient measuring the sensor response rate under the indicated conditions. Using τ_c as the deviation from nominal in the temperature sensor, (5.20) and (5.21) can be consolidated in the orthodox manner: The state ordering in what follows is $x = \text{vec}(\tau_m, \tau_c)$.

The coefficients of the local panel models are given as a function of insolation in [Sch80]. For a single panel and two insolation levels, the coefficients and the nominal operating points are as displayed in Table 5.3. (The sensor model is not given in the reference.) The nominal panel temperature at high insolation is 17°F higher than that for low insolation: $\chi = (1017, 1000)\otimes\mathbf{1}$.

The sensors at the site provide two types of measurements:

- a direct measurement of T_c in noise,
- an insolation measurement.

For the purposes of this example, the former will be sampled every 0.5 min in white Gaussian noise of standard deviation 2°F:

$$y[k] = H\chi[k] + n[k],$$

where $H = \mathbf{e}_2'$ and $R_x = 4$.

As noted above, the insolation can change suddenly and frequently. Figure 5.9 shows sample functions of $\{T_m\}$ and $\{T_c\}$ for the insolation path: $\phi_t = \mathbf{e}_1 I_{[0,1)\cup[6,9)} + \mathbf{e}_2 I_{[1,6)\cup[9,11)}$. Both temperatures were initialized at 1,017°. After the cloud comes over the heliostats at $t = 1$, T_m moves rapidly toward 1,000°F with T_c moving at a much slower rate. A 6° temperature difference between the sensor and the panel develops in 5 min. When the sun returns, the sensor output has significant lag.

In the hybrid model, the modal-state is an indicator of the set point, and the base-state is derived from a linearization about this nominal point. A change in $\{\phi_t\}$ is associated with both a change in local dynamics and a discontinuity in the base-state (though not in the plant state). Further, $\{y[k]\}$ is an explicit function of $\{\chi_t\}$ not $\{x_t\}$. In Figure 5.9, the mode-dependent behavior is evident. But an estimator that

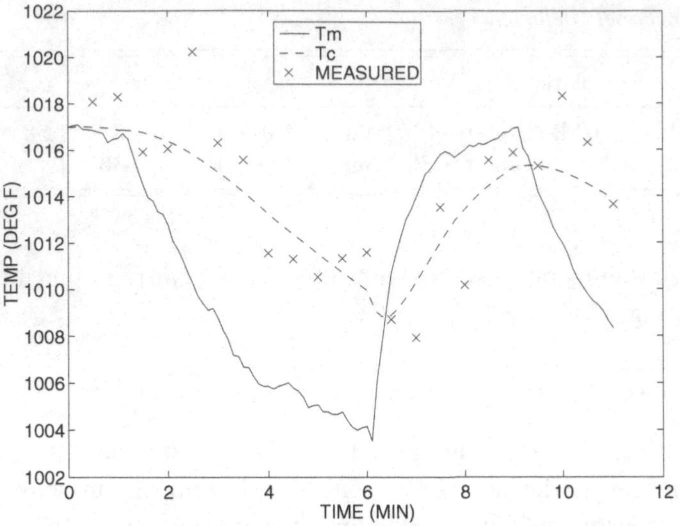

Figure 5.9. Measurements of panel temperature along with the actual temperatures.

depends only on $\{y[k]\}$ would have a difficult time distinguishing the current mode. The temperature measurements are marked on the figure with an x. Determining $\{T_m\}$ from such an abbreviated data set is a difficult task indeed.

Using data from the site, the modal dynamics can be particularized. A simple Markov model matching the sample means of the insolation sojourns during a period of partial clouds is

$$Q = \begin{bmatrix} -0.25 & 0.25 \\ 0.50 & -0.50 \end{bmatrix}$$

(mean-sojourn in sun is 4 min, while it is 2 min in cloud).

To improve performance, the insolation sensor gives a direct measurement of $\{\phi_t\}$. In the application, the insolation is not uniformly distributed across the panel: The geometry of the sun–mirror panel is such that the distribution of the solar flux across a panel is irregular with possible hot spots. Insolation sensors only measure conditions in a local neighborhood, and to determine the effective insolation across the panel, several were arrayed across the panel. The highest reading from the group was used as a measure of panel insolation. This inclination toward high insolation readings can be accommodated with an asymmetrical $\mathbf{P}$. It will be assumed that the insolation sensor is sampled every 0.1 min.

Suppose we wish to design a remote estimator of panel temperature. In what follows, three algorithms will be contrasted. The first, PME(IG) uses the most discriminating modal measurement suite. The discernibility matrix for the insolation

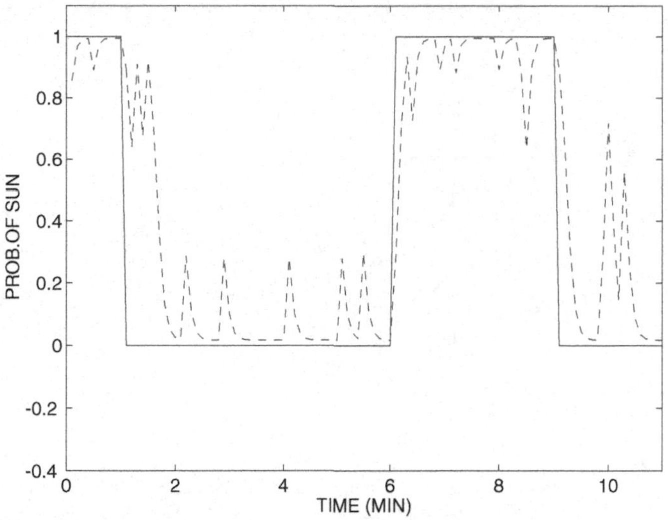

Figure 5.10. The probability of sun using PME(IG).

sensor is

$$\mathbf{P}_1 = \begin{bmatrix} -0.85 & 0.20 \\ 0.15 & -0.80 \end{bmatrix}.$$

This sensor distinguishes conditions quite well but has a bias toward high insolation. Figure 5.10 shows a plot of $\{\phi_1\}$ (sojourns in the sun), the modal measurements, and $\{\hat{\phi}_1\}$ as computed by PME(IG). The good quality of the data is reflected in a expeditious response to modal transitions ($\sim$1 min) and a generally high confidence in the modal identification.

The second estimator, PME(IP) uses a less accurate insolation sensor:

$$\mathbf{P}_2 = \begin{bmatrix} -0.65 & 0.40 \\ 0.35 & -0.60 \end{bmatrix}.$$

Again there is a bias toward high insolation, but the precision of the classifier is such that the modal-state cannot be identified with confidence.

The third estimator, labeled EKF, uses the conventional Kalman algorithm with pseudonoise. Suppose that Q is a symmetric chain with mean sojourns of 2 min. The stationary distribution of the chain is $\hat{\phi}_t = 0.51$. To account for the discontinuity in the base-state, the plant noise in the EKF is increased by

$$\chi V(\mathbf{e}_1)\chi' = d\langle x,x\rangle_t/dt.$$

In the EKF there are no modal measurements and the cross moments in P_{XX} are

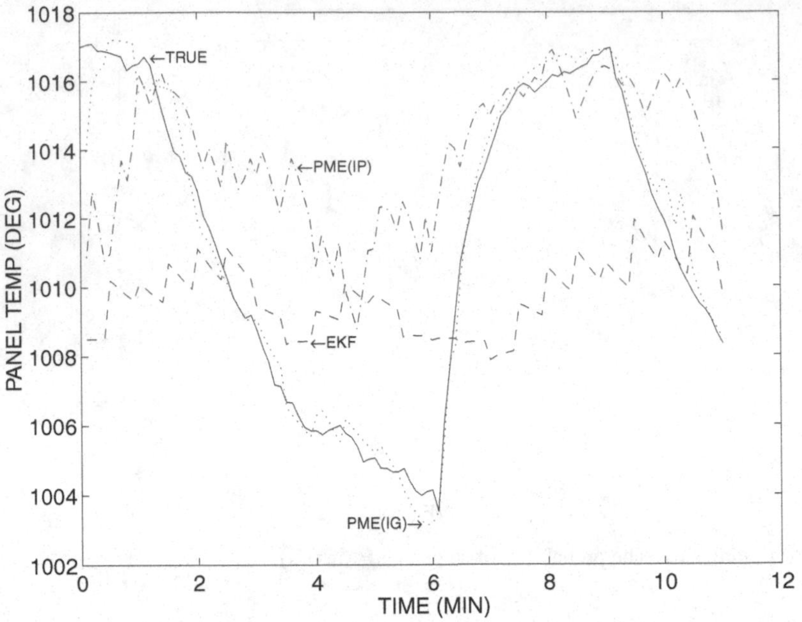

Figure 5.11. Temperature estimates from PME(IG), PME(IP), and EKF.

neglected:

$$\frac{d}{dt}\hat{x}_t = \hat{A}_t\hat{x}_t$$

and

$$\frac{d}{dt}P_{xx} = \hat{A}P_{xx} + P_{xx}\hat{A}' + \hat{R}_\chi + \chi V(\mathbf{e}_1)\chi'$$

subject to appropriate initial conditions and update equations.

Figure 5.11 shows the response of PME(IG), PME(IP), and EKF in this application. The PMEs initialize themselves in the first minute. During the first cloudy period, both of the PMEs follow the metal temperature, but PME(IG) is far quicker. PME(IP) mistakes the modal condition at ~5 min, and this leads to a premature increase in the temperature estimate. Still, given the inability of PME(IP) to resolve the modal condition, this response is surprisingly good.

EKF uses the average dynamics for extrapolation and bases its innovations estimate on the average panel temperature. It is both slow to respond to modal changes and languid in its reaction. The temperature variation in T_m is 16°F; for PME(IG) it is 16°F; for PME(IP) it is 9°F; for EKF it is only 3°F.

The three estimators studied in the example all compute estimates of their own uncertainty. Uncertainty can be expressed in different ways: One of them is the computed standard deviation of the error in the panel temperature base-state, $\{x_1\}$.

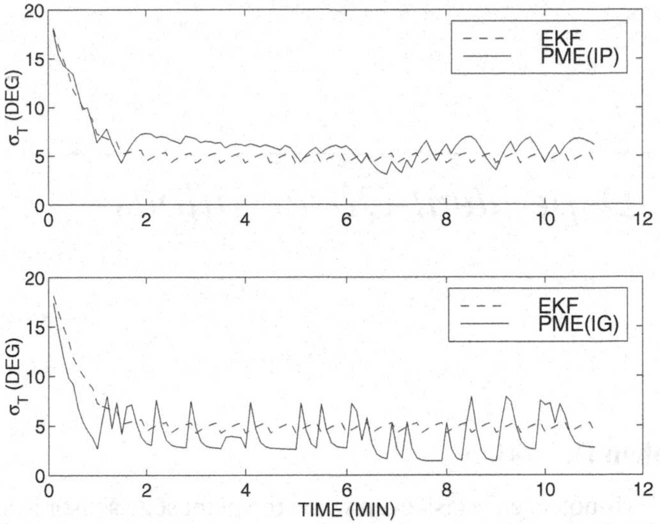

Figure 5.12. Computed standard errors for PME(IG), PME(IP), and EKF.

Figure 5.12 shows a plot of $\sqrt{P_{xx}}(1, 1) = \sigma_T$. The behavior of $\{\sigma_T;$ EKF$\}$ is what would be expected. From the initial value of 20°F it drops to about 5°F, increasing between temperature measurements. Although the temperature sensor noise is only 2°F, $\{\sigma_T;$ EKF$\}$ is larger because the measurement is indirect. Note that the actual temperature deviation can be closer to 6°F.

The uncertainty in the two PMEs is mode dependent. The computed $\{\sigma_T;$ PME(IP)$\}$ tends to be about 6°F, a conservative value. During a brief period during which PME(IP) becomes confident of its modal identification (at ~ 7 min), $\{\sigma_T;$ PME(IP)$\}$ drops to a value of 4°F. In PME(IG) the mode dependence is clearest because modal identification is more accurate. The standard error, $\{\sigma_T;$ PME(IG)$\}$, is seen to move from about 2°F to about 8°F whenever an unexpected modal observation is received. This is to be expected since an observation suggesting a modal change causes the PME to suspect there might be a larger base-state error.

6

Mode-Dependent Observations

6.1 Problem Definition

Zygostate estimation is not an easy task even when the plant state sensor is modeled as in (1.17):

plant state measurement: time-discrete

$$y[k] = H\chi[k] + n[k]. \qquad (1.17)$$

In some applications, the problem is more perverse because the coefficients of the plant state sensor depend on the modal-state. Regime dependence may occur because the linearization is different for different operating points or because the intensity of the sensor noise is thus dependent. This latter circumstance occurs, for example, when the plant output is intentionally jammed. In this chapter, we will explore the utility of the PME in this application. To keep the presentation as concise as possible, the focus will be on linear jump systems (LJS) ($\chi = 0$) without endogenous actuation.

We begin with a more flexible plant state observation model:

plant state measurement: time-discrete; variable quality

$$y[k] = \sum_i (H_i x[k] + D_i n[k])\phi_i, \qquad (6.1)$$

where $\{n[k]\}$ is a white Gaussian sequence and the covariance of $D_i n[k]$ is $R_x(i)$. Though the measurement model is similar in form and motivation to (1.17), the zygostate estimation problem is now complicated by the fact that we are not as sure of what we are measuring.

In (6.1) both the gain and the noise intensity are random processes. The measurement gain, H, has considerable impact on the base-state update in the Kalman filter. The measurement residual (the innovations increment) at time $t = (k+1)T$ is $r[k+1] = y[k+1] - \hat{y}[k+1]^-$. The Kalman update is

$$\Delta\hat{x}[k+1] = \gamma_x r[k+1], \tag{6.2}$$

$$\Delta P_{xx}[k+1] = -\gamma_x R_{yy}[k+1]\gamma_x', \tag{6.3}$$

where the covariance of the residual is $R_{yy}[k+1] = HP_{xx}[k+1]^- H' + R_x$ and $\gamma_x = P_{xx}[k+1]^- H' D_{yy}[k+1]$. It is evident that mistaking the value of H causes the filter to assign the correction preferentially in the wrong direction.

The influence of a mistake in assigning an intensity to the sensor noise is more subtle. A good estimator attempts to smooth the white observation noise by averaging. As R_x increases, the filter gain increases the interval over which averaging takes place. But what should be done when the intensity of the observation noise is not known? If the noise is larger than it is thought to be, the Kalman gain will be too high and the filter will not smooth the noise. However, if the model noise is larger than the actual noise, not all of the information in the observation will be used to make the correction.

In this chapter, we will utilize the PME to create a finite-dimensional algorithm for integrating these modified plant state observations with the conventional modal-state observations. This requires a generalization of the PME as carried out by Boyd in [Boy96]. The same group of canonical moments required in earlier applications reappears. The explicit change in the algorithm appears in the update equation that accommodates a plant state measurement. The PME is recursive, and the influence of the new $\{y[k]\}$ update propagates through the estimation process.

6.2 The PME

The PME fuses the two innovation processes into the required zygostate estimate. The plant state observation model given in (6.1) modifies the calculation of the base-state innovations process. The observation residual, $r[k+1]$, is the difference between the observation and its expectation:

$$r[k+1] = y[k+1] - \hat{y}[k+1]^-.$$

In previous chapters, the plant state residual was written

$$r[k+1] = y[k+1] - H\hat{x}[k+1]^-.$$

This form will not suffice for this analysis. Instead we must use

$$r[k+1] = y[k+1] - \sum_i H_i R_{x\phi_i}^-. \tag{6.4}$$

The residual process is white Gaussian with positive covariance $R_{yy} = D_{yy}^{-1}$:

$$R_{yy} = \sum_i H_i \left(R^-_{xx\phi_i} - R^-_{x\phi_i} R^-_{\phi_i x}\right) H_i' + R_x(i)\hat{\phi}_i^- . \tag{6.5}$$

Define a gain function γ_x by

$$\gamma_x = \sum_i P^-_{x(x\phi_i)} H_i' D_{yy}. \tag{6.6}$$

In these terms, the PME is written:

the PME: mode dependence in the plant state sensor

Between observations:

$$\frac{d}{dt}\hat{\phi}_t = Q'\hat{\phi}_t,$$

$$\frac{d}{dt}\hat{x}_t = \sum_i A_i R_{x\phi_i},$$

$$\frac{d}{dt}P_{x\phi} = \sum_i A_i P_{(x\phi_i)\phi} + P_{x\phi} Q,$$

$$\frac{d}{dt}P_{xx} = \sum_i \left(A_i P_{(x\phi_i)x} + (\cdot)' + R_\chi(i)\hat{\phi}_i\right),$$

$$\frac{d}{dt}P_{xx\phi_m} = \sum_i \left(A_i P_{(x\phi_i)x\phi_m} + (\cdot)' + P_{xx\phi_i} Q_{im} R_\chi(i) P_{\phi_i\phi_m}\right).$$

At a modal observation:

$$\hat{\phi}^+ = \hat{\phi}^- * \Delta\vartheta,$$

$$\Delta\hat{x} = P_{x\phi}\Delta\vartheta,$$

$$\Delta P_{x\phi} = -\Delta\hat{x}\,\Delta\hat{\phi}' + \sum_k P_{x\phi\phi_k}\Delta\vartheta_k,$$

$$\Delta P_{xx} = -\Delta\hat{x}\,\Delta\hat{x}' + \sum_k P_{xx\phi_k}\Delta\vartheta_k,$$

$$\Delta P_{xx\phi_m} = -\Delta\hat{\phi}_m \Delta\hat{x}\,\Delta\hat{x}' - \Delta\hat{\phi}_m P^+_{xx} - \Delta\hat{x}\, P^+_{\phi_m x} - P^+_{x\phi_m}\Delta\hat{x}' + \sum_k P_{xx\phi_m\phi_k}\Delta\vartheta_k.$$

At a base-state observation:

$$\Delta\hat{x} = \gamma_x \Delta\nu_x,$$

$$\Delta P_{x\phi} = -\gamma_x \sum_i H_i P_{x\phi_i}(\mathbf{e}_i - \hat{\phi})',$$

$$\Delta P_{xx} = \gamma_x R_{yy}\gamma_x' - \gamma_x \sum_i H_i(P_{xx\phi_i} + \hat{\phi}_i P_{xx}) - (\cdot)',$$

$$\Delta P_{xx\phi_m} = -\gamma_x \sum_i H_i(P_{xx\phi_i\phi_m} + \hat{\phi}_i P_{xx\phi_m}) + (\cdot)'.$$

Where comparable, the PME mimics the Kalman update with the substitutions

$$\sum_i H\hat{x}\phi_i \mapsto \sum_i H_i R^-_{x\phi_i},$$

$$P^-_{xx} H' D_{yy} \mapsto \sum_i P^-_{x(x\phi_i)} H_i' D_{yy},$$

$$H P^-_{xx} H' + R_x \mapsto \sum_i H_i(R^-_{xx\phi_i} - R^-_{x\phi_i} R^-_{\phi_i x})H_i' + R_x(i)\hat{\phi}_i^- .$$

The first of these replacements is reasonable: If $\{\phi_t\}$ is not known, $x\phi_i$ is replaced by its expectation. The second is harder to devine but plausible. The equation for the covariance of the measurement residual is the result of a direct calculation.

The choice of γ_x is surprising in one respect. The covariance of the measurement residual, R_{yy}, is an aggregation over the modal-state space. The factor $\sum_i P^-_{x(x\phi_i)} H_i$ is another aggregate over the same space. Intuition would suggest

$$\gamma_x = \sum_i P^-_{x(x\phi_i)} H_i' (H_i(R^-_{xx\phi_i} - R^-_{x\phi_i} R^-_{\phi_i x})H_i' + R_x(i)\hat{\phi}_i^-)^{-1},$$

but this composition is not correct; the aggregations take place separately.

6.3 Modal Estimation Using the PME

The effectiveness of the PME depends upon its ability to fuse disparate measurement sequences to create a zygostate estimate. Even when there is considerable error in $\{\hat{\phi}_t\}$, the error cross moments are used to improve $\{\hat{x}_t\}$. The changes in the PME created by (6.1) are made explicit in the $\{y[k]\}$ update, but they are already latent in the $\{z_t\}$ update. And it is from $\{\mathcal{Z}_t\}$ that the mode is identified. To illustrate this aspect of the the estimation problem, let us focus on the modal subfilter. Suppose an imager

Table 6.1. *Imager error parameters.*

Mode	UDE	NNE	PE
$\mathbf{e}_1$	0.1	0.1	0.2
$\mathbf{e}_2$	0.9	0.0	0.0

is used in a tracker following a maneuvering target in the plane. The imager places the orientation of the target in bins of width 30° at a rate of 10 frames/s ($\lambda = 10$). The target has the three turning modes encountered in earlier applications:

$$\Phi_t \in \{0.2 \text{ r/s}, \alpha_t = \mathbf{e}_1; 0 \text{ r/s}, \alpha_t = \mathbf{e}_2; -0.2 \text{ r/s}, \alpha_t = \mathbf{e}_3\}.$$

The imager quality is satisfactory during normal operation, but the target uses countermeasures from time to time. During intervals of jamming, the image is so degraded that it can easily be placed in any of the angular bins. Let $\{r_t\}$ be an indicator of the countermeasure mode: $r_t = \mathbf{e}_1$ if jamming is off; $r_t = \mathbf{e}_2$ if jamming is on. The error parameters of the imager are given in Table 6.1.

Even in normal operation, this imager is not as good as imager IG in Table 4.1. It makes the correct classification only 60% of the time. During intervals when the countermeasures are active, the imager appears to be essentially useless; classification errors occur 90% of the time. Figure 6.1 shows a sample of the output

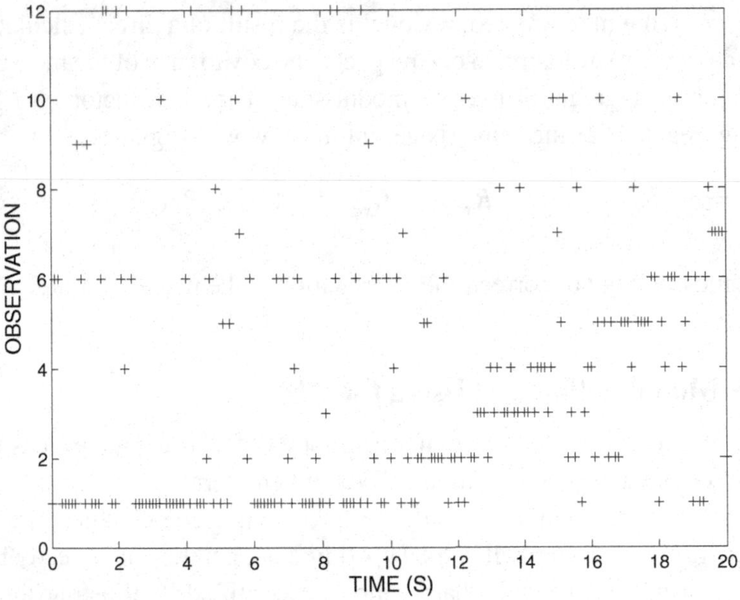

Figure 6.1. Imager observations during periods of jamming.

of the imager for a simple turn. The target is coasting on the interval $t \in [0, 10)$ and is located in bin 1; $\rho_t = \mathbf{e}_1$. At $t = 10$ s it begins to turn, but it does not cross the nearest angular bin boundary until $t = 11$ s. It continues to turn until $t = 20$ s, at which time $\rho_t = \mathbf{e}_7$.

This angular motion is clearly visible in Figure 6.1. What may not be so clear is that on the intervals from [5,6) and [15,16), the target jams the imager: $r_t = \mathbf{e}_2$; $t \in [5, 6) \cup [15, 16)$. Particularly in the latter interval, the presence of jamming is masked by the chance occurrence of observations in orientation bin $\mathbf{e}_3$ and its contiguous bins. Since such measurements correspond to permissible target motions, these measurements could be taken to be unjammed indications of the modal state.

To delineate the PME, the modal dynamics must be specified. The dimension of ϕ_t is $12 \times 3 \times 2 = 72$. Suppose the mean sojourn times are

- 1.2 s in an angular bin during a turn,
- 10 s in a turn,
- 20 s in a coast,
- 1 s in a jamming mode,
- 10 s in an unjammed mode.

From this, Q can be deduced. The turn rate is actually slower than the PME expects. This type of model mismatch commonly occurs in applications. Figure 6.2 shows the probability of jamming as computed by the PME. The identification of

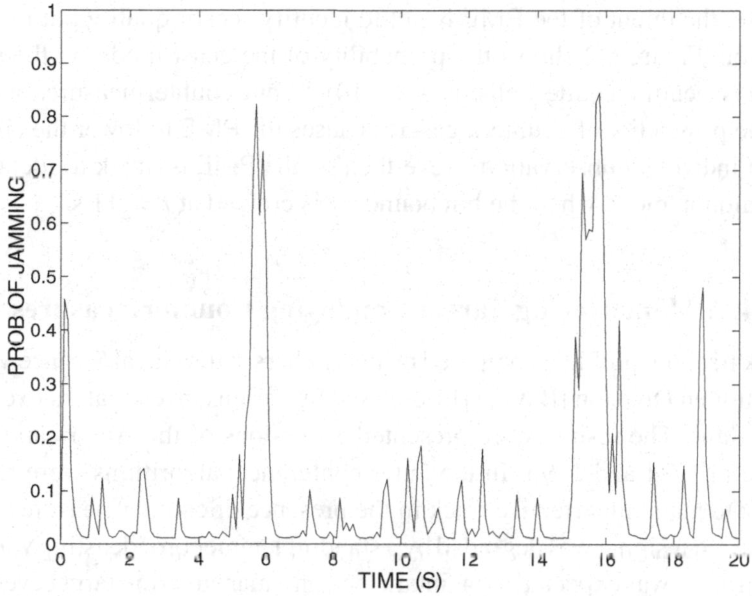

Figure 6.2. Probability of jamming on the sample trajectory.

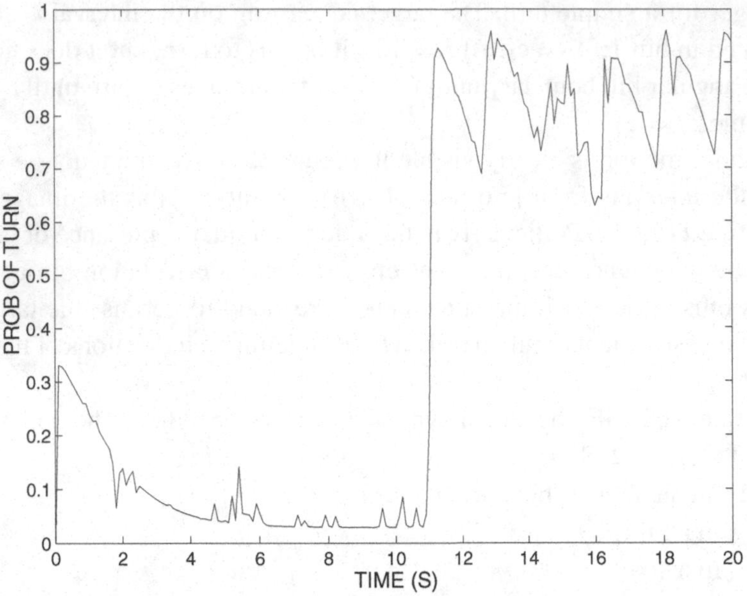

Figure 6.3. Probability of turn on the sample trajectory.

the jamming intervals is surprisingly good given the ambiguity of the observation. Countermeasures are identified expeditiously and with confidence. Artifacts in the second jamming interval delayed recognition of a return to the normal mode of observation.

Of course, the intent of the PME is not to identify sensor quality but rather the target motion. Figure 6.3 shows the probability of the coast mode on the sample path. Coast is identified quite well on $t \in [0, 10)$ despite countermeasures at around $t = 5$ s. The possibility of countermeasures causes the PME to lower the credence given to an individual observation. Nevertheless, the PME is quick to identify the change in motion mode when the bin boundary is crossed at $t = 11$ s.

6.4 A Maneuvering Target Employing Countermeasures

Benchmark tracking problems prepared by researchers at the Naval Surface Warfare Center, Dahlgren Division [BWC94] were used by various investigators to exercise their algorithms. The results were presented at sessions of the American Control Conference in 1994 and 1995. In the latter conference, algorithms were rated on their ability to maintain effective track in the presence of countermeasures. In one instance, the radar signal was degraded by a standoff jammer broadcasting wideband noise. The tracker was expected to maintain lock on a maneuvering target even while the jammer cycled on and off.

Wideband jamming creates more noise in the plant state measurement. During intervals of jamming, the filter gain should be reduced to prevent excessive volatility in the estimates. Unfortunately, an EKF does not know how to weight the observations because it is not sure of its measurement mode. If the jamming is manifest in the modal observation, the dual path architecture of the PME can be used to resolve the modal ambiguity and assist in the base-state estimation.

To elucidate the response of the PME in this environment, return to one of the problems described in Chapter 4. An antiship missile is launched from an aircraft at a range of about 80 km and falls to an altitude of 780 m. After engine ignition, it approaches the ship at a speed of 335 m/s with constant altitude. As the missile nears the ship it performs a series of evasive 7 g jinks, coasts for 10 seconds, then makes a final 3 g turn toward the ship. The path and its planar projection are shown in Figure 6.4. Figure 4.3 shows part of this path after detection.

The planar motion model for the missile is

$$d\begin{bmatrix} X \\ Y \\ V_x \\ V_y \end{bmatrix} = \begin{bmatrix} 0 & 0 & 1 & 0 \\ 0 & 0 & 0 & 1 \\ 0 & 0 & 0 & -\Phi \\ 0 & 0 & \Phi & 0 \end{bmatrix}\begin{bmatrix} X \\ Y \\ V_x \\ V_y \end{bmatrix} dt + \begin{bmatrix} 0 & 0 \\ 0 & 0 \\ 1 & 0 \\ 0 & 1 \end{bmatrix} d\begin{bmatrix} w_x \\ w_y \end{bmatrix}, \tag{6.7}$$

where $\{X, Y\}$ are position coordinates, and $\{V_x, V_y\}$ are associated velocities. The target is subject to both a wideband omnidirectional acceleration represented by

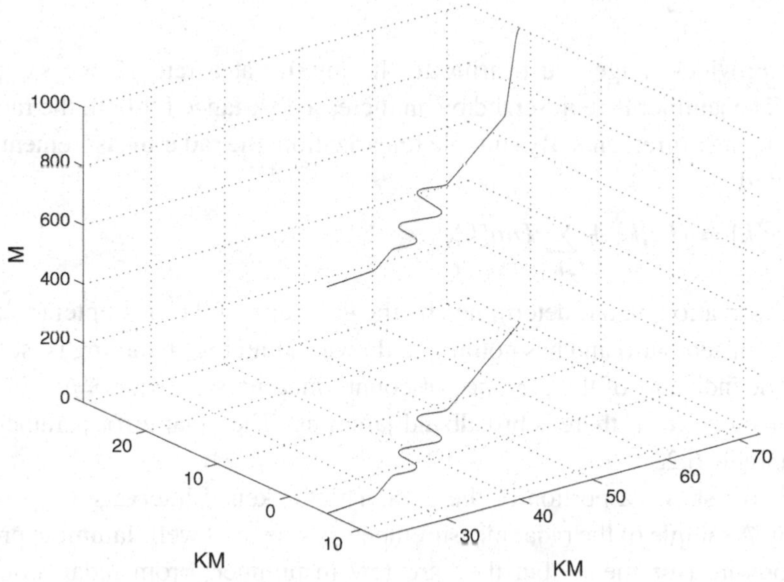

Figure 6.4. Trajectory of a missile with its shadow in the $Z = 0$ plane.

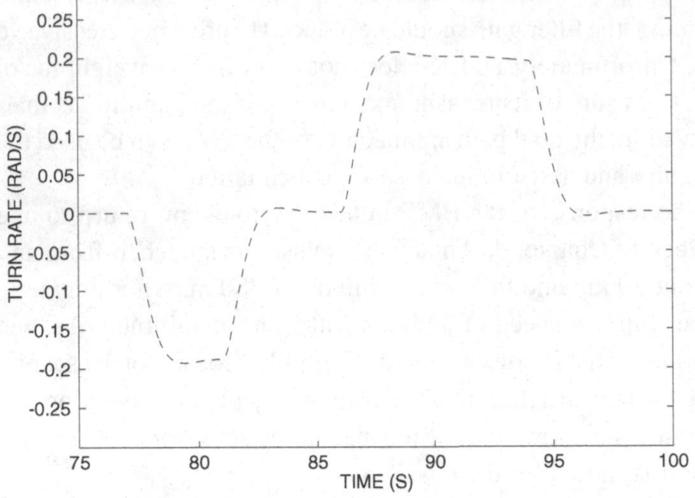

Figure 6.5. Turn rate profile for the test trajectory.

the $\mathcal{F}_t$-Brownian motion $\{w_x, w_y\}$ and the maneuver acceleration represented by the turn rate process $\{\Phi_t\}$.

Figure 6.5 shows the turn rate profile of the benchmark path from the beginning of the jinking phase to impact. The piecewise constant turn rate model is a good approximation to target motion, but it does not reproduce the turns exactly.

With the usual modal identification, (6.7) can be written:

$$dx_t = \sum_i A_i x_t \alpha_i \, dt + C dw_t.$$

A radar provides range and bearing to the missile at a rate of one sample per second. The jammer is represented by an increased variance for both the range and the bearing measurements. By suitable linearization, the radar measurement can be represented

$$y[k] = Hx[k] + \sum_i D_i n[k] \phi_i. \tag{6.8}$$

The observation gain is determined by the geometry of the encounter (calculated on the estimated path) and has nothing to do with jamming. Jamming is isolated in $R_x(i)$. The indicator of the presence of countermeasures is $\{r_t\}$: $r_t = \mathbf{e}_1$ if normal conditions; $r_t = \mathbf{e}_2$ if there is broadband jamming. The radar error parameters are given in Table 6.2.

Figure 6.6 shows a portion of the path. The thickened intervals are periods of jamming. A sample of the radar measurements is shown as well. Jamming produces anomalous measurements, but they are few in number. From radar alone, it is difficult to isolate jamming sojourns in a timely manner.

Table 6.2. *Radar errors.*

Mode	Range	Bearing
$\mathbf{e}_1$	40 m	0.1°
$\mathbf{e}_2$	400 m	1°

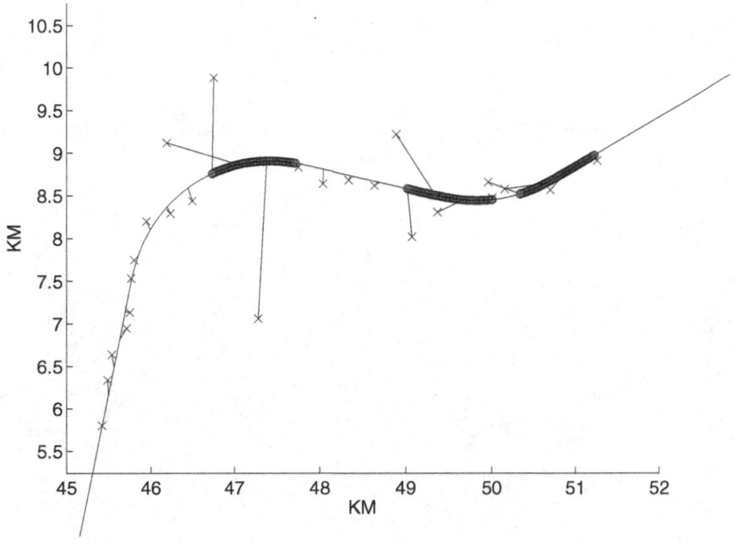

Figure 6.6. Target path with intervals of jamming.

The EKF developed in Chapter 4 can be used in this application too. The wide-band disturbances on the path are small: $W = 1$. However, to account for jamming, pseudonoise should be added. Figure 6.7 shows the error ellipses for EKF(W=25) on a sample path (not the sample shown in Figure 6.6). It is clear that EKF(W=25) is confused by the maneuvers, and the jamming compounds the confusion. The acceleration model predicts an omnidirectional acceleration of constant intensity. The filter is ill prepared for the turn preceded by jamming. The error ellipses of EKF(W=25) are larger than those of EKF(W=1), but the target position is rarely close to the one-σ ellipses.

EKF(W=25) does not expect jammed radar measurements. The large, unmodeled error in the measurements shows up in the filter output as big jumps in the estimated position. The filter gain is too low for the turn, but it is too high to effectively filter the measurement noise. The jumps induce errors in the estimated target velocity as is evident when the estimated position tacks away from the true trajectory.

In contrast to EKF(W=25), the PME acknowledges not only the existence of target maneuvers, but also the possibility of jamming. The jammed imager will

Table 6.3. *Imager error parameters.*

Mode	UDE	NNE	PE
$\mathbf{e}_1$	0.05	0.05	0.2
$\mathbf{e}_2$	0.8	0.0	0.0

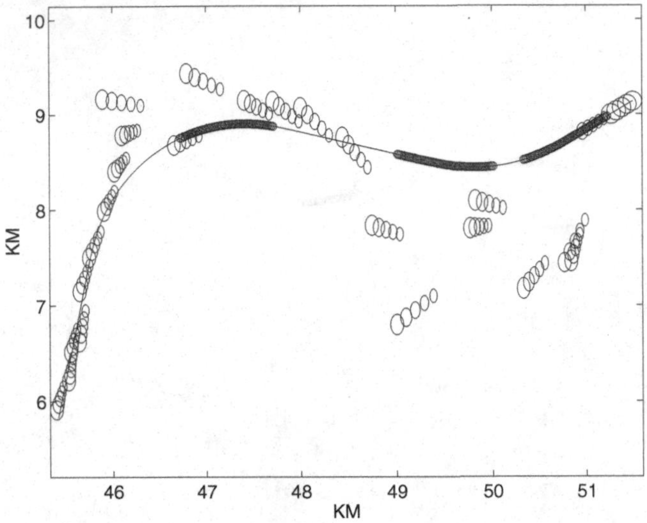

Figure 6.7. Performance of EKF(W=25) on a path with jamming.

have a large probability of uniformly distributed random error, reducing its overall effectiveness in identifying a turn. For this trajectory, the jamming intervals are particularly inopportune: They mask the beginning of a maneuver mode change in all three cases.

In the PME it is necessary to delineate a modal-state that embraces maneuver mode, missile orientation, and jamming mode. Partition the turn rates in the usual way:

$$\Phi_t \in \{0.2 \text{ r/s}, \alpha_t = \mathbf{e}_1; 0 \text{ r/s}, \alpha_t = \mathbf{e}_2; -0.2 \text{ r/s}, \alpha_t = \mathbf{e}_3\}.$$

The imager quantization is $2\pi/L$. The three modal primitives, α_t, ρ_t, and r_t, may now be combined to give a modal-state: $\phi_t = \alpha_t \otimes \rho_t \otimes r_t$.

The modal dynamics follow from the sojourn times in each mode:

- 5 s in a turn,
- 10 s in a coast,
- 3 s in a jamming mode,
- 3 s in an unjammed mode,

and the imager quality is given in Table 6.3.

The PME is simplified when H is constant. If $H_i \equiv H$, then

$$\sum_i H_i (R^-_{xx\phi_i} - R^-_{x\phi_i} R^-_{\phi_i x}) H_i' = H P^-_{xx} H',$$

$$\sum_i P^-_{x(x\phi_i)} H_i D_{yy} = P^-_{xx} H' D_{yy}.$$

Also, $R_{yy} = H P^-_{xx} H' + \hat{R}^-_x$. As a consequence, the plant state measurement update is that used in Chapter 4:

At a base-state observation:

$$\Delta\hat{x} = \gamma_x \Delta\nu_x,$$

$$\Delta P_{x\phi} = -\gamma_x H P_{x\phi},$$

$$\Delta P_{xx} = -\gamma_x R_{yy} \gamma_x',$$

$$\Delta P_{xx\phi_m} = -\gamma_x H P_{xx\phi_m} - P_{xx\phi_m} H' \gamma_x'.$$

Figure 6.8 shows the tracking performance of the PME. The error ellipses are larger during intervals of jamming, indicating the filter's decreased confidence in its own estimates. At the same time, the filter gain is moderated by its model of the large observation errors during jamming: The large jumps in estimated target position evident in Figure 6.7 are avoided. The error ellipses for the PME almost always enclose the true target position.

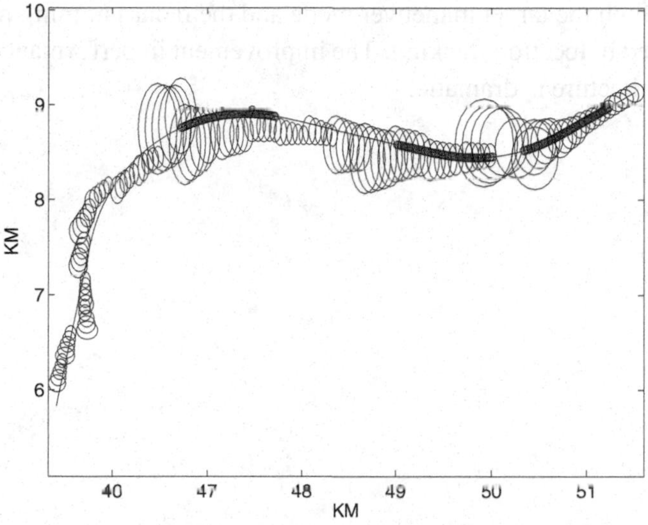

Figure 6.8. Performance of PME on a path with jamming.

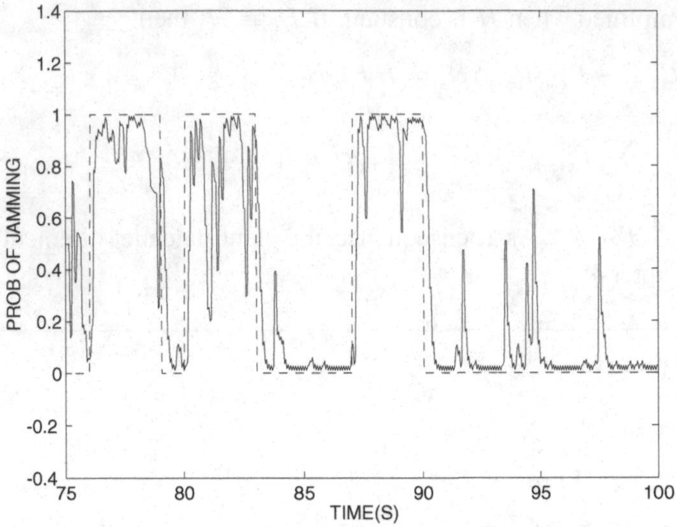

Figure 6.9. Computed probability of jamming and the turn jamming state.

Figure 6.9 shows the computed probability of jamming. Despite its unpropitious placement, the onset of each interval of countermeasures is recognized within a second or so. The motion mode is identified almost as well as it was in the absence of jamming.

EKF(W=25) has difficulty tracking a maneuvering target because the conventional approach to estimation does not induce the filter to change its dynamics with the maneuver and the measurement mode. In contrast, the PME accommodates both random target turns and the intervals of degraded target measurements. The image observations allow both the target maneuver mode and the radar jamming mode to be identified and used in location tracking. The improvement in performance using the dual-sensor architecture is dramatic.

7

Control of Hybrid Systems

7.1 Feedback Regulation of Hybrid Systems

In the applications of hybrid estimation we have explored thus far, we have ignored the influence of the endogenous actuating signal. For example, although a hostile aircraft is controlled, the control action is not generated at the tracker. In other applications that involve an endogenous control, the effect of the actuating signal on estimation is passive. A prototype of this passivity is the Kalman filter for an LJS. The base-state dynamics are given in Chapter 1 (1.8):

$$dx_t = \sum_i ((A_i x_t + B_i u_t)\, dt + C_i\, dw_t)\phi_i. \tag{7.1}$$

The presence of a control signal results in a simple modification of the $\mathcal{G}_t^{\phi}$-conditional mean of the state. The actuating signal appears *only* in the extrapolation equation for $\{\hat{x}_t\}$ in the form of a term $\sum_i B_i u_t \phi_i$. The base-state error covariance is independent of $\{u_t\}$: The control has no impact on the quality of the estimate.

The independence of $\{P_{xx}\}$ and $\{u_t\}$ has important implications for the design of a feedback control algorithm. To illustrate, let us review some results on the regulation of an LJS on the time interval $t \in [0, T]$:

Kalman filter: time-continuous state, time-discrete measurements, endogenous actuating signal

Between observations:

$$\frac{d}{dt}\hat{x}_t = \sum_i (A_i \hat{x}_t + B_i u_t)\phi_i, \tag{7.2}$$

$$\frac{d}{dt}P_{xx} = \sum_i (A_i P_{xx} + P_{xx} A_i' + R_\chi(i))\phi_i. \tag{7.3}$$

At an observation:

$$\Delta\hat{x}[k+1] = \gamma_x r[k+1], \tag{7.4}$$

$$\Delta P_{xx}[k+1] = -\gamma_x R_{yy}[k+1]\gamma_x'. \tag{7.5}$$

In an LJS, the set points are zero: χ and υ are both zero matrices, and the base-state is continuous at modal changes.

The process $\{u_t\}$ is necessarily adapted to the filtration generated by the measurements. Different measurement architectures lead therefore to different controllers. Recall that $\mathcal{F}$ is the underlying σ-algebra on $(\Omega, \mathcal{P})$ with respect to which all of the relevant plant processes are defined. There are several subfiltrations that appear in applications:

- $\{\mathcal{F}_t\}$ is generated by $\{x_t, \phi_t, y_t, z_t\}$: complete knowledge of all plant processes up to the present.
- $\{\mathcal{G}_t\}$ is generated by $\{y_t, z_t\}$: complete knowledge of all plant outputs up to the present.
- $\{\mathcal{G}_t^{\phi}\}$ is generated by $\{y_t, \phi_t\}$: complete knowledge of all base-state observations and the modal path up to the present.
- $\{\mathcal{G}_t^{\phi\mathrm{T}}\}$ is generated by $\{y_t\}$ and $\{\phi_\tau;\, \tau \in [0, \mathrm{T}]\}$: complete knowledge of the base-state observations up to the present and the complete (past and future) modal path.
- $\{\mathcal{X}_t^{\phi}\}$ is generated by $\{x_t, \phi_t\}$: complete knowledge of the zygostate path up to the present.
- $\{\mathcal{X}_t^{\phi\mathrm{T}}\}$ is generated by $\{x_t\}$ and $\{\phi_\tau;\, \tau \in [0, \mathrm{T}]\}$: complete knowledge of the base-state path up to the present and the complete modal path.
- $\{\mathcal{X}_t^{z}\}$ is generated by $\{x_t, z_t\}$: complete knowledge of the base-state and modal observation up to the present.

The regulation problem can be posed within each of these filtrations. For example, the filtration $\{\mathcal{G}_t^{\phi\mathrm{T}}\}$ differs from $\{\mathcal{G}_t^{\phi}\}$ insofar as the former knows the future modal path while the latter can only speculate. In the Kalman filter, the future modal path is superfluous. Equations (7.2)–(7.5) are properly called the $\mathcal{G}_t^{\phi}$-Kalman filter (or equivalently the $\mathcal{G}_t^{\phi\mathrm{T}}$-Kalman filter). The development of the PME permits no measurements of the modal future though they would be useful if available. In regulation, the modal future is of considerable utility.

To illustrate the analytical design of a feedback controller for a hybrid system, consider first the case where the plant is noise free: $\{R_\chi(i) = 0;\, i \in \boldsymbol{S}\}$. Let us find that $\mathcal{X}_t^{\phi\mathrm{T}}$-adapted actuating signal that minimizes the quadratic performance index

$$J(u) = \int_{[0,\mathrm{T}]} (x_t' R x_t + u_t' S u_t)\, dt, \tag{7.6}$$

where $R \geq 0$ and $S > 0$ are symmetric matrices.

The solution to this optimization problem is well known [AM90, Chapter 2]. The $\mathcal{X}_t^{\phi\mathrm{T}}$-quadratic regulator (called the *LQ-regulator*) is given by the following construction:

$$u_t = -\sum_i S^{-1} B_i' L_t x_t \phi_i, \tag{7.7}$$

where $\{L_t\}$ satisfies

$$\frac{d}{dt} L_t = (-A_i' L_t - L_t A_i + L_t B_i S^{-1} B_i' L_t - R)\mathbf{e}_i' \phi_t \tag{7.8}$$

with $L_\mathrm{T} = 0$. Note that L_t depends on the modal path, that is, the sequence of modes taken by the plant on the time interval of interest.

Some properties of the LQ-regulator are evident:

- u_t is linear in the base-state;
- the control gain depends on the current control matrix, $\sum_i B_i \phi_i$;
- the control gain depends on the future values of the modal process through $\{L_t\}$.

The properties of the $\mathcal{X}_t^{\phi\mathrm{T}}$-regulator are much studied, and this feedback controller has been proposed for many applications. If the $R_\chi(i)$; $i \in S$ are not zero, $J(u)$ is random. In this case, $E[J(u)]$ is often used as a performance index. The $\mathcal{X}_t^{\phi\mathrm{T}}$-regulator is optimal even in this case.

To make the problem somewhat more realistic, remove the perfect measurement of the base-state. Suppose, instead of $\{x_t\}$, only $\{y_t\}$ is available. The performance index remains the expectation of the quadratic functional:

$$J(u) = E\left[\int_{[0,\mathrm{T}]} (x_t' R x_t + u_t' S u_t)\, dt\right]. \tag{7.9}$$

The $\mathcal{G}_t^{\phi\mathrm{T}}$-quadratic regulator is that which minimizes $J(u)$. This problem has a surprisingly simple solution. The $\mathcal{G}_t^{\phi\mathrm{T}}$-quadratic regulator is given by [AM90, Section 8.2]:

$$u_t = -\sum_i S^{-1} B_i' L_t \hat{x}_t \phi_i, \tag{7.10}$$

where $\{L_t\}$ is given in (7.8).

The $\mathcal{G}_t^{\phi\mathrm{T}}$-quadratic regulator is composed of

- the mean of the base state generated by the $\mathcal{G}_t^{\phi}$-Kalman filter,
- the gain of the $\mathcal{X}_t^{\phi\mathrm{T}}$-quadratic regulator.

This is to say that if x_t is replaced in the $\mathcal{X}_t^{\phi\mathrm{T}}$-regulator by its $\mathcal{G}_t^{\phi}$-mean, the $\mathcal{G}_t^{\phi\mathrm{T}}$-regulator results. The $\mathcal{G}_t^{\phi\mathrm{T}}$-quadratic regulator is an instance of the *certainty equivalence principle*: An unknown quantity (x_t) can be replaced by its mean ($\hat{x}_t$) in an algorithm derived on the basis of flawless knowledge of the quantity. It is also an instance of the *separation principle*: State estimation can be disengaged from regulator design with each algorithm derived independently and the results conjoined.

To increase the realism of the model on another level, let us take from the regulator its knowledge of the modal future and pose the problem in $\{\mathcal{X}_t^\phi\}$. Estimation is of no consequence since the zygostate is known, but the future is murky. Again use the performance index $J(u)$:

$$J(u) = E\left[\int_{[0,\mathsf{T}]} (x_t' R x_t + u_t' S u_t)\, dt\right].$$

Even with flawless measurements, certain problems immediately present themselves. The $\mathcal{X}_t^{\phi\mathsf{T}}$-quadratic regulator cannot be implemented because the required gain is explicitly dependent on the modal future. Analysis of this problem leads to the $\mathcal{X}_t^\phi$-quadratic regulator [Swo69a]:

$$u_t = -\sum_i S^{-1} B_i' L_i x_t \phi_i, \tag{7.11}$$

where the $\{L_i;\, i \in \mathbf{S}\}$ satisfy

$$\frac{d}{dt} L_i = -A_i' L_i - L_i A_i + L_i B_i S^{-1} B_i' L_i - R - \sum_j Q_{ij} L_j \tag{7.12}$$

with $L_i(\mathsf{T}) = 0;\, i \in \mathbf{S}$. Comparing (7.7) with (7.11), it is evident that the $\mathcal{X}_t^{\phi\mathsf{T}}$-quadratic regulator shares many similarities with the $\mathcal{X}_t^\phi$- quadratic regulator. Both are linear in x_t' with a gain that is influenced by the modal-state. They differ, however, in the calculation of the $\{L_t\}$ factor. The former can use its knowledge of the modal future to prepare itself for the modal changes it knows will occur. The latter has only a stochastic model with which to predict the future.

The issues of stability of the closed-loop $\mathcal{X}_t^\phi$-system are more subtle than those encountered in the $\mathcal{X}_t^{\phi\mathsf{T}}$-system. The closed-loop $\mathcal{X}_t^{\phi\mathsf{T}}$-quadratic optimal system is linear with time variable coefficients and the notions of stability are applied in the usual way. The closed-loop $\mathcal{X}_t^\phi$-quadratic optimal system is different insofar as the modal path is not predetermined. In fact, the closed-loop system can be unstable in certain regimes (have closed-loop poles of the regime-specific system in the right half plane) even though the overall system is quadratic optimal [Swo69b]. Issues of stability are discussed in much more detail in [FLJC92], [FLF95], and [Mar90] for $\mathcal{X}_t^\phi$-closed-loop systems.

We have employed a quadratic performance index to construct the feedback regulator. In the $\mathcal{X}_t^\phi$-case, this leads to control gains that are indexed by the operating mode. This form of feedback has been used in applications without carrying along the associated optimization. Equation (7.11) is an example of *gain scheduling* in which "linear design methods are applied to the linearized model at each operating point in order to arrive at a set of feedback control laws that perform satisfactorily when the system is operated near the respective operating points"[Rug91]. This more myopic view of regulation localizes the control design problem to the current regime and seeks an acceptable regime-specific controller for that operating condition.

When the sample paths of $\{\phi_t\}$ are not constant, the adequacy of a controller using gain scheduling is more difficult to determine. For example, if $\{A_i, B_i; i \in S\}$ is a family of controllable models, there is a set of feedback regulators with gains $\{K_i; i \in S\}$ such that the regime-specific closed-loop poles are well located no matter the regime:

$$u_t = -\sum_i K_i x_t \phi_i. \tag{7.13}$$

Indeed, we could select the $\{K_i\}$ such that the closed-loop poles did not change on the modal path: When $\{\phi_t\}$ makes the transition $\mathbf{e}_i \mapsto \mathbf{e}_j$, $K_i \mapsto K_j$ so as to keep the poles of $A_j - B_j K_j$ in the same place as those of $A_i - B_i K_i$. This particular $\mathcal{X}_t^\phi$-regulator may not make the transfer function invariant on the path, but this choice of pole locations avoids the concerns engendered by the local instabilities that may arise using the $\mathcal{X}_t^\phi$-quadratic regulator.

The $\mathcal{G}_t^\phi$-quadratic regulator of the LJS system also satisfies the certainty equivalence principle [SA77]. When there are base-state discontinuities at modal transitions, the $\mathcal{G}_t^\phi$-quadratic regulator takes the form of a linear controller with a bias [Swo82], [DE98]:

$$u_t = -\sum_i (K_i \hat{x}_t + M_i)\phi_i.$$

The bias, $\sum_i M_i \phi_i$, situates the base-state at a point favorable to the anticipated discontinuity.

When the modal-state measurements are noisy the problem is considerably more complex even when the base-state is measured perfectly. Suppose $\{x_t\}$ is measured but the modal state is inferred from $\{z_t\}$:

$$dz_t = h'\phi_t \, dt + d\eta_t, \tag{7.14}$$

where $\{\eta_t\}$ is a Brownian motion. Since only the modal-state is uncertain, it can be inferred using (2.22). We can adapt the gain scheduling approach presented earlier to create an $\mathcal{X}_t^z$-regulator: Replace the gain $\sum_i K_i \phi_i$ with its mean $\hat{K}_t$:

$$u_t = -\sum_i K_i \hat{\phi}_i x_t. \tag{7.15}$$

The advantage of (7.15) is that only $\{\hat{\phi}_t\}$ need be computed on-line.

The $\mathcal{X}_t^z$-quadratic control problem has been often investigated and has proven to be intractable. One approach that has achieved some success is useful when the modal measurement is accurate (i.e., $\hat{\phi}_t \approx \phi_t$) except for short delays after a modal transition. Let $\{\phi_t^a\}$ be a process on the modal-state space such that $E[(\phi_t^a - \phi_t)'(\phi_t^a - \phi_t) \mid \mathcal{G}_t]$ is small. For example, ϕ_t^a could be the most likely mode. Then $\{\phi_t^a\}$ could replace $\{\phi_t\}$ in the $\mathcal{X}_t^{\phi\mathrm{T}}$-regulator, (7.11). Or better, the regulator could be modified to account for the delay in identifying the modal transitions (see [SC86] and [DE98]).

One attempt to find the solution to the more general problem is described in [LKM85]: The finite-state modal process is replaced with a diffusion and $\{z_t\}$ is as given in (7.14). Even if the number of relevant $\mathcal{Z}_t$-moments of ϕ_t is truncated arbitrarily, the design problem is not tractable. A representation for the $\mathcal{X}_t^z$-quadratic control was proposed in the reference:

$$u_t = -\sum_i S^{-1}\hat{B}'L_t x_t. \tag{7.16}$$

Unfortunately, the moments upon which $\{L_t\}$ depends cannot be generated without considerably simplifying the problem (e.g., using an EKF to compute them).

The problems encountered in [LKM85] are compounded if $\{x_t\}$ is not observed. One cause of these difficulties is that the actuating signal may now be used to influence the ability of the regulator to determine the regime itself: The actuating signal can be used to *probe* the plant and enhance modal estimation. At a fundamental level, the quadratic regulator has twin roles: It regulates the base-state and it reduces the zygostate uncertainty. When this twofold task is acknowledged, controller design is called a problem in *dual control* [Swo66a, Swo66b].

In the above implementations, the controller is passive with respect to estimation, and the dual nature of the problem is either moot (e.g., because the modal-state is measured) or ignored. Since the influence of the actuating signal estimation on estimation accuracy is minimal, the focus can be placed on regulation. The $\mathcal{G}_t$-quadratic optimal control problems are commonly formulated in such a way that the conditional mean estimate of the base-state (called the *information state*) plays the role of the base-state in the $\mathcal{F}_t$-optimization problem. The index $J(u)$ is related to the quality of the base-state estimate plus a quadratic form in the information state and control. Since the latter is independent of the control algorithm, the optimization problem becomes an orthodox LQ-problem in the information state.

We will not address the dual nature of the $\mathcal{G}_t$-feedback regulator in this chapter except perhaps peripherally. A good illustration of the issues arising in nondual hybrid control appears in [EB96]. First note that a perfunctory generalization of the $\mathcal{F}_t$-quadratic regulator to a system with noisy zygostate measurements would be

$$u_t = -\sum_i K_i \hat{\phi}_i \hat{x}_t.$$

It is shown in [EB96], however, that at least for the $\mathcal{Y}_t$-quadratic-optimal problem, the controller has a form somewhat different. The $\mathcal{Y}_t$-quadratic regulator problem is intractable, but by ignoring the probing aspects (seeking what is called the *open-loop-optimal-feedback* (OLOF) control), it is shown that an attractive feedback regulator can be written

$$u_t = -\sum_i K_i(\mathcal{Y}_t) R_{x\phi_i}. \tag{7.17}$$

The gain factor, $K_i(\mathcal{Y}_t)$, is quite complicated and involves moments of $\{\phi_t\}$.

The second factor in (7.17) is perhaps of more interest to us. It is one of the canonical moments that is computed as part of the PME. This suggests that the geometry of the zygostate is useful even for nonprobing control. Although computing $\hat{\phi}_t$ and $\hat{x}_t$ is difficult using conventional algorithms, finding a reasonable approximation to $R_{x\phi}$ is even more so.

In this chapter we will use the PME to illuminate some of the issues that arise in the regulation of hybrid systems. This will be done with controls of a restricted sort and only for LJS. The construction of the PME is such that full probing is not achieved in the feedback system. It will be seen, however, that the second mixed cross-zygostate moment is important for effective control.

7.2 The PME

Let us consider the regulation of an LJS with time-discrete base-state measurements and classificational measurements of the modal-state:

$$dz_t = \lambda \mathbf{P} \phi_t \, dt + d\eta_t. \tag{7.18}$$

Unless the delays are short (e.g., $\{\phi_t\}$ is measured in small white noise in [DBE96]), the estimation problem is difficult. Finite-dimensional approximations to the conditional probabilities are proposed and exploited in [LBV91].

The PME for a linear jump system is:

the PME: LJS with control

Between observations:

$$\begin{aligned}
\frac{d}{dt}\hat{\phi}_t &= Q'\hat{\phi}_t, \\
\frac{d}{dt}\hat{x}_t &= \sum_i (A_i R_{x\phi_i} + B_i u\hat{\phi}_i), \\
\frac{d}{dt}P_{x\phi} &= \sum_i (A_i P_{(x\phi_i)\phi} + B_i u P_{\phi_i\phi}) + P_{x\phi}Q, \\
\frac{d}{dt}P_{xx} &= \sum_i ((A_i P_{(x\phi_i)x} + B_i u P_{\phi_i x}) + (\cdot)' + R_\chi(i)\hat{\phi}_i), \\
\frac{d}{dt}P_{xx\phi_m} &= \sum_i ((A_i P_{(x\phi_i)x\phi_m} + B_i u P_{\phi_i x\phi_m}) + (\cdot)' + P_{xx\phi_i} Q_{im}).
\end{aligned}$$

At a modal observation:

$$\begin{aligned}
\hat{\phi}^+ &= \hat{\phi}^- * \Delta\vartheta, \\
\Delta\hat{x} &= P_{x\phi}\Delta\vartheta, \\
\Delta P_{x\phi} &= -\Delta\hat{x}\Delta\hat{\phi}' + \sum_k P_{x\phi\phi_k}\Delta\vartheta_k, \\
\Delta P_{xx} &= -\Delta\hat{x}\Delta\hat{x}' + \sum_k P_{xx\phi_k}\Delta\vartheta_k, \\
\Delta P_{xx\phi_m} &= -\Delta\hat{\phi}_m\Delta\hat{x}\Delta\hat{x}' - \Delta\hat{\phi}_m P^+_{xx} - \Delta\hat{x}P^+_{\phi_m x} - P^+_{x\phi_m}\Delta\hat{x}' \\
&\quad + \sum_k P_{xx\phi_m\phi_k}\Delta\vartheta_k.
\end{aligned}$$

At a base-state observation:

$$\begin{aligned}
\Delta\hat{x} &= \gamma_x\Delta\nu_x, \\
\Delta P_{x\phi} &= -\gamma_x H P_{x\phi}, \\
\Delta P_{xx} &= -\gamma_x R_{yy}\gamma_x', \\
\Delta P_{xx\phi_m} &= -\gamma_x H P_{xx\phi_m} - P_{xx\phi_m} H'\gamma_x.
\end{aligned}$$

Its development is presented in Appendix 1.

This PME, which we denote by PME(B), is similar to the Kalman filter. The term $\sum_i (A_i \hat{x}_t + B_i u_t)\phi_i$ in the Kalman filter is replaced by $\sum_i (A_i R_{x\phi_i} + B_i u_t \hat{\phi}_i)$, a not surprising change given that ϕ_t is not now known and is correlated with x_t. If all of the B_i were equal to the constant B we would have

$$\sum_i (A_i R_{x\phi_i} + B_i u_t \hat{\phi}_i) = \sum_i A_i R_{x\phi_i} + B u_t.$$

The contribution of the actuating signal is added to the uncontrolled drift of $\{\hat{x}_t\}$.

The calculation of $\{P_{xx}\}$ in PME(B) is similar to that performed in the Kalman filter (e.g., the term of $\sum_i A_i P_{xx}\phi_i$ in the Kalman filter is replaced by the moment $\sum_i A_i P_{(x\phi_i)x}$). This change in the PME accounts for the correlation of the modal×base-state error and is avoided in the Kalman filter because of the assumed perfect measurement of $\{\phi_t\}$. More peculiar is the fact that $\{P_{xx}\}$ (and hence the gain in PME(B)) depends upon $\{u_t\}$. The inclusion of a control extends the equation for the base-state error covariance with the terms

$$\frac{d}{dt} P_{xx} = \cdot + \sum_i (B_i u_t P_{\phi_i x} + P_{x\phi_i} u_t' B_i').$$

In contrast with the Kalman filter, PME(B) can exert influence over its own uncertainty. But this influence is limited. If B_i is a constant, $\sum_i B_i P_{\phi_i x} = 0$: If the control matrix is independent of mode, the control has no influence on the base-state error covariance. Further, the control-dependent terms in $\{P_{xx}\}$ are effective only to the degree that the cross-error moment $P_{x\phi}$ is large: When the regime is identified with confidence, the control has little influence on $\{P_{xx}\}$. Consequently, a regulator based upon the PME cannot be said to be a *dual* controller. Any heightening of the probing function is secondary to its role as a base-state regulator.

The influence of $\{u_t\}$ on the other canonical moments is similar:

$$\frac{d}{dt} P_{x\phi} = \cdot + \sum_i B_i u_t P_{\phi_i \phi},$$

$$\frac{d}{dt} P_{xx\phi_m} = \cdot + \sum_i (B_i u_t P_{\phi_i x\phi_m} + P_{x\phi_i\phi_m} u_t' B_i').$$

Again, if the B_i $(i \in S)$ are constant,

$$\sum_i B_i u_t P_{\phi_i \phi} = 0$$

and

$$\sum_i B_i u_t P_{\phi_i x\phi_m} = 0.$$

Also, if the regime is known with confidence, $P_{x\phi_i\phi_m} \approx 0$ and $P_{\phi_i\phi} \approx 0$. So the influence of control on the higher moments is significant only during the intramodal transients and only when the regime-specific control matrices are different.

The weak dependence of the higher moments on $\{u_t\}$ suggests a way to simplify the PME. Let PME($\hat{B}$) use the LJS algorithm presented in Section 4.4 with the addition of $\hat{B}_t u_t$ to the extrapolation equation of $\{\hat{x}_t\}$. Ignore $\{u_t\}$ in the other moment equations. This is an algorithm that retains the geometry of the motion without the complication of the actuating signal in moments where $\{u_t\}$ seems to have less influence. As the estimator's confidence in its estimate improves, the influence of the actuating signal wanes anyway: As $\hat{\phi}_t \approx \mathbf{e}_i$, the PME(B) becomes identical to the PME($\hat{B}$). In the next section, we will look at an example that contrasts these algorithms.

7.3 An RPV Subject to Subsystem Failure

To illustrate the usefulness of the PME in a feedback control context, consider an elementary scenario. A remotely piloted vehicle (RPV) is detected at a range of approximately 2.5 km. It is desired that the RPV be directed to its station (located at coordinates (0,0,0)). It is first returned to a neighborhood of the station at a constant altitude (the capture region) with handover to a landing controller in the terminal phase. During the constant-altitude portion of this operation, the RPV will be thought of as moving in the X–Y plane (X and Y are location coordinates referenced to the station) with elementary and uncoupled motion dynamics: $\frac{d^2}{dt^2}X = b_x u_x +$ noise, and similarly for Y; $b_x = 1$ is the actuator gain in the X direction (respectively $b_y = 1$ is the actuator gain in the Y direction), and $\{u_x\}$ is the actuating signal in the X direction (respectively $\{u_y\}$ in the Y direction). The noise is unit Gaussian white noise and is independent in direction. This nominal mode of operation will be labeled $\phi_t = \mathbf{e}_1$.

Without compensation, the X–Y-dynamics are unattractive: four open-loop poles at the origin. The usual state space description of the system would be four dimensional: $x_t = \mathrm{vec}[X, Y, V_x, V_y]$, where V_x is the X component of the velocity, and similarly for V_y. The base-state model of the RPV is controllable using these coordinates, and it is easily seen that the four closed-loop poles can be placed arbitrarily with linear state feedback. Specifically, if $u_t = -K_1 x_t$ the closed-loop poles can be placed on a circle with $\omega_n = 0.2$ and with damping ratios $\xi = 0.5$ and 0.75:

$$K_1 = \begin{bmatrix} 0.04 & 0.20 & 0 & 0 \\ 0 & 0 & 0.04 & 0.30 \end{bmatrix}.$$

As introductory textbooks often point out, linear feedback can be used to convert an unstable system to one that has satisfactory performance.

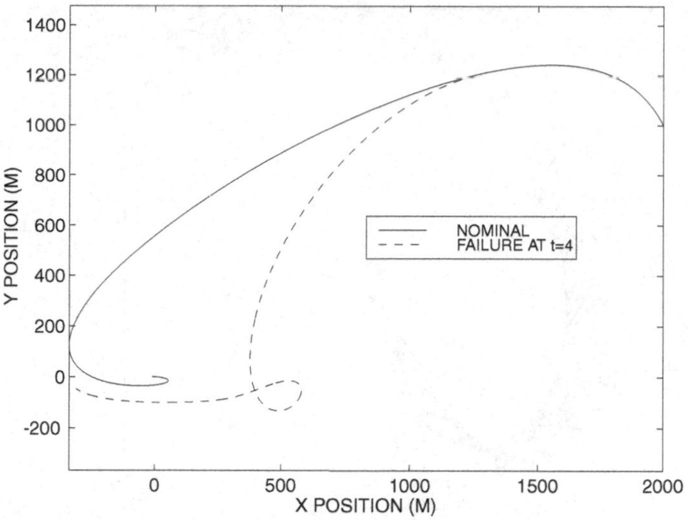

Figure 7.1. The nominal path of the RPV and the effect of a failure at $t = 4$ s.

Unfortunately, this RPV is subject to two kinds of subsystem failure. The first (labeled $\phi_t = \mathbf{e}_2$) causes the RPV to turn at a rate of 0.2 rad/s (about 6 g at the speed of the RPV), and it simultaneously reduces the gain of the X-actuator by a factor of ten ($b_x = 0.1$). The second (labeled $\phi_t = \mathbf{e}_3$) causes the RPV to turn in the opposite direction at the same rate, and it reduces the gain of the Y-actuator by a factor of ten ($b_y = 0.1$).

Figure 7.1 shows the effect of the subsystem failure on the performance of the nominal regulator, $u_t = -K_1 x_t$. From detection at $(X, Y) = (2, 1)$ km and velocity $(V_x, V_y) = (-100, 200)$ m/s, the regulator corrects the initial velocity and brings the RPV toward the station. The solid curve shows the response of the RPV in the nominal mode: $\phi_t = \mathbf{e}_1$. As the RPV approaches the station, it overshoots – primarily in X. This is corrected and the RPV nears the capture region. The path shown is for the first 80 s after detection.

Suppose there is a subsystem failure: $\phi_t \mapsto \mathbf{e}_2$ at $t = 4$ s. The X-gain is suddenly reduced and the vehicle begins to turn; the RPV begins to head south. The failure causes the pirouette shown in Figure 7.1. The controller corrects this error and moves toward the station, but the RPV enters the capture region with a disadvantageous state; the west velocity is far too high. Again only 80 s of the path is shown.

It is easily determined that the closed-loop system in the degraded mode is still stable using the gain K_1, but the closed-loop poles in the $\mathbf{e}_2$ regime move to $\omega_n = 0.28$, $\xi = 0.53$ and $\omega_n = 0.044$, $\xi = 0.19$. The former pair are well damped and fast. It is the latter pair that gives the motion its peculiar character. They are less well damped and their natural frequency has been reduced by a factor of five.

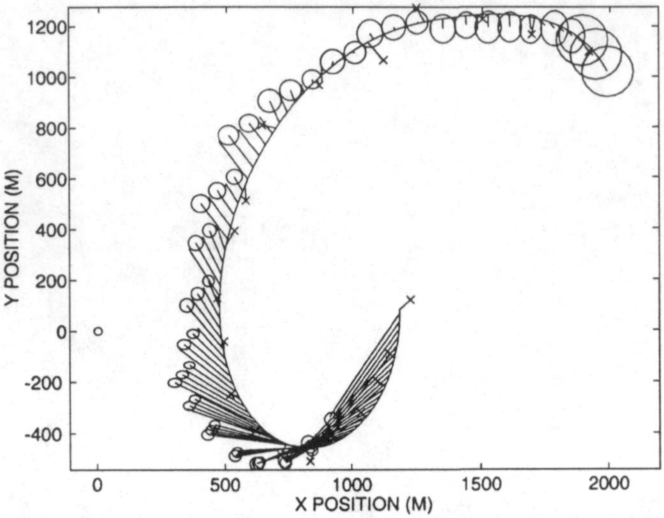

Figure 7.2. A sample path of the controlled RPV using EKF(W=1).

Figure 7.1 shows the motion with perfect and continuous location and velocity sensors. Suppose, however, that the location of the RPV is actually measured in range and bearing with a radar located at the station. The radar has a sample rate of 1 dwell/s, a standard deviation in the (X, Y)-range measurement of 40 m, and a standard deviation in (X, Y)-bearing of 50 mr (50 m at a range of 1 km). An EKF (labeled EKF(W=1)) for the nominal system can be constructed that will generate estimates of the motion state. The estimates generated by the EKF will certainly be good before a failure since the motion has a LGM representation. It is plausible to assume that after the initial transient, $\hat{x}_t \approx x'_t$ because the radar noise is small. However, a one-second update is significant in this application because of the smoothing in the EKF. Figure 7.2 shows a sample path of the first 20 s of RPV motion using the nominal control: $u'_t = -K_1\hat{x}_t$. This figure has a feather plot with position shown every 0.2 s. The radar measurements (marked with an "x") are shown as well.

Before the subsystem failure, the *certainty-equivalent* regulator performs well (compare with Figure 7.1). After the failure, the closed-loop response begins to deteriorate. The flight path has a pirouette like that seen in Figure 7.1, but the size of the loop is now larger. It is interesting to note that EKF(W=1) exhibits the phenomenon of excess error even in the regulation context: The error in the location exceeds the error in the radar fixes. This is common in EKFs when used in a multimode environment; the estimate is worse than the data. The closed-loop system is still stable, but the lags and the misinterpretation of the data are such that the EKF-based regulator is unacceptable.

To improve performance suppose there is a sensor onboard the RPV that measures the operating mode and transmits this information to the controller ten times a second. The accuracy of this sensor is not good. The modal sensor gives a correct classification of the motion mode only 60% of the time:

$$\mathcal{P}(\Delta z_t = \mathbf{e}_i \,|\, \phi_t = \mathbf{e}_i) = 0.6,$$

with the errors uniform in the remaining bins. From this the discernibility matrix, **P**, can be constructed. The dynamics of the modal process are delineated by the elements of the Q matrix. Suppose the chain is symmetric about $\phi_t = \mathbf{e}_1$ with mean time to failure of 4 s. Repair is not possible: The mean time to repair is selected to be 30 s and this is longer than the simulation interval.

The base-state dynamics are

$$d\begin{bmatrix} X \\ Y \\ V_x \\ V_y \end{bmatrix} = \begin{bmatrix} 0 & 0 & 1 & 0 \\ 0 & 0 & 0 & 1 \\ 0 & 0 & 0 & -\Phi \\ 0 & 0 & \Phi & 0 \end{bmatrix} \begin{bmatrix} X \\ Y \\ V_x \\ V_y \end{bmatrix} dt + \begin{bmatrix} 0 & 0 \\ 0 & 0 \\ b_x & 0 \\ 0 & b_y \end{bmatrix} u_t \, dt$$

$$+ \begin{bmatrix} 0 & 0 \\ 0 & 0 \\ 1 & 0 \\ 0 & 1 \end{bmatrix} d\begin{bmatrix} w_x \\ w_y \end{bmatrix}.$$

In normal operation ($\Phi_t \equiv 0$), there are two accelerations:

- an endogenous acceleration $\{u_t'\}$ weighted by the matrix $\mathbf{e}_2 \otimes B_i$,
- a wideband, omnidirectional, exogenous acceleration represented by $\{w_t\}$.

The subsystem failure manifests itself in an abrupt change in the system gains and a turning motion. Specifically, if a failure of type one occurs ($\phi_t = \mathbf{e}_2$), the gain in the X direction decreases ($B_2 = \text{diag}(0.1, 1)$), and the aircraft starts to turn ($\Phi_t = 0.2$ r/s). Alternatively, if failure of type two occurs ($\phi_t = \mathbf{e}_3$), the gain in the Y direction decreases ($B_3 = \text{diag}(1, 0.1)$), and the platform starts to turn in the other direction ($\Phi_t = -0.2$ r/s). The static stability of the matrices A_2 and A_3 show little improvement over A_1. In open loop, each failure mode has a pair of poles at the origin and a complex conjugate pair on the imaginary axis. After a failure, the X–Y dynamics cannot be separated because the turn introduces coupling in the X and the Y directions.

We have reviewed several algorithms in which the $\mathcal{F}_t$-regulator takes the form

$$u_t = -\sum_i K_i \phi_i x_t. \tag{7.19}$$

The gains $\{K_i; i \in S\}$ may be given by the solution of a set of ordinary differential equations. Or perhaps "the controller parameters are calculated at a number of operating conditions using some suitable design method. The controller is thus tuned or calibrated for each operating condition" [AW95, Chapter 11]. This latter method of selection may involve robustness [BB96, Shi96] or stability [FLC96]. In any event, such a regulator presupposes that the zygostate can be measured without error.

In this problem, we will restrict attention to a variant on (7.17),

$$u_t = -\sum_i K_i R_{x\phi_i}. \tag{7.20}$$

The $\{K_i\}$ can be selected to suit the application. Since the RPV is controllable in every regime, there exists a set of feedback gains $\{K_i; i \in S\}$ such that $\boldsymbol{A}_i = A_i - B_i K_i$ has the same poles for all i. In the case of perfect state feedback ($\{x_t\}$ and $\{\phi_t\}$ are observed) and $u_t = -\sum_i K_i \phi_i x_t$, the location of the static closed-loop poles would be constant over the full control interval.

Using the radar with modal sensor augmentation, the coefficients of PME(B) can be determined. Figure 7.3 shows the performance of the regulator $u_t = -\sum_i K_i R_{x\phi_i}$. The flight path of the RPV is displayed as the solid curve. The PME(B) generates estimates of location along with guiding the RPV. The feather plot and the one σ-uncertainty ellipses are shown as well. For clarity, the ellipses are shown every 0.4 s. Not only is the track better using PME(B) but the envelope of the uncertainly curves represents the actual path error quite well. This is in contrast

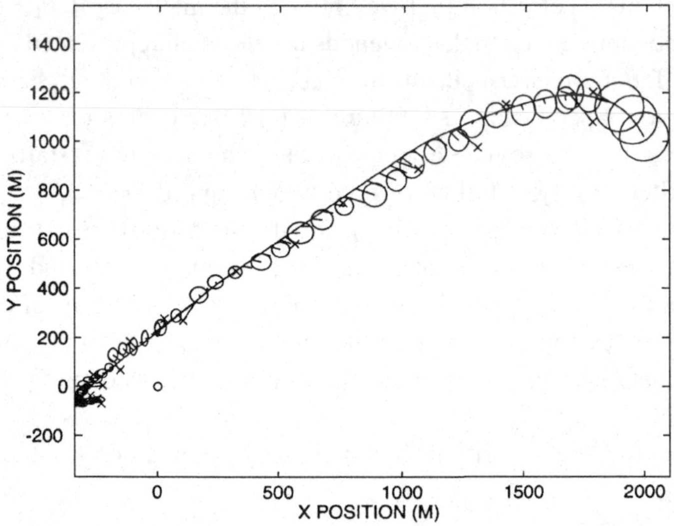

Figure 7.3. A sample path of the controlled RPV using PME(B).

with the EKF(W=1)-based regulator for which the putative one-σ ellipses differ from the actual path by as much as 15 standard deviations.

The error ellipses generated by PME(B) are relatively slow to decrease in amplitude. This is due to the poor modal sensor and the residual effect of modal estimation errors. At the end of the path, the PME(B)-based regulator is certain of its location. The guidance to the capture region is accomplished with small influence from the failure.

In lieu of the pole-invariant regulator, PME(B), an *averaging* analogue can be used (labeled PME($\hat{B}$)). The regulator is still $u_t = -\sum_i K_i R_{x\phi_i}$, but $R_{x\phi}$ is computed from the PME using

$$d\hat{x}_t = \sum_i (A_i - \hat{B}_t K_i) R_{x\phi_i}\, dt,$$

where $\hat{B}_t$ is short for $\sum_i \hat{\phi}_i B_i$. The control signal is neglected in computing the higher moments. During transients, this regulator does not maintain the filter poles in their desired locations. The poles of the estimator return to their design values whenever the estimation quality is good ($\hat{\phi}_t \approx \phi_t$). The performance of the regulator based upon PME($\hat{B}$) is shown in Figure 7.4.

The track of the PME($\hat{B}$)-regulator is shown along that of the PME(B)-regulator in an expanded window near the capture region in Figure 7.5. The EKF(W=1)-regulator is worse than either of these and its track does not enter the window shown during the simulation interval. The PME(B)-regulator takes a more direct path to the station than does the PME($\hat{B}$)-regulator. The latter also tends to loop out during

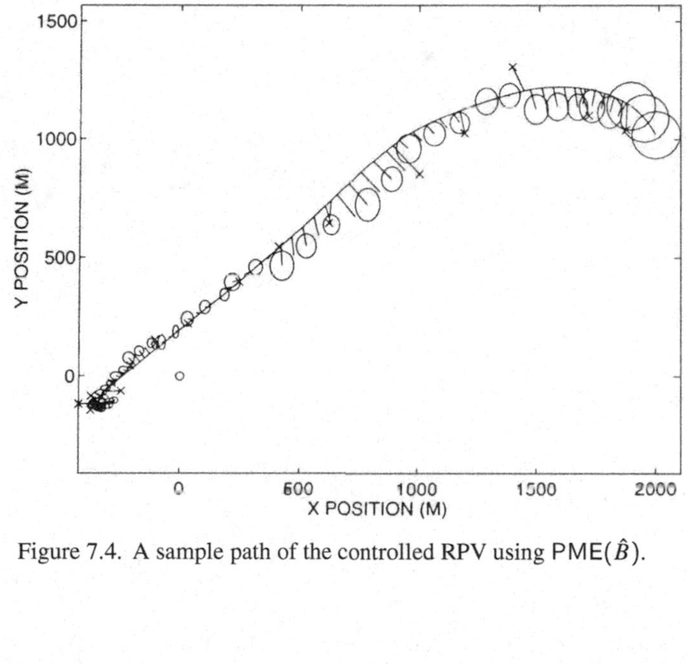

Figure 7.4. A sample path of the controlled RPV using PME($\hat{B}$).

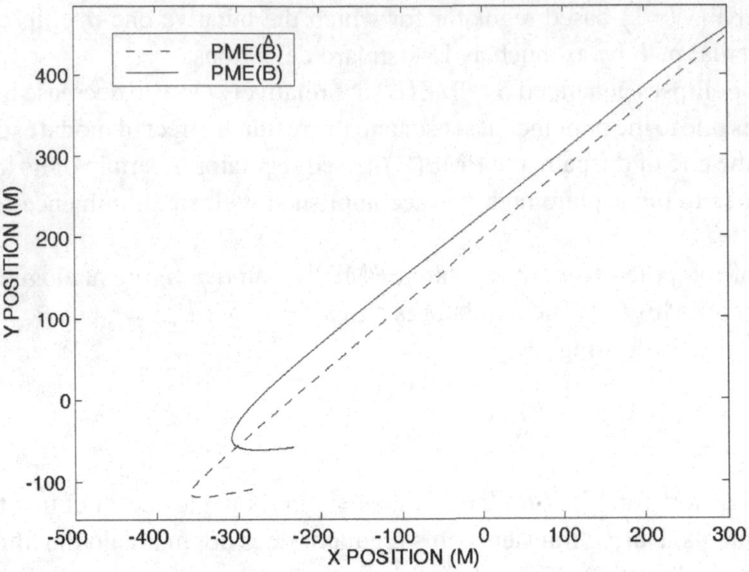

Figure 7.5. The closed-loop response of PME(B) and PME($\hat{B}$) near the capture region.

the initialization transient when the regulator is turning the RPV toward the station. Despite its more direct approach to the station, the PME(B)-regulator overshoots less than the PME($\hat{B}$)-regulator, and for this portion of the scenario, it ends nearer the center of the capture region.

8

Target Recognition and Prediction

8.1 Problem Statement

State estimation and control is made difficult in a hybrid system by the multiplicative nonlinearities in the equation of base-state evolution. The PME fuses complementary data streams using the dual path architecture shown in Figure 4.1 in a finite-dimensional algorithm for approximating the $\mathcal{G}_t$-error moments useful in a broad range of applications.

In earlier chapters we have used the PME to estimate the base-state of a moving platform (called here the *target*). In these applications, a plausible motion model for the target was known a priori. There are situations in which this important information is lacking, for example in cases where the target must be identified while simultaneously tracking it and predicting its future motion. Identification in this context is called *automatic target recognition* (ATR). Prediction can take many forms, but we will focus on predicting the location where the target intersects a boundary in state space.

Uncertainties in target identification compound those already present in target location at time of detection and the accretion of disturbances along the path. Model-based path-following algorithms utilize a formal model to represent target evolution, and the selection of a tracking algorithm is based upon the target dynamics as articulated in the model. Because the tracking algorithm is tuned to a particular dynamic class, it is advantageous to know the proper class in advance, or if unknown, to identify it as soon as possible. Often the tracking and identification aspects of this problem are treated separately: ATR is accomplished using one sensor (e.g., from a picture of the target generated by an imager), and location estimation using another (e.g., a radar). High level tracking/ATR fusion architecture could involve:

- Using the ATR to identify the target type and an EKF to track and predict using the model identified by the ATR.

- Tracking/prediction with an EKF using some kind of average target dynamics of the target aggregate. ATR would be handled as a separate (or subsidiary) problem.

In the former case, an additional level of uncertainty is introduced by the variability, over time, of target identification. The ostensible target might be identified as being in one class at time $t = t_1$ only to be reclassified at a later time. In the latter case, the dynamic model used to generate the estimates, while constant, is not specific to any actual target. The method of average dynamics is common, and it must be said that an EKF derived in this way performs surprisingly well given that the model is necessarily poor. This seems to be yet another example of the principle that poor models do not prevent good estimation if the SNR is high.

With imaging sensor/processor outputs coming into common use, the spatially extended features of the target should be processed, along with their more prosaic brethren, to achieve simultaneous target recognition and position estimation/prediction [MSG95, TRK92]. As we have seen, the PME generates estimates of all elements of the comprehensive state vector including the mean position and velocity of the target (along with suitable uncertainty measures). This chapter investigates a problem in which the target dynamics are not known a priori. This situation arises when targets of different classes may appear in an engagement. Target recognition is achieved both from direct measurements (e.g., the shape of the target in the image) and from the angular motion as determined from image data. Because of subtle biases in the models, the latter may not, in itself, be adequate for ATR.

8.2 Recognition and Tracking a Maneuvering Target

Synthesis of a model-based tracker begins with a quantitative representation of target motion. The simple two-dimensional motion model used here is given in (4.1):

$$d\begin{bmatrix} X \\ Y \\ V_x \\ V_y \end{bmatrix} = \begin{bmatrix} 0 & 0 & 1 & 0 \\ 0 & 0 & 0 & 1 \\ 0 & 0 & 0 & -\Phi \\ 0 & 0 & \Phi & 0 \end{bmatrix} \begin{bmatrix} X \\ Y \\ V_x \\ V_y \end{bmatrix} dt + \begin{bmatrix} 0 & 0 \\ 0 & 0 \\ 1 & 0 \\ 0 & 1 \end{bmatrix} d\begin{bmatrix} w_x \\ w_y \end{bmatrix}, \tag{8.1}$$

where $\{X, Y\}$ are position coordinates, and $\{V_x, V_y\}$ are associated velocities. The target is subject to both a wideband omnidirectional acceleration represented by the $\mathcal{F}_t$-Brownian motion $\{w_x, w_y\}$ and a maneuver acceleration represented by the turn rate process $\{\Phi_t\}$.

For this analysis, it will be assumed that the targets distinguish themselves by their agility. The maneuver indicator process, $\{\alpha_t\}$, is an $\mathcal{F}_t$-Markov process on the canonical unit vectors: $\alpha_t = \mathbf{e}_i$ if $\Phi_t = a_i$; $i \in \boldsymbol{K}$. Equation (8.1) can be written as a LJS:

$$dx_t = \sum_i A_i x_t \alpha_i \, dt + C dw_t,$$

where the definition of $\{A_i; i \in \boldsymbol{K}\}$ is evident from the context. The motion model (8.1) is a hybrid, nonlinear, stochastic differential equation in which the maneuvers create a family of motion modes within an LGM framework.

In contrast to the situation studied in Chapter 4, suppose that there are R different kinds of targets. Denote the R-dimensional target-class indicator vector by r_t: $r_t = \mathbf{e}_i$ if the target is of the ith class. Different targets have different capabilities. Suppose the tempo of maneuvers is a target class dependent Markov chain: If $r_t = \mathbf{e}_i$, $\{\alpha_t\}$ is a Markov chain with $K \times K$-transition rate matrix Q^i_α. Again divide the orientation range space of the target into L equally spaced bins, and let the orientation indicator process be $\{\rho_t\}$. The angular bin sequence will also be represented with a Markov process for a specific turn rate $\times$ target class: If $\alpha_t \otimes r_t = \mathbf{e}_i$, $\{\rho_t\}$ is Markov with transition rate matrix Q^i_ρ. Target type is constant: $r_t \equiv r_0$. The comprehensive maneuver-state of the target is given by $\phi_t = \alpha_t \otimes r_t \otimes \rho_t$. The $KLR \times KLR$-dimensional transition rate matrix, Q, for the Markov process $\{\phi_t\}$ can be produced using elementary methods if the primitive processes have no common jumps. The definition of A_i can be directly extended to ϕ_t and the dynamics $\{\phi_t\}$ are given by

$$d\phi_t = Q'\phi_t \, dt + dm_t, \tag{8.2}$$

where $\{m_t\}$ is an $\mathcal{F}_t$-martingale.

In tracking/recognition, the modal measurement will be processed to yield both orientation and the ostensible target class. We postulate, as before, two sensors:

- The point sensor (the radar) produces a sequence of range-bearing measurements: $y[k] = Hx[k] + n[k]$.
- The modal classifier places the measured orientation $\times$ type into one of LR bins at a rate λ frames/s:

$$dz_t = \lambda \mathbf{P}\phi \, dt + d\eta_t.$$

The modal-state measurement sequence, $\{z_t\}$, is a counting process of dimension LR, the ith component of which is the number of times on $[0, t]$ the imager has placed the target orientation/type in bin i. The quality of the imager is determined both by the frame rate, λ, and by the fidelity of the processing of a single data frame, the latter embodied in an $LR \times LR$ discernibility matrix $\mathbf{P}$.

The PME for this problem is as given in Chapter 4. The modal observation contains both target type and orientation. Let $\boldsymbol{\lambda}_t = \lambda\mathbf{P}(r_t \otimes \rho_t)$, and let $\{\vartheta_t\}$ be a piecewise constant process with increments

$$\Delta\boldsymbol{\vartheta} = \lambda \mathbf{P}'(\hat{\boldsymbol{\lambda}}_t^{-1} * \Delta z_t).$$

If the RL-dimensional vector $\boldsymbol{\lambda}_t$ is repeated K times, $\vartheta_t = \mathbf{1}_K \otimes \boldsymbol{\vartheta}_t$ is formed. From $\hat{\phi}_t$, it is a direct calculation to find $\hat{\alpha}_t$ and $\hat{r}_t$ (the $\mathcal{G}_t$-probability of target class):

the PME: continuous base-state; time-discrete measurements

Between observations:

$$\frac{d}{dt}\hat{\phi}_t = Q'\hat{\phi}_t,$$

$$\frac{d}{dt}\hat{x}_t = \sum_i A_i R_{x\phi_i},$$

$$\frac{d}{dt}P_{x\phi} = \sum_i A_i P_{(x\phi_i)\phi} + P_{x\phi}Q,$$

$$\frac{d}{dt}P_{xx} = \sum_i \left(A_i P_{(x\phi_i)x} + (\cdot)'\right) + R_\chi,$$

$$\frac{d}{dt}P_{xx\phi_m} = \sum_i \left(A_i P_{(x\phi_i)x\phi_m} + (\cdot)' + P_{xx\phi_i}Q_{i,m}\right).$$

At a modal observation:

$$\hat{\phi}^+ = \hat{\phi}^- * \Delta\vartheta,$$

$$\Delta\hat{x} = P_{x\phi}\Delta\vartheta,$$

$$\Delta P_{x\phi} = -\Delta\hat{x}\Delta\hat{\phi}' + \sum_k P_{x\phi\phi_k}\Delta\vartheta_k,$$

$$\Delta P_{xx} = -\Delta\hat{x}\Delta\hat{x}' + \sum_k P_{xx\phi_k}\Delta\vartheta_k,$$

$$\Delta P_{xx\phi_m} = -\Delta\hat{\phi}_m\Delta\hat{x}\Delta\hat{x}' - \Delta\hat{\phi}_m P_{xx}^+ - \Delta\hat{x}P_{\phi_m x}^+ - P_{x\phi_m}^+\Delta\hat{x}' + \sum_k P_{xx\phi_m\phi_k}\Delta\vartheta_k.$$

At a base-state observation:

$$\begin{aligned}\Delta\hat{x} &= \gamma_x \Delta\nu_x,\\ \Delta P_{x\phi} &= -\gamma_x H P_{x\phi},\\ \Delta P_{xx} &= -\gamma_x R_{yy} \gamma_x',\\ \Delta P_{xx\phi m} &= -\gamma_x H P_{xx\phi_m} - P_{xx\phi_m} H' \gamma_x.\end{aligned}$$

8.3 Automatic Target Recognition

To illustrate the versatility of the PME in a problem of tracking and identification, return to the antiship missile scenario studied earlier. This time suppose that there are two possible target types: $r_t \in \{\mathbf{e}_1, \mathbf{e}_2\}$. The targets have distinguishing turn dynamics and shapes [Kuh92].

Target(Nom): This is the nominal antiship missile and is close to that studied in Chapter 4. The set of permissible turn rates is given by

$$\Phi_t \in \{0.2 \text{ r/s}, \alpha_t = \mathbf{e}_1; 0 \text{ r/s}, \alpha_t = \mathbf{e}_2; -0.2 \text{ r/s}, \alpha_t = \mathbf{e}_3\}.$$

The mean sojourn time in a turn is 4 s (5 s in Chapter 4), with each turn being followed by a coast. There is symmetry around coast, and the initial maneuver modes are equally likely. Coast has a mean duration of 5 s.

Target(Agl): The second target is more agile than Target(Nom). The set of turn rates is

$$\Phi_t \in \{0.4 \text{ r/s}, \alpha_t = \mathbf{e}_1; 0 \text{ r/s}, \alpha_t = \mathbf{e}_2; -0.4 \text{ r/s}, \alpha_t = \mathbf{e}_3\}.$$

The mean sojourn time in each of the maneuver states is 2 s. A coast has mean duration 5 s.

Although the mean angular increment for both targets is the same, 0.8 r, Target(Agl) executes the turn in half the time. The target types will be assumed to be equally likely.

The sensor suite consists of the radar used earlier (errors are Gaussian with standard deviation 40 m in range and 1.75 mr in bearing) and a collocated imager. The nominal radar interdwell time is 1 s. The radar-exclusive EKF selected for comparison is found by neglecting the maneuvers (labeled EKF($W = 1$) or EKF(1)). The target types are not distinguishable by EKF(1). They differ in their turn dynamics but these are ignored by the EKF: The dependence of the target model on type is moot in EKF(1). The initial covariance is as in Chapter 4 (diagonal with standard deviation in position (100 m) and velocity (20 m/s)). EKF(1) is initialized

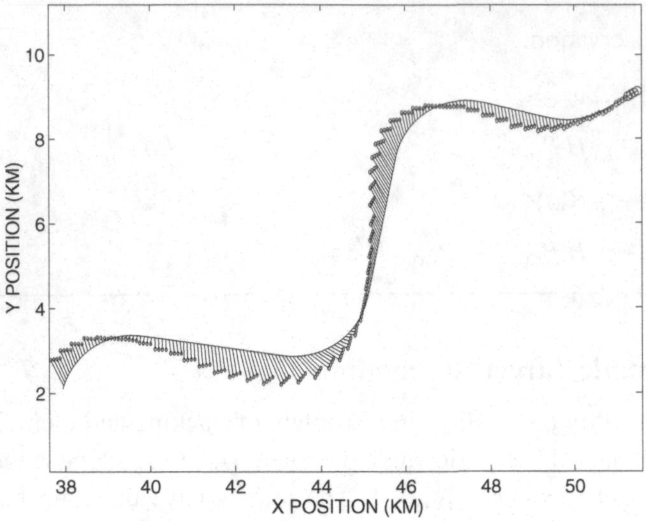

Figure 8.1. Performance of EKF(1) on a jinking path.

on the true target motion at the time of detection. The target motion is most closely described as coming from target class Target(Nom). A sample showing the performance of EKF(1) on a portion of the path is shown in Figure 8.1. This extends and combines data from Figures 4.3 and 4.5.

Now complement the radar with an imager. The PME must accomplish two tasks simultaneously: target identification and tracking. It will do this on the basis of a direct measurement of type and also on the estimated turn dynamics. It will be assumed that the imager operates at $\lambda = 10$ frames/s. In orientation, the imager is of the same quality as that given in Table 5.2, but smaller orientation bins will be used than those used there: $L = 24$ (15° bins). Target classification is also achieved from shape or spectral analysis. Each frame is classified as coming from one of the two target classes. Specifically, suppose the type signature is weak at the tracking range: $\mathcal{P}$(target is classified as j | target is type j) $=$ 0.55 (labeled target fidelity (TF)). Table 8.1 gives the attributes of the classifier. The orientation

Table 8.1. *Modal sensor parameters.*

Error Type	Probability
UDE	0.1
NNE	0.1
PE	0.3
TF	0.55

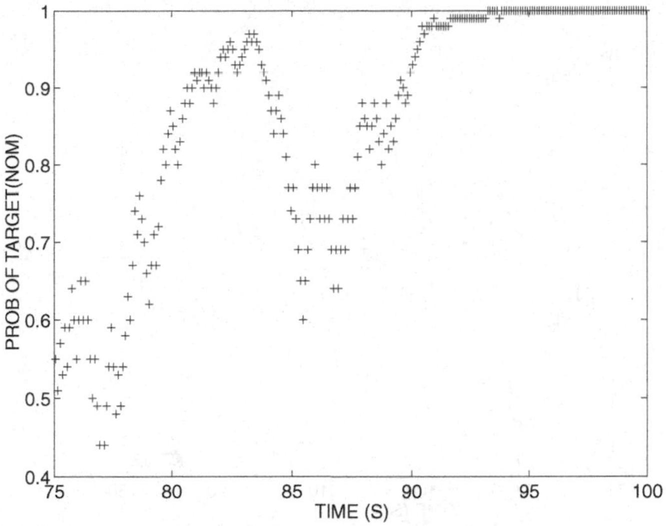

Figure 8.2. Target identification using PME(0.55, 10).

errors are mutually exclusive, and if it is assumed that the target type classification is independent of the orientation classification, **P** can be produced. From this the PME can be deduced. We denote this PME by PME(TF= 0.55, λ=10), or just PME(0.55,10).

The modal classifier is not of particularly good quality, correctly classifying target orientation only 50% of the time in orientation and 55% of the time in type. The standard deviation of orientation error is 75°. This error is so big as to suggest the imager would be of little use in tracking. Figure 8.2 shows the $\mathcal{G}_t$-$\mathcal{P}$(Target(Nom)) as computed by PME(0.55,10). PME(0.55,10) is puzzled by the first turn since the path matches neither dynamic model well and has an anomalous acceleration before jinking. When a turn is suspected, the PME tends to think the target is the more agile one (e.g., at ~87 s). Ultimately the target is identified, and since $r_t = \mathbf{e}_1$ is absorbing, volatility in $\mathcal{G}_t$-$\mathcal{P}$(Target(Nom)) diminishes.

Figure 8.3 shows the radial error for both algorithms on a sample path. With the perfect initialization, the EKF(1) is superior to PME(0.55,10) into the first turn. In contrast to the errors in the EKF (800 m), PME(0.55,10) keeps errors during the turn to less than 150 m (75 m after a radar update). It is close to the performance achieved by PME(IG) in Chapter 4: Finer orientation bins help but poorer image quality and multiple target classes hinder tracking.

Figure 8.4 contrasts the performance of EKF(1) and PME(0.55,10). The poor performance of the EKF is typical. PME(0.55,10) does a much better job, and if Figure 8.4 is inspected carefully, it is seen that the area of the error ellipses of PME(0.55,10) decreases as confidence in target identification increases.

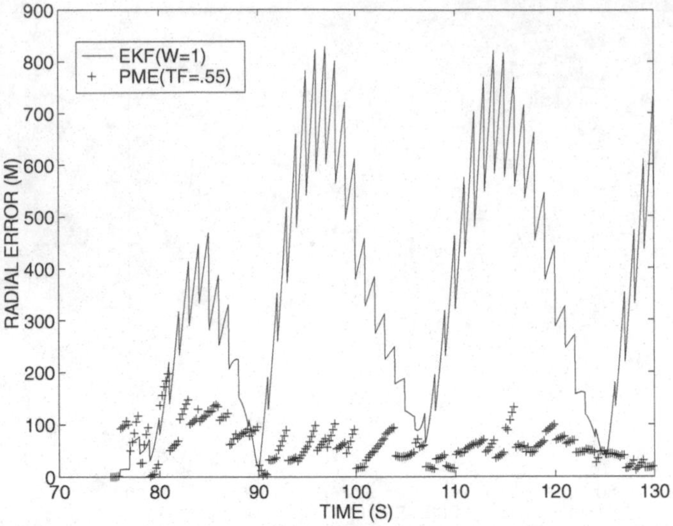

Figure 8.3. Mean radial tracking error for EKF(1) and PME(0.55,10).

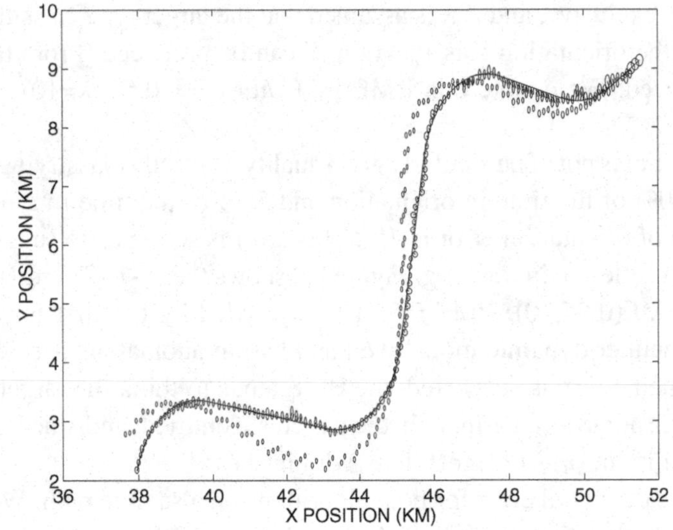

Figure 8.4. Tracking performance of EKF(1) and PME(0.55,10).

To better understand the issues engendered by ATR, let us generalize the previous scenario. Instead of two possible targets there are now three ($r_t \in \{\mathbf{e}_1, \mathbf{e}_2, \mathbf{e}_3\}$):

- Target(Nom), $r_t = \mathbf{e}_1$;
- Target(Agl), $r_t = \mathbf{e}_2$;
- Target(Lan), $r_t = \mathbf{e}_3$.

The third target, Target(Lan), has a more languid motion than do either of the

Table 8.2. *PME frame rate and error parameters.*

PME	UDE	NNE	PE	TF	λ
PME(0.55,10)	0.1	0.1	0.3	0.55	10
PME(0.75,10)	0.1	0.1	0.3	0.75	10
PME(0.50,10)	0.1	0.1	0.3	0.5	10
PME(0.55,1)	0.1	0.1	0.3	0.55	1

others. The set of turn rates is

$$\Phi_t \in \{0.1 \text{ r/s}, \alpha_t = \mathbf{e}_1;\ 0 \text{ r/s}, \alpha_t = \mathbf{e}_2;\ -0.1 \text{ r/s}, \alpha_t = \mathbf{e}_3\}.$$

The mean sojourn time in each of the maneuver states is extended to 8 s with the mean duration in coast again 5 s.

Each target has the same mean turn amplitude (46°) and the same mean sojourn time in coast. The targets are distinguished by their turn rates when they turn and by the probability that they are in a particular motion mode; for example, since the coast durations are the same for all of the targets, Target(Agl) will spend more time in coast mode than either of its fellows. Not every target is present in every encounter. In one case, the tracker-ATR must choose between Target(Nom) and Target(Agl), and in another between Target(Nom) and Target(Lan). At the beginning of an engagement, the PME knows the pertinent target types and assumes them to be equally likely.

Again we will look at an encounter involving a target modeled most closely by Target(Nom). The same radar will be used along with a collocated imager. As before, the imager operates at a nominal rate of $\lambda = 10$ frames/s with 15° orientation bins. The imagers are listed in Table 8.2. They all have the same quality for classifying orientation, but they differ in target recognition, and in one case, the frame rate is slower than nominal. The radar-exclusive algorithm selected for comparison is again EKF(1).

8.3.1 Engagement Nom-Agl

In this first engagement, suppose the trackers are not sure whether the target is of class Target(Nom) or of class Target(Agl). Consider first the nominal track/recognition algorithm, PME(0.55,10). Image resolution is not particularly good. Figure 8.4 shows the path of the target along with the one-σ error ellipses. Figure 8.5 shows the probability that the target is in the coast mode as generated by PME(0.75,10) through PME(0.50,10): the $\mathcal{G}_t$-$\mathcal{P}(\alpha_t = \mathbf{e}_2)$ for the PMEs over the range from good target type indication to no type indication at all. Target detection occurs at $t = 75$

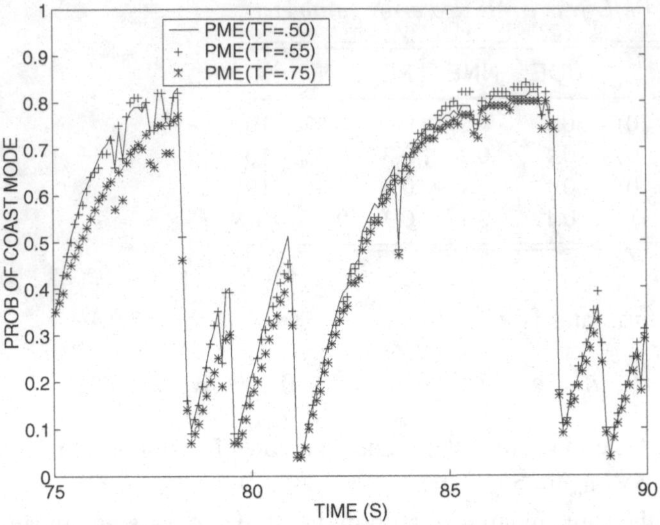

Figure 8.5. Coast mode identification using PME(0.75,10), PME(0.55,10), and PME (0.50,10).

in the midst of a short coast. All of the PMEs move from their initial modal uncertainty toward favoring coast. The coast actually ends at $t = 77.3$, but none of the PMEs are immediately aware of this. The initial orientation is centered in an angular bin, and the orientation angle of the target must traverse the bin to recognize the transition in $\{\alpha_t\}$. The PME with the most accurate target recognition processor is most conservative in declaring the target is coasting; Target(Nom) spends less time in coast than does Target(Agl). Figure 8.5 shows only the time interval [75, 90] s because all three of the PMEs show the same estimates of the modal path after $t = 90$ s. Note that the PMEs detect certain parts of a turn quite accurately – when the orientation crosses a bin boundary, the target is turning. Identifying a coast is inherently more problematic. The coast mode manifests itself in the absence of angular bin crossings, an equivocal signature. This causes all of the PMEs to think the coast mode lasts longer than it really does, and this tends to favor that target with the longest sojourns in coast – Target(Nom) in this case.

A sample path of the target recognition response for PME(0.75,10) through PME(0.50,10) is displayed in Figure 8.6: $\mathcal{G}_t$-$\mathcal{P}$(Target(Nom)). Beginning at maximum uncertainty, PME(0.75,10) quickly moves toward $\hat{\alpha}_t = \mathbf{e}_1$. This is not surprising given the accurate classification processing in PME(0.75,10). Less accurate type measurements cause PME(0.55,10) and PME(0.50,10) to be considerably more hesitant to declare Target(Nom). The coast sojourn (at ~85 s) causes PME(0.55,10) (but not PME(0.50,10)) to become less sure of its target selection. The estimate generated by PME(0.50,10) is less volatile than that generated by PME(0.55,10).

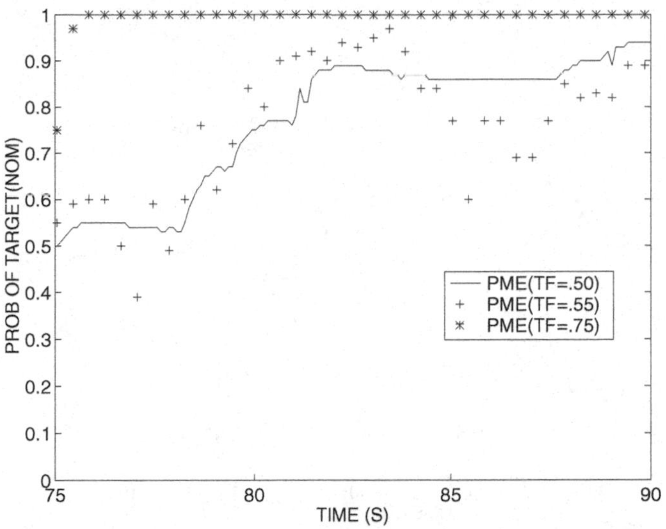

Figure 8.6. Target recognition using PME(0.75,10), PME(0.55,10), and PME(0.50,10).

This is not surprising since recognition in the PMEs is achieved by processing orientation estimates and this has a smoothing effect.

Accurate shape processing is helpful in target identification, but its efficacy in tracking is considerably less. The Markov maneuver model is a coarse representation of the sojourn times in a bin or in a motion mode. The lifetimes of the modal process are much more regular than the motion model suggests, and without making this information explicit in the PMEs, the link between the modal process and the target type is attenuated. The Markov model is useful, however, for quantifying the acceleration geometry. The contrast between the tracking performance of PMEs is not great. Figure 8.7 shows the radial tracking error for PME(0.75,10) and PME(0.50,10) (the extremes of recognition fidelity) on the portion of the path during which the PMEs are identifying the target: $t \in [75, 100]$ s. The radar samples are independent, but the imager samples are the same. Before target identification is conclusive, the radial error associated with PME(0.50,10) tends to be smaller than that for PME(0.75,10), at least in the extremes. Single samples should not be over interpreted, but it is evident that target recognition is not definitively linked to tracking error.

The image rate in the PMEs considered thus far has been 10 frames/s. If this creates an excessive computational burden, the frame rate can be reduced. Figure 8.8 shows the target and the error ellipses calculated by PME(0.55,1), which uses the nominal imager/radar with the frame rate of the imager reduced by a factor of 10. Comparing these ellipses to those for PME(0.55,10) (see Figure 8.4), we see that the PME is less certain of its estimates when the sample rate is slower. The

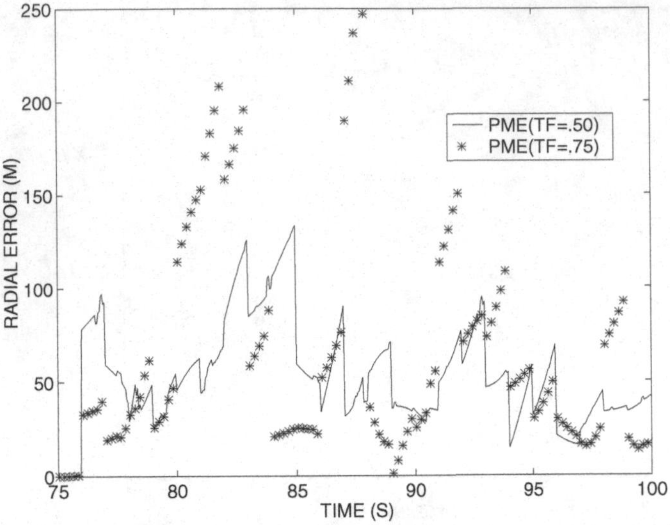

Figure 8.7. Radial tracking error for PME(0.75,10) and PME(0.50,10).

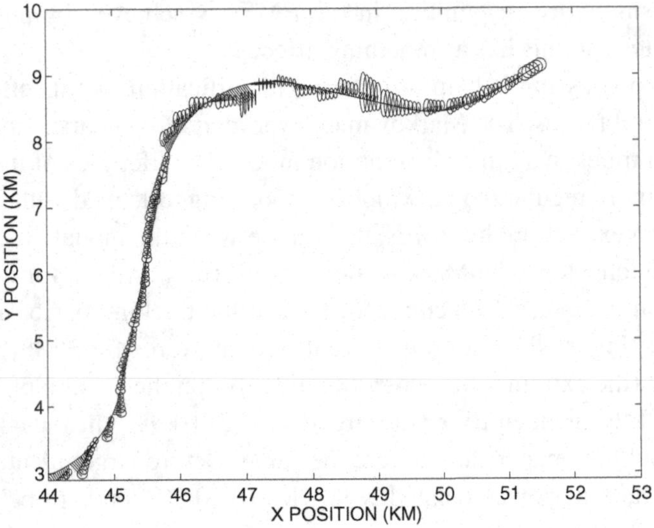

Figure 8.8. Tracking performance for PME(0.55,1).

ellipses grow significantly during intrasample intervals, and the error grows during transitions in maneuver mode. Still, in contrast to EKF(1), the true path is seldom far beyond the envelope of the one-σ error ellipses.

Target recognition degrades with the slower frame rate. This is caused both by the paucity of direct measurements and by the difficulty in determining the motion mode at the lower frame rate. Figure 8.9 shows a sample of target recognition for

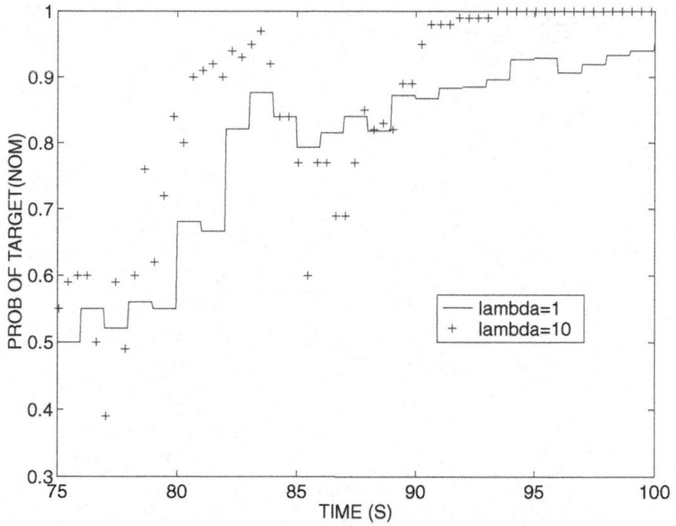

Figure 8.9. Target recognition for PME(0.55,10) and PME(0.55,1).

PME(0.55,1) superimposed on the previously displayed plot of PME(0.55,10). A higher frame rate leads to quicker recognition, though the ATR does tend to be more volatile (see the coast sojourn at ~87 s). The larger error circles appearing in Figure 8.8 are associated with the intervals in which the confidence in target identification is low. Once the target type is known, PME(0.55,1) proceeds to maintain its track, but the tracking errors grow at the beginning of the right turn.

8.3.2 Engagement Nom-Lan

For reasons discussed earlier, the PME is such that the target identification tends to be biased toward the tamest target in a group. In the preceding scenario, the tame target was the actual target, Target(Nom). Consider the alternative situation in which the permissible targets are Target(Nom) and Target(Lan). The trajectory, again associated with Target(Nom), is the flight path shown in Figure 8.1. The gains of PME(0.75,10) through PME(0.50, 10) must be recomputed to reflect this changed target environment. Figure 8.10 shows the motion mode estimate (probability of coast) over the early part of the engagement: $t \in [75, 90]$ s. Modal estimation is no more difficult in the Target(Nom)/Target(Lan) engagement if the target classification data are accurate: The curves associated with PME(0.75,10) are essentially the same in Figures 8.5 and 8.10. When the target type measurements are of low quality, modal identification degrades. PME(0.55,10) and PME(0.5, 10) again differ little, but they differ considerably from the similar Target(Nom)/Target(Agl) situation. This is particularly noticeable during the interval of coast, at ~85 s. The difference

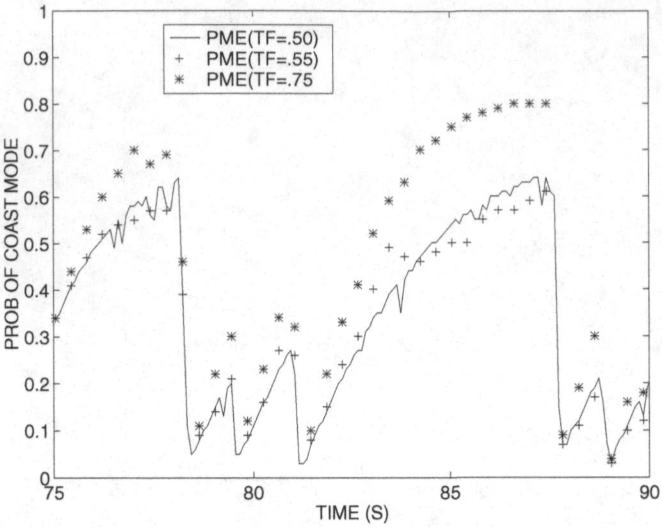

Figure 8.10. Coast mode identification using PME(0.75,10), PME(0.55,10), and PME (0.50,10).

between the PMEs is much less when the target is turning, because of the relatively unambiguous signature of this motion mode.

Target recognition from motion reconstruction is more difficult in the Target (Nom)/Target(Lan) engagement. Figure 8.11 shows the computed probability of Target(Nom) for PME(0.55,10) and PME(0.50,10) during the initial 25 s of the

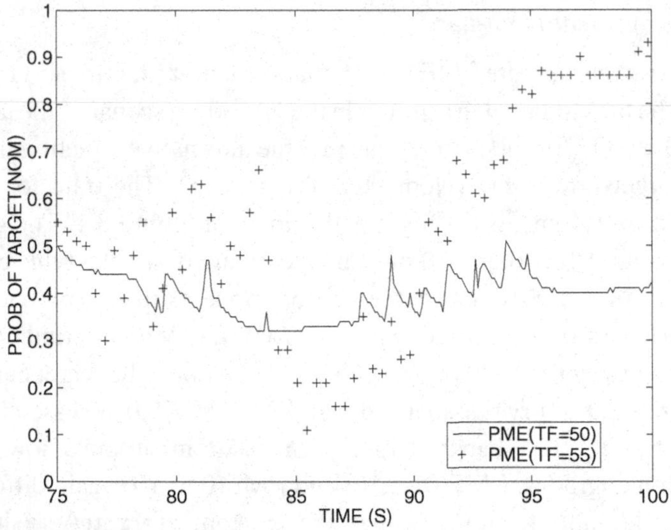

Figure 8.11. Target recognition for PME(0.55,10) and PME(0.50,10).

encounter. The response of PME(0.75,10) is not shown because it differs little from that shown in Figure 8.6: In a high SNR environment, target identification is based primarily on the direct type measurement. Not surprisingly, PME(0.55,10) is able to identify the target, though it takes 5 s more to do it conclusively than in the earlier scenario. What is surprising is the response of PME(0.50,10). Ultimately the target will be identified since $r_t = \mathbf{e}_1$ is an absorbing state. But with PME(0.50,10), $\mathcal{G}_t$-$\mathcal{P}$(Target(Nom)) shows no appreciable motion in that direction on the interval shown. The Markov motion model, by itself, is insufficient in this case to differentiate the targets in a timely manner.

In the Target(Nom)/Target(Lan) engagement, it is more difficult for PME (0.55,10) to identify the target, and this is reflected in the size of its error ellipses (see Figure 8.12). Comparing this plot with Figure 8.4, PME(0.55,10) tracks about as well as it did in the Target(Nom)/Target(Agl) scenario. Until PME(0.55,10) can classify the target with confidence, its error ellipses are increased to make the uncertainty associated with the proper motion model explicit (see the first left turn). Nonetheless the target trajectory is well contained by the envelope of the error ellipses.

Clearly then, the PME is a more accurate tracker than is an EKF. If path following is the only requirement, target recognition may not be needed for high quality performance. Without direct target type measurements, the PME does not provide an accurate indication of the target class. This is primarily due to the fact that the maneuver and the orientation processes are not truly Markov processes. A maneuver

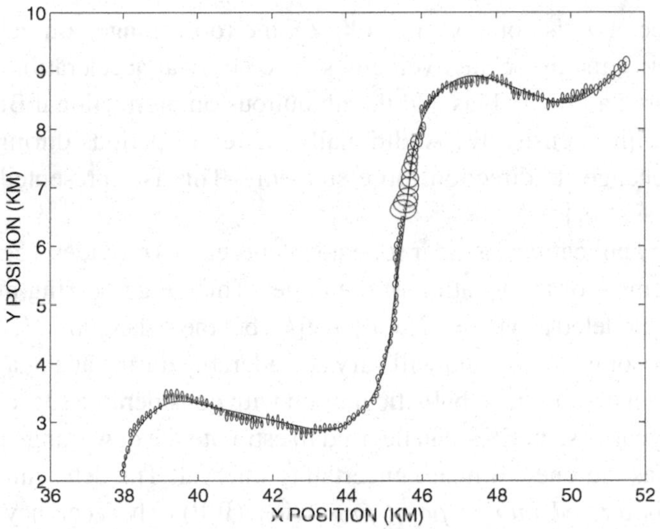

Figure 8.12. Performance of PME(0.55,10) in the Target(Nom)/Target(Lan) engagement.

model that more clearly identifies the modal residence time distributions could be expected to improve performance. (See the discussion of this issue in Chapter 4.) It seems, however, to be more efficient to use more detailed shape analysis to acheive target recognition rather than make the motion model more specific.

8.4 Path Prediction

Although the need to track agile targets is evident, in some applications it is also necessary to predict their future motion. For example, in theater air defense systems a diverse set of sensors is employed to identify threats at launch, predict their path, and intercept them at an opportune time. Because of the predictable motion dynamics of *ballistic* targets, good performance is achieved against them. For such targets, conventional techniques, with small modifications, can be employed for both estimation and prediction. A more difficult situation arises when the targets are *smart* (i.e., maneuvering) [Tin95].

Synthesis of a predictor begins with a quantitative representation of target motion. Suppose the target is a missile following a near-ballistic path in the plane (flat earth; constant gravity, g). A planar model is given by

$$d\begin{bmatrix} X \\ Z \\ V_x \\ V_z \end{bmatrix} = \begin{bmatrix} 0 & 0 & 1 & 0 \\ 0 & 0 & 0 & 1 \\ 0 & 0 & 0 & -\Phi \\ 0 & 0 & \Phi & 0 \end{bmatrix} \begin{bmatrix} X \\ Z \\ V_x \\ V_z \end{bmatrix} dt - \begin{bmatrix} 0 \\ 0 \\ 0 \\ g \end{bmatrix} dt + \begin{bmatrix} 0 & 0 \\ 0 & 0 \\ 1 & 0 \\ 0 & 1 \end{bmatrix} d\begin{bmatrix} w_x \\ w_y \end{bmatrix}, \tag{8.3}$$

where the state vector x_t is composed of (X, Z), the (downrange, altitude) coordinates, and (V_x, V_z), the associated velocities. The nominal acceleration process consists of two parts: a gravity bias and the ubiquitous omnidirectional Brownian excitation, $\{w_t\}$, with intensity W. Additionally, there are periods during which the missile may change its direction more suddenly. This is represented by the "maneuver" process, $\{\Phi_t\}$.

In most tracking applications, a radar (or equivalent sensor) provides a center-of-reflection measurement of the location of the target. This range-bearing measurement can again be modeled as $y[k] = Hx[k] + n[k]$, but the radar *gain*, H, depends upon the target–sensor geometry and will vary considerably during an engagement.

The EKF has been used to track ballistic motion with considerable success. With such unadorned dynamics, an EKF can be used to estimate the downrange position at impact as well as the one-σ impact uncertainty interval. The determination of this alert region is termed *impact point prediction* (IPP). The accuracy of this extrapolation has important implications. The assets in most danger from the missile are those at or near the impact point. Assets within a reasonably chosen uncertainty interval around the projected impact should be on alert status and should take action to mitigate the effect of the threat.

To determine the impact region explicitly, suppose at time t_0 the state of the EKF is $(\hat{x}_{t_0}, P_{xx}(t_0))$. The missile will impact when its altitude is zero (when $Z_t = 0$). The time of impact is random, but we can estimate *time-to-go* (t_g) by extrapolating along a ballistic path forward from $\hat{x}_{t_0}$ to $\hat{Z}_t = 0$. The *impact point* (X_{imp}) is the downrange position at this time. Thus the estimate of time-to-go is simply

$$t_g = \left(\hat{V}_z + \sqrt{(\hat{V}_z^2 + 2g\hat{Z})}\right) \Big/ g.$$

During this time, the missile will move downrange to the point

$$X_{\text{imp}} = \hat{X}_{t_0} + t_g \hat{V}_x.$$

This extrapolation to impact is derived from a parabolic approximation to the missile path.

In IPP, not only must the mean location of impact be determined but the uncertainty region as well. To determine the latter, note that a one-σ interval can be placed about X_{imp} by computing the error covariance at the time of impact: $P_{xx}(t_g)$. Beginning at $P_{xx}(t_0)$, we can find the uncertainty region by integrating

$$\frac{d}{dt} P_{xx} = A P_{xx} + P_{xx} A' + \mathbf{E}_2 \otimes W.$$

The downrange coordinate is X_t. The variance of the downrange impact point is

$$(P_{xx}(t_g))_{11} = \sigma_X^2(t_g).$$

Direct calculation yields

$$\sigma_X^2(t_g) = \mathbf{e}_1'(\mathbf{\Phi}(t_g, 0) P_{xx}(t_0) \mathbf{\Phi}(t_g, 0)')\mathbf{e}_1 + W t_g^3,$$

where the transition matrix for the ballistic motion is

$$\mathbf{\Phi}(t_g, 0) = \begin{bmatrix} 1 & 0 & t_g & 0 \\ 0 & 1 & 0 & t_g \\ 0 & 0 & 1 & 0 \\ 0 & 0 & 0 & 1 \end{bmatrix}.$$

The extrapolation logic for the PME is considerably more complicated than was the case for the EKF. Suppose, that at time t_0, the PME is at state $\{\hat{\psi}_{t_0}, \hat{x}_{t_0}, P_{xx}(t_0), P_{x\phi}(t_0), P_{xx\phi}(t_0)\}$. From these initial conditions, t_g must be estimated by integrating the expected motion forward to impact:

PME: impact point prediction

Between observations:

$$\frac{d}{dt}\hat{\phi}_t = Q'\hat{\phi}_t,$$

$$\frac{d}{dt}\hat{x}_t = \sum_i A_i R_{x\phi_i},$$

$$\frac{d}{dt}P_{x\phi} = \sum_i A_i P_{(x\phi_i)\phi} + P_{x\phi}Q,$$

$$\frac{d}{dt}P_{xx} = \sum_i (A_i P_{(x\phi_i)x} + (\cdot)') + R_X,$$

$$\frac{d}{dt}P_{xx\phi_m} = \sum_i (A_i P_{(x\phi_i)x\phi_m} + (\cdot)' + P_{xx\phi_i}Q_{im}),$$

with impact at $\hat{Z}_t = 0$ and time t_g. The variance of the downrange impact point is again $\sigma_X^2(t_g)$.

8.5 A Missile Test in Australia

During October and November 1995, in a cooperative experiment, the U.S. Ballistic Missile Defense Organization (BMDO) and the Australian Defence Science and Technology Organization (DSTO) conducted a series of multistage rocket launches at the Woomera Test Range in South Australia [CSB96b]. The primary purpose of the experiment was to launch scientific payloads, but the tests also provided a means to exercise some algorithms that had been proposed for long-range detection and tracking. The Innovative Science and Technology Experimentation Facility (ISTEF) of BMDO furnished tracking sensors that provided precision range and angle–angle track data during the trials. GPS data were also available for post launch analysis. The totality of data generated from this diverse grouping of sensors permitted tracking algorithms to be evaluated in a realistic environment. The demonstration verified the utility of infrared (IR) and radar data fusion for following a nearly ballistic trajectory and showed that track data and video could be transmitted in real time to a remote site for processing and interpretation. The experiment did not involve a maneuver-capable target. It is useful, however, to abstract and generalize the experiment and show that when the target has maneuver capability, the PME is a useful alternative to the current system based upon an EKF.

At Woomera, the payload had a boost phase. After separation, the payload flew a near-ballistic trajectory until impact. The Woomera sensor suite included both a radar and an IR sensor. Location measurements were from the Woomera Adour C-band radar (for range and bearing) and the ISTEF optical tracking mount equipped with a suite of telescopes and sensors, which provided precision angle track data. The angular data from the IR sensor were much more accurate than those from the radar. The radar and optical sensors were located 27 km downrange and together acted as an equivalent point-target sensor, which we label the IR/radar. The origin of the (X, Z)-coordinate system is placed at the sensor. From the raw Woomera data it was determined that the nominal IR/radar errors have standard deviations of 40 m in range and 0.4 mr in bearing (at 150 km this implies a cross line-of-sight error of about 70 m). The nominal IR/radar interdwell interval is 0.1 s. Despite these highly accurate measurements, the IR/radar tracker was unable to maintain lock on the early and relatively benign portion of the path.

To complicate the experiment, suppose that the missile has maneuver capability that is utilized to make tracking and prediction difficult near impact. Specifically, the missile accelerates from launch (at $t = 0$ s) for 44 s at 5.8 g. Payload motion at separation is represented by a 3 s, 3 g turn down. The target flies a ballistic path on the complement of these two intervals until $t = 450$ s. At this time the missile executes a 5 s, 3 g pull up for evasive purposes. The missile then flies a ballistic path until impact. Table 8.3 summarizes the acceleration modes. Detection occurs 50 s after launch. This initialization gives each tracker 22 s to reach quiescent operation before the separation event occurs. The nominal path of the payload is shown in Figure 8.13. The evasive maneuver is barely visible in the figure because of its short duration and the high speed of the payload. Until the terminal maneuver, the path is essentially that flown in the Woomera experiment.

To test the algorithms that follow, each is used on a single sample of the target path. Suppose the trackers begin with initial standard deviations of 100 m in the position coordinates and 20 m/s in the velocities. Because of the small intensity of the omnidirectional accelerations, $W = \mathbf{I}$. The tracker at Woomera was an EKF. With these parameters, the nominal EKF (labeled $\mathsf{EKF}(W = 1, \lambda = 10)$, or $\mathsf{EKF}(1,10)$) can be produced.

Table 8.3. *Motion modes.*

Time	Status
0	launch
(0, 44)	boost @ 5.8 g
(72, 75)	separation @ 3 g
(450, 455)	evasive maneuver @ 3 g
528	impact

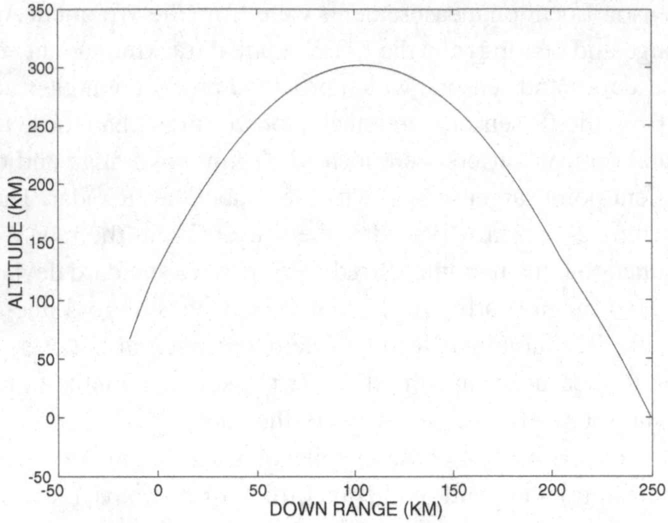

Figure 8.13. The path of the maneuvering missile.

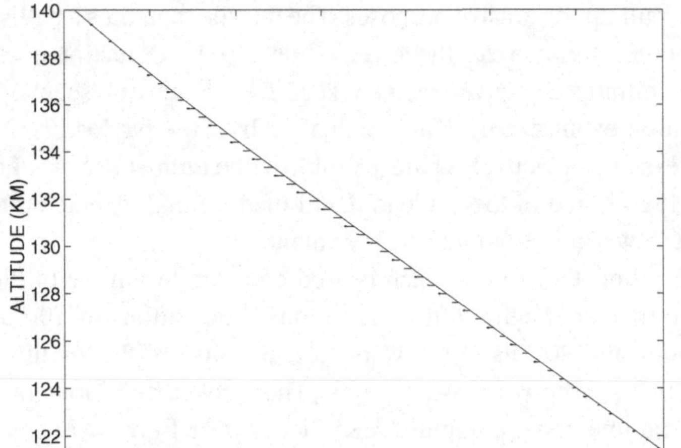

Figure 8.14. Performance of EKF(1,10) near the reentry maneuver.

The reentry maneuver begins slightly below (203,140) km. Figure 8.14 shows an expanded feather plot in this region. Before the maneuver begins, EKF(1,10) has located the target and has little error in its velocity estimate. EKF(1,10) does not expect the pull up, however, and falls far behind when it occurs: The error exceeds 200 m and is primarily in the downrange direction. The error ellipses are barely visible in the figure.

The failure of EKF(1,10) to follow the reentry maneuver is explained in part by its inability to compute its own error covariance correctly. Going into the pull up, both the error and the error ellipses are small. At this time, the error ellipses enclose the target path and give a good indication of the uncertainty region for this specific tracker. The situation changes after the maneuver. EKF(1,10) lags the turn but is oblivious of the fact. For EKF(1,10) the uncertainly ellipses do not give a valid indication of tracking fidelity; the true path is several standard deviations away from the estimate. The maneuver-induced correction takes about 10 s (not shown), and this for a turn that lasts only 5 s.

Suppose now that the IR/radar measurements are augmented with those from a modal sensor. Although the two turn events induce different target turn rates, suppose a low dimension model is used in the design of the PME: Let

$$\Phi_t \in \{15 \text{ mr/s}, \alpha_t = \mathbf{e}_1;\ 0 \text{ r/s}, \alpha_t = \mathbf{e}_2;\ -15 \text{ mr/s}, \alpha_t = \mathbf{e}_3\}.$$

This coarse parsing of the turn rates reduces the complexity of the equations for the PME gains, and such simplifications should be sought whenever possible. A similar simplification is used to quantify the maneuver tempo: The mean sojourn time in a ballistic phase will be 50 s, while that in a turning mode will be 5 s. This single tempo is crude and, more importantly, does not alert the PME to regions of likely evasion. Each turn will be followed by a ballistic segment. From a nonmaneuvering mode, it will be supposed that turns in either direction are equally likely. From this description, the maneuver dynamics (i.e., Q) can be deduced.

The quality of the modal sensor is delineated by the frame rate, λ, and the discernibility matrix, $\mathbf{P}$. First suppose that the modal sensor is operating at the rate of the IR/radar: $\lambda = 10$ frames/s. The discernibility matrix characterizes the fidelity of the modal sensor/processor for a single frame. In this study, the features in an image are assumed to lead directly to a classification of turn rate – target orientation is not measured at this range. The sensor is of good quality, correctly classifying the maneuver mode 80% of the time with the errors uniformly distributed over the complementary bins. From this, PME($W = 1, \lambda = 10$) (or, more simply, PME(10)) can be deduced. A summary comparison of the tracking algorithms is given in Table 8.4.

Table 8.4. *Tracker parameters.*

Algorithm	Radar Rate	Modal Rate	Radar Error	Modal Error
EKF(1,10)	10/s	—	40 m/.4 mr	—
EKF(1,1)	1/s	—	40 m/.4 mr	—
PME(10)	10/s	10/s	40 m/.4 mr	20%
PME(1)	1/s	10/s	40 m/.4 mr	20%

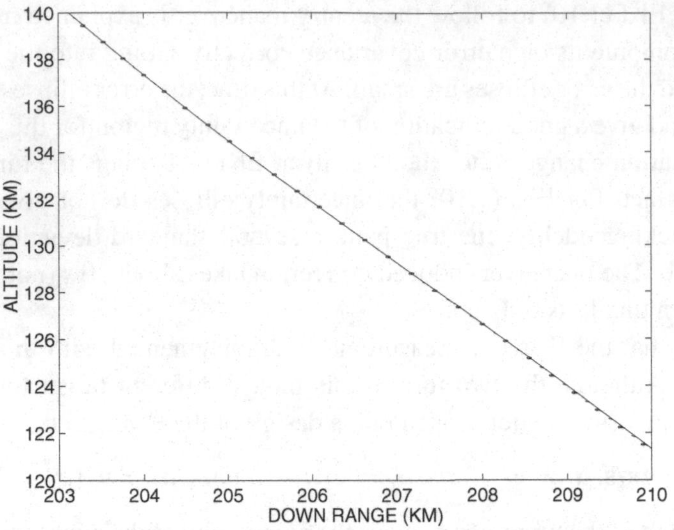

Figure 8.15. Performance of PME(10) near the reentry maneuver.

The performance of PME(10) is given in Figure 8.15 and is far superior to that of EKF(1,10). At the scale of the figure, the errors of PME(10) are barely visible. To contrast the two trackers in more detail, Figure 8.16 shows the radial error for both EKF(1,10) and PME(10). Both trackers do well on ballistic segments with errors typically below 20 m and excursions to 40 m (and 60 m for the EKF). This

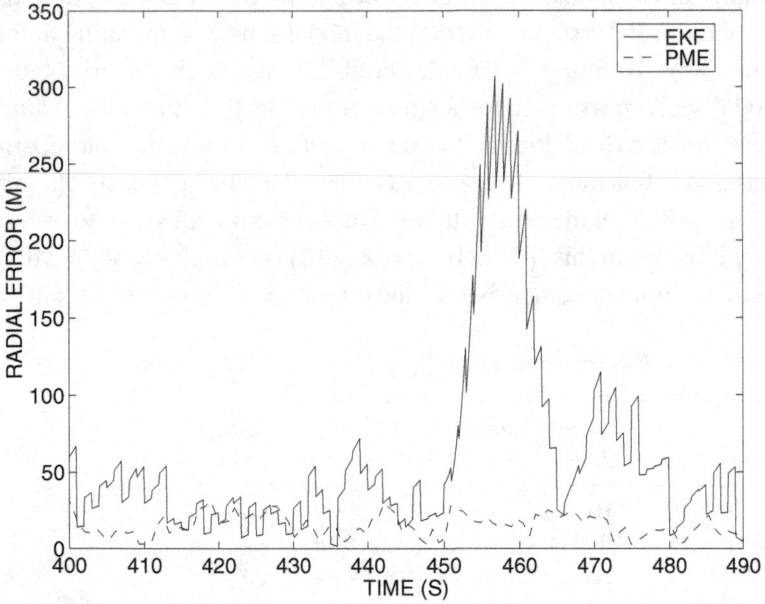

Figure 8.16. Radial tracking error for EKF(1,10) and PME(10).

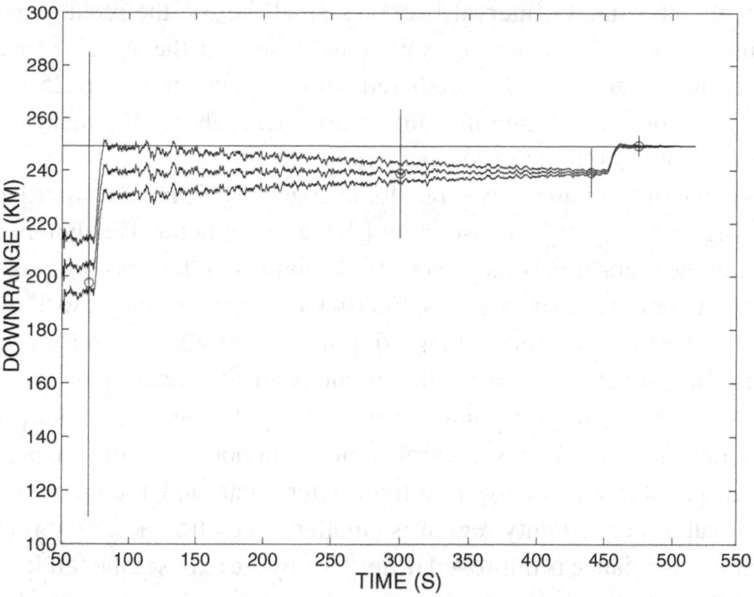

Figure 8.17. Impact point prediction for EKF(1,10) (continuous) and PME(10) (four samples) together with 1-σ error bounds.

is reasonable given the IR/radar accuracy. The PME performs slightly better than the EKF during ballistic motion, which is mildly surprising given the fact that the modal sensor merely confirms the preconception of the EKF. During the reentry turn, the radial error of EKF(1,10) explodes to approximately 300 m while that of PME(10) remains around 30 m. This is a performance improvement of a factor of 10 with a similar improvement at separation.

Figure 8.17 shows a plot of X_{imp} and the associated uncertainty interval as computed by EKF(1,10) as a function of time since launch. (The local variability in the uncertainty interval is due to the errors in the IR/radar measurement.) After detection, EKF(1,10) predicts an impact at about 205 km downrange. By placing a one-σ region about the expected impact point, EKF(1,10) predicts with confidence that impact will occur within the interval in [195,215] km. But the actual impact point is 250 km. This lies about four standard deviations from X_{imp}; this is a major error since the actual impact point would naturally be considered to be outside the area of influence of the target in this engagement.

Impact prediction does not change much between detection and separation. After separation the impact point expected by EKF(1,10) suddenly shifts – it is closer to 240 km. As time moves on, the uncertainty region narrows due to the fact that the time-to-go is getting shorter: $\{P_{xx}\}$ is about constant and the integration time is growing ever smaller. While the absolute prediction error is less than it was before separation, the uncertainty interval becomes small before the reentry maneuver, engendering again a false sense of security in those near the actual impact point. After the reentry maneuver, the predicted impact point moves to 250 km with the uncertainty interval becoming quite short. Here EKF(1,10) makes accurate calculations because no more turns take place.

It is easy to intuit the qualitative results in Figure 8.17, but the figure affords a clear warning regarding myopic use of an EKF as a predictor. The dynamic model that underlies the algorithm is incorrect (it is oblivious of the possibility of maneuvers), but with frequent measurements, the EKF can track the target well over most of its path. However, its errors are magnified when the EKF is used for prediction; there is insufficient data to disabuse the predictor of its model-induced biases. Its inability to predict the impact point – it was off by 40 km – is not surprising. A ballistic extrapolation is the most natural choice after boost. It is the misplaced confidence in its prediction that is so troubling. After separation, the predicted impact is adjusted, but the uncertainty region is smaller; prediction accuracy normalized by the computed variance is improved little. This figure exposes the fallacy of using the EKF to activate fixed-point defense systems; the wrong ones may be alerted and the right ones may be told they are outside the attack envelope.

Calculation of X_{imp} for the PME is more difficult. It cannot easily be done as a running function of time as was the case for EKF(1,10). For PME(10), Figure 8.17 shows the value of X_{imp} along with the one-σ uncertainty interval for four times: $t = 70$ s, just before separation; $t = 300$ s, after separation; $t = 444$ s, before the reentry maneuver; and $t = 480$ s, after the final maneuver. The value of X_{imp} for PME(10) differs little from that computed by EKF(1,10). Each extrapolation took place in a ballistic interval, and because of symmetry, the PME extrapolation has the same character as the EKF. The one-σ uncertainty interval is far more representative of the actual impact point than is that computed from the EKF. At every point, even those immediately preceding a turn, the true impact point is in the alert region. Use of the PME is warranted in this problem if only because of the improvement in IPP.

One of the objectives of the Woomera experiment was to determine if it is possible to transmit tracking data in real time to a remote processing site. To reduce the communications burden, suppose that the Woomera IR/radar is again used, but in an infrequent measurement mode, with an IR/radar interdwell time of 1 s. (The communication rate has been reduced by a factor of 10.) With this measurement specification, the congruous EKF (labeled EKF(1,1)) can be derived. Figure 8.18 shows the feather plot of the path following error of EKF(1,1) near the reentry maneuver. The lower sampling rate causes the error to increase (compare Figure 8.5). The error is again primarily downrange, but in contrast with

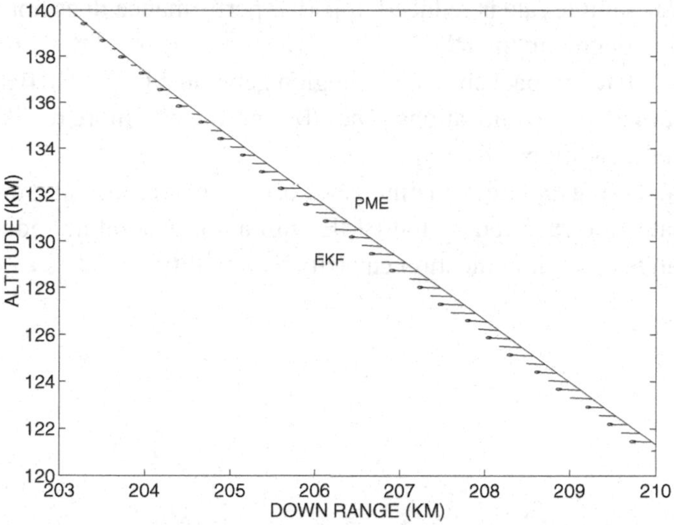

Figure 8.18. Performance of EKF(1,1) and PME(1) near the reentry maneuver (the PME points are barely visible at this scale).

EKF(1,10), the tracker is very slow to recover from the measurement-induced transient.

The Woomera IR/radar operating at this same rate can be used in the PME to create a tracker labeled PME(1). The errors for PME(1) are so small that the feather plot shows little at the scales indicated. Figure 8.18 shows both EKF(1,1) and PME(1) near the reentry maneuver. EKF(1,1) is not so sure of its estimates as was EKF(1,10), but the actual curve is still far from the one-σ confidence region. The error ellipses for PME(1) are bigger but still enclose the true path (not visible on this scale).

Extrapolation to impact can be done with both EKF(1,1) and PME(1). The results differ little from Figure 8.17. Suffice it to say the primary determinant of the extrapolation uncertainty interval in both trackers is the exogenous acceleration during the interval to impact, and performance is not very sensitive to the current tracking error moments.

It is clear that the PME permits a more efficient use of the primary sensor in this application. The effectiveness of the PME is due to its ability to use the modal information both for direct updates and for adjustments to the tracker gains and time constants (see also [CSB96a] and [CS95]). It is difficult to isolate the primary influence of the modal sensor. However, it can be seen that:

- The tracking error after maneuvers can be reduced with a modal sensor as an adjunct to the base-state sensor.

- If the radar sample rate is reduced, tracking performance does not degrade to the same degree in the PME.
- The extrapolated impact uncertainty region generated by the EKF may lead to untoward resource allocations. The PME produces a more realistic view of terminal conditions.
- When processing and track maintenance occurs at a remote site, fewer raw data updates are required by the PME, with a corresponding reduction in the demands on communication capacity and quality.

9

Hybrid Estimation Using Measure Changes

9.1 Change of Measure

In earlier chapters of this book, we studied state estimation and regulation with an emphasis on time-continuous hybrid plants with a mix of time-continuous and time-discrete observations. The plant input/state-output representation is such that each state category is associated with an observation process and filtration: $\{y_t\}$ and $\{\mathcal{Y}_t\}$ for the base-state, and $\{z_t\}$ and $\{\mathcal{Z}_t\}$ for the modal-state. Low level data fusion generates the filtration $\mathcal{G}_t = \mathcal{Y}_t \vee \mathcal{Z}_t$. An engineer seeks practical algorithms for approximating the $\mathcal{G}_t$-regime probabilities along with those $\mathcal{G}_t$-moments (including cross moments) important in the application. The primary tool in those chapters was the polymorphic estimator in its sundry realizations. The PME generates serviceable approximations to relevant $\mathcal{G}_t$-conditional moments. Although $\hat{\phi}_t$ is the vector of conditional regime probabilities, the PME does not provide the $\mathcal{G}_t$-distribution function of the zygostate.

The PME is premised on the assumption that the modal measurement is a good one; at least $\{z_t\}$ is the best measurement available as regards the regime. The modal measurement is not perfect to be sure, but it is good enough that base-state $\times$ modal-state cross moments depend incrementally upon $\{z_t\}$ alone. This hierarchical processing structure has its rationale in the way in which the engineer assembles the sensor suite. For example, the initial design of the panel temperature regulator for the solar central receiver (see Chapter 5) did not include insolation sensors. They were added only after it became apparent that acceptable control of the receiver panel was impossible without them; they were an expensive supplement. In the same way, the sensor suite used in the example of image-enhanced tracking may have included an imager, but the imager was originally intended for target detection and/or recognition. In both examples, conventional sensor fusion took place at a high level (path fusion) because of the narrow role assigned to the modal sensors. The PME brings improved performance because it exploits the synergies inherent in early data integration. However, even the PME neglects the base-state innovations in the calculation of the canonical moments and this leads to suboptimal results.

Perhaps surprisingly, the PME shows itself to advantage even when the regime measurement is of fairly low quality. This is due to the fact that the PME enjoys a knowledge of the plant-state geometry and it computes the error cross moments required to utilize this geometry in its estimates. Despite its success in problems where the measurement quality and frequency make the assumptions underlying the PME difficult to defend, there are situations in which the innovations exclusivity tries a designer's intuition. For example, it might be obvious to an engineer that the base-state observations could help reduce the frequency of aliasing errors in $\{z_t\}$. By ignoring $\{\mathcal{Y}_t\}$ in generating $\{\hat{\phi}_t\}$, lower quality estimates necessarily result.

Exact generation of the $\mathcal{Y}_t$-zygostate moments is not possible in any practical sense. Unless the modal measurements are flawless (and frequent), the same statement applies to the $\mathcal{G}_t$-moments. In this chapter, we will look at an approximation to $\mathcal{G}_t$-estimation in the context of a time-discrete plant and observation. It is useful to recall that the IMM is a time-discrete $\mathcal{Y}_t$-estimator of the zygostate when there are no base-state discontinuities induced by the modal transitions. The IMM generates a usable approximation to $\{\hat{x}_t\}$ in applications where high accuracy calculation of $\{\hat{\phi}_t\}$ and the higher cross moments is not required.

In this chapter we will use Bayes' formula to generate $\mathcal{G}[k]$-estimates, and we will use pruning instead of merging to manage the level of algorithmic complexity. As in the referenced studies of the MM-estimators, attention will be focused on plants without endogenous excitation ($u[k] \equiv 0$), but set point discontinuities will be included in the model.

The model of zygostate evolution is (see (1.29) and (1.32))

$$x[k] = \sum_j (A_j x[k-1] + C_j w[k])\mathbf{e}_j'\phi[k-1] - \chi\Delta\phi[k-1], \tag{9.1}$$

$$\phi[k] = \Pi\phi[k-1] + m[k], \tag{9.2}$$

where $\{C_i; i \in \mathbf{S}\}$ are nonsingular with

$$C_i'C_i = R_\chi(i) = D_\chi(i)^{-1} > 0.$$

The regime transition probabilities are given by Π. In the intersample interval $[(k-1)T, kT)$ the base-state evolves according to the dynamics of the regime at time $(k-1)T$. If there is a modal transition, the base-state will experience the discontinuity $-\chi\Delta\phi[k-1]$.

The measurement model is essentially that given in (1.17) and (1.54) with a regime-dependent SNR in the base-state channel (see Chapter 1):

$$y[k] = \sum_i H_i \mathbf{e}_i' \phi[k] \chi[k] + n[k], \tag{9.3}$$

$$z[k] = \mathbf{P}\phi[k] + \eta[k], \tag{9.4}$$

where $\{n[k]\}$ is a Gaussian white noise sequence with positive covariance $F'F = R_x = D_x^{-1}$. As in Chapter 6, the plant state observation gain is mode dependent.

The development that follows is simpler if we modify the primary filtration. The composite sequence

$$\{w[0], \ldots, w[k], m[0], \ldots, m[k], y[0], \ldots, y[k-1], z[0], \ldots, z[k-1]\}$$

along with the vectors $x[0]$ and $\phi[0]$ generates $\mathcal{F}$[k]. The initial plant states are independent and have the probability distributions $x[0] \in \mathbf{N}(\hat{x}[0], P_{xx}[0])$, and $\phi[0]$ is distributed according to $\hat{\phi}[0]$. In this construction, the observation filtration, $\mathcal{G}[k] = \mathcal{Y}[k] \vee \mathcal{Z}[k]$, is not contained in $\mathcal{F}[k]$ but rather in $\mathcal{F}[k+1]$. The exogenous processes in (9.1)–(9.4) are $\mathcal{F}[k]$-martingale increments (or $\mathcal{F}[k-1]$-martingale increments), and the detailed characteristics of their paths are given in Chapter 1.

In contrast to the moment-based approach used to frame the PME, we will seek an implementable approximation to the $\mathcal{G}[k]$-conditional distribution of the zygostate. At time $t = kT$, suppose the base-state has the $\mathcal{G}[k]$-conditional vector probability density, $p[k] = [p^i[k]]$:

$$p^i[k](z)\,dz = \mathcal{P}(x[k] \in [z, z+dz], \phi[k] = \mathbf{e}_i | \mathcal{G}[k]). \tag{9.5}$$

From $p[k]$, the requisite moments can be obtained directly. We seek a mapping from $\{p[k], y[k+1], z[k+1]\}$ to $p[k+1]$. This mapping can be stated most concisely when phrased in terms of a set of unnormalized densities, $q[k](z) = [q^i[k](z)]$, derived using a *change-of-measure* (COM) approach. The resulting algorithm will be called the *COM-estimator*.

To many engineers, the COM calculations that follow will lack intuitive appeal, but they are actually those common in applications. To provide a simple illustration of the COM technique, let $(\Omega, \mathcal{F}, \mathcal{P})$ be a probability space on which $\Omega = [0, 1]$, $\mathcal{F}$ is the Borel field, and $\mathcal{P}$ is a probability measure on [0, 1] with density p_x. On this space, $x = \omega$ is a random variable. If f is a well-behaved function of x, its expectation is easily computed:

$$E[f] = \int_\Omega f(\omega)\,d\mathcal{P}(\omega) = \int_{[0,1]} f(u) p_x(u)\,du. \tag{9.6}$$

There is another way of performing this calculation. Let $(\Omega, \mathcal{F}, \bar{\mathcal{P}})$ be another probability space (same sample space and events but a different probability measure, $\bar{\mathcal{P}}$). Let $\bar{\mathcal{P}}$ be length-measure: The probability of an element of $\mathcal{F}$ is its length. The random variable $x = \omega$ is $\bar{\mathcal{P}}$-uniformly distributed on [0, 1] and so the probability density of x is identically one on [0, 1]. The probability measures on the event space are different, and the moment properties of $x(\omega)$ are different as well. The $\mathcal{P}$-probability of an event like $x \in [u, u + du]$ is $p_x(u)du$ whereas the $\bar{\mathcal{P}}$-probability of this identical event is du: Even though the sample spaces and the random variables are duplicates, the probabilities of relevant events are different.

The derivative of $\mathcal{P}$ with respect to $\bar{\mathcal{P}}$, denoted $\bar{\Lambda}$, is in this case equal to p_x:

$$\frac{d\mathcal{P}}{d\bar{\mathcal{P}}}(u) = \bar{\Lambda}(u) = p_x(u).$$

To distinguish it from $\mathcal{P}$-expectation, denote expectation with respect to $\bar{\mathcal{P}}$ by $\bar{E}$. Let $\mathcal{G}$ be a sub-σ-field of $\mathcal{F}$. The conditional Bayes' Theorem [EAM95, Theorem 3.2] relates the operator E to the operator $\bar{E}$:

$$E[f \mid \mathcal{G}] = \frac{\bar{E}[f\bar{\Lambda} \mid \mathcal{G}]}{\bar{E}[\bar{\Lambda} \mid \mathcal{G}]}. \tag{9.7}$$

Equation (9.7) shows that an expectation with respect to $\mathcal{P}$ can be performed as an expectation with respect to a different measure $\bar{\mathcal{P}}$ if the derivative of the measure is included properly in the calculation. For example, if $\mathcal{G}$ is the trivial σ-field, $E[f] = \bar{E}[f\bar{\Lambda}]/\bar{E}[\bar{\Lambda}]$. Since $\bar{E}[\bar{\Lambda}] = \int_0^1 \bar{\Lambda}(u)\,du = 1$ and $\bar{E}[f\bar{\Lambda}] = \bar{E}[fp_x]$, Equations (9.6) and (9.7) are identical.

In this illustrative example, there would be little inclination to use COM for the calculation, but this is not always so. Return to the hybrid estimation problem. On the original event space and filtration, $(\Omega, \mathcal{F}; \mathcal{F}[k])$, consider a new probability measure, $\bar{\mathcal{P}}$. With respect to $\bar{\mathcal{P}}$, there are several independent, vector-valued, random sequences: $\{w[k]\}$ and $\{y[k]\}$ are Gaussian white noise sequences ($w[k]$ and $y[k]$ are $\mathbf{N}(0, \mathbf{I})$); $\{z[k]\}$ is an independent, identically distributed (iid) sequence that is uniformly distributed across $\{\mathbf{e}_1, \ldots, \mathbf{e}_S\}$; $\{m[k]\}$ is a sequence of martingale increments. This change to $\bar{\mathcal{P}}$ reflects no change for $\{w[k]\}$ and $\{m[k]\}$, but the character of $\{y[k]\}$ and $\{z[k]\}$ is considerably different. Although the processes $\{y[k]\}$ and $\{z[k]\}$ will retain the name "observation processes" even in the new probability space, they are both $\bar{\mathcal{P}}$–iid sequences and convey little pertinent information.

The $\mathcal{G}[k]$-zygostate estimation problem under $\bar{\mathcal{P}}$ is easy to solve but not particularly interesting since $\{y[k]\}$ and $\{z[k]\}$ are independent of the zygostate. To use a COM-estimator to compute the $\mathcal{P}$-moments of interest, we must first delineate the relationship between $\bar{\mathcal{P}}$ and $\mathcal{P}$. Denote the density function of a unit Gaussian random vector by $\Phi(u) = \mathbf{N}_u(0, \mathbf{I})$. For $l \in \{0, 1, \ldots\}$, define a local likelihood function

$$\bar{\lambda}[l] = \frac{S}{\Phi(y[l])} z[l]'\mathbf{P}\phi[l]|F|^{-1}\Phi\left(F^{-1}\left(y[l] - \sum_i H_i \mathbf{e}_i' \phi[l](x[l] + \chi_i)\right)\right) \tag{9.8}$$

and let $\bar{\Lambda}[k]$ be a continuing product of the $\bar{\lambda}[l]$:

$$\bar{\Lambda}[k] = \prod_0^k \bar{\lambda}[l]. \tag{9.9}$$

Denote the derivative of $\mathcal{P}$ with respect to $\bar{\mathcal{P}}$ by $\bar{\Lambda}$. It will first be shown that the restriction of $\bar{\Lambda}$ to $\mathcal{F}[k] \vee \mathcal{G}[k]$ is $\bar{\Lambda}[k]$; that is, $\mathcal{P}$ can be found from $\bar{\mathcal{P}}$ by multiplication by $\bar{\Lambda}[k](z[k], y[k])$:

$$\left.\frac{\partial \mathcal{P}}{\partial \bar{\mathcal{P}}}\right|_{\mathcal{F}[k]\vee\mathcal{G}[k]} = \bar{\lambda}[k](z[k], y[k]). \tag{9.10}$$

To see this, observe that the probability of an event is the expectation of the indicator of that event (e.g., $\mathcal{P}(y[k] \in A \mid \mathcal{F}[k]) = E[I(y[k] \in A) \mid \mathcal{F}[k]]$). Consider the compound event $I(y[k] \le t)I(z[k] = \mathbf{e}_i)$:

$$\mathcal{P}(y[k] \le t, z[k] = \mathbf{e}_i \mid \mathcal{F}[k]) = E[I(y[k] \le t)I(z[k] = \mathbf{e}_i) \mid \mathcal{F}[k]].$$

To find the probability, the $\mathcal{P}$-expectation of the indicator function $I(y[k] \le t)$ $I(z[k] = \mathbf{e}_i)$ must be calculated conditioned on $\mathcal{F}[k]$. It should be far simpler if we use $\bar{E}$ since the observation processes have such an elementary structure under $\bar{\mathcal{P}}$:

$$\mathcal{P}(y[k] \le t, z[k] = \mathbf{e}_i \mid \mathcal{F}[k]) = \frac{\bar{E}[\bar{\Lambda}[k] I(y[k] \le t) I(z[k] = \mathbf{e}_i) \mid \mathcal{F}[k]]}{\bar{E}[\bar{\Lambda}[k] \mid \mathcal{F}[k]]}.$$

Many of the factors in $\bar{\Lambda}[k]$ are $\mathcal{F}[k]$-adapted ($\bar{\Lambda}[k-1]$ is $\mathcal{F}[k]$-adapted) and common to numerator and denominator. These factors can be canceled to yield

$$\mathcal{P}(y[k] \le t, z[k] = \mathbf{e}_i \mid \mathcal{F}[k]) = \frac{\bar{E}[\bar{\lambda}[k] I(y[k] \le t) I(z[k] = \mathbf{e}_i) \mid \mathcal{F}[k]]}{\bar{E}[\bar{\lambda}[k] \mid \mathcal{F}[k]]}. \tag{9.11}$$

Look first at the denominator of (9.11):

$$\bar{E}[\bar{\lambda}[k] \mid \mathcal{F}[k]] = \bar{E}\left[\frac{S}{\Phi(y[k])} z[k]'\mathbf{P}\phi[k]|F|^{-1} \times \Phi\left(F^{-1}\left(y[k] - \sum_i H_i \mathbf{e}_i' \phi[k](x[k] + \chi_i)\right)\right) \,\middle|\, \mathcal{F}[k]\right].$$

But $z[k]$ is $\bar{\mathcal{P}}$–iid and independent of $y[k]$. So $\bar{E}[\bar{\lambda}[k] \,|\, \mathcal{F}[k]]$ can be separated into factors that depend upon the $y[k]$ and $z[k]$ individually. First,

$$S\bar{E}[z[k]'\mathbf{P}\phi[k] \,|\, \mathcal{F}[k]] = S(\mathbf{1}'\phi[k]/S) = 1.$$

Second,

$$\bar{E}\left[\frac{|F|^{-1}\Phi(F^{-1}(y[k] - \sum_i H_i\mathbf{e}_i'\phi[k](x[k] + \chi_i)))}{\Phi(y[k])}\middle| \mathcal{F}[k]\right]$$
$$= \int_\Omega \left(\Phi\left(F^{-1}\left(u - \sum_i H_i\mathbf{e}_i'\phi[k](x[k] + \chi_i)\right)\right)\middle/ \Phi(u)\right)$$
$$\times |F|^{-1}\Phi(u)\, du.$$

Let $\zeta = F^{-1}(u - \sum_i H_i\mathbf{e}_i'\phi[k](x[k] + \chi_i))$. The Jacobian of the transformation is $|F|$. Substituting $|F|d\zeta = du$ we obtain

$$\bar{E}\left[\frac{|F|^{-1}\Phi(F^{-1}(y[k] - \sum_i H_i\mathbf{e}_i'\phi[k](x[k] + \chi_i)))}{\Phi(y[k])}\middle| \mathcal{F}[k]\right]$$
$$= \int_\Omega \Phi(\zeta)\, d\zeta = 1$$

and the value of $\bar{E}[\bar{\lambda}[k] \,|\, \mathcal{F}[k]]$ is one as well:

$$\bar{E}[\bar{\lambda}[k] \,|\, \mathcal{F}[k]] = 1. \tag{9.12}$$

With the denominator equal to one, only $\bar{E}[\bar{\lambda}[k]I(y[k] \le t)I(z[k] = \mathbf{e}_i) \,|\, \mathcal{F}[k]]$ need be evaluated:

$$\mathcal{P}(y[k] \le t, z[k] = \mathbf{e}_i \,|\, \mathcal{F}[k]) = \bar{E}[\bar{\lambda}[k]I(y[k] \le t)I(z[k] = \mathbf{e}_i) \,|\, \mathcal{F}[k]].$$

Again, separate the factors involving $y[k]$ from those involving $z[k]$. First, for the modal-state measurements, we get

$$S\bar{E}[z[k]'\mathbf{P}\phi[k]I(z[k] = \mathbf{e}_i) \,|\, \mathcal{F}[k]] = (\mathbf{P}\phi[k])_i$$

or

$$\mathcal{P}(z[k] = \mathbf{e}_i \,|\, \mathcal{F}[k]) = (\mathbf{P}\phi[k])_i. \tag{9.13}$$

Second, for the base-state measurements we have

$$\mathcal{P}(y[k] \le t \,|\, \mathcal{F}[k])$$
$$= \bar{E}\left[I(y[k] \le t)|F|^{-1}\frac{\Phi(F^{-1}(y[k] - \sum_i H_i\mathbf{e}_i'\phi[k](x[k] + \chi_i)))}{\Phi(y[k])}\middle| \mathcal{F}[k]\right]$$

$$= \int_{\Omega} I(u \leq t) \frac{\Phi(F^{-1}(u - \sum_i H_i \mathbf{e}_i' \phi[k](x[k] + \chi_i)))}{\Phi(u)} |F|^{-1} \Phi(u)\, du$$

$$= \int_{\Omega} I\left(F\zeta + \sum_i H_i \mathbf{e}_i' \phi[k](x[k] + \chi_i) \leq t \right) \Phi(\zeta)\, d\zeta$$

So $y[k]$ is an $\mathbf{N}(\sum_i H_i \mathbf{e}_i' \phi[k](x[k] + \chi_i), R_x)$ random variable under $\mathcal{P}$, and the $\mathcal{P}$-observation model is that given in (9.3) and (9.4).

Suppose f is a scalar-valued function of the base-state. As before, its expectation could be computed using either $\mathcal{P}$ or $\bar{\mathcal{P}}$:

$$E[\mathbf{e}_i' \phi[k+1] f(x[k+1]) \,|\, \mathcal{G}[k+1]]$$
$$= \frac{\bar{E}[\mathbf{e}_i' \phi[k+1] \bar{\Lambda}[k+1] f(x[k+1]) \,|\, \mathcal{G}[k+1]]}{\bar{E}[\bar{\Lambda}[k+1] \,|\, \mathcal{G}[k+1]]}. \tag{9.14}$$

The denominator has nothing to do with f and is simply a normalizing factor. The numerator, $\bar{E}[\mathbf{e}_i' \phi[k+1] \bar{\Lambda}[k+1] f(x[k+1]) \,|\, \mathcal{G}[k+1]]$, will be thought of as the $\mathcal{G}[k+1]$-conditional expectation of $f(x[k+1])$ with respect to an *unnormalized* $\mathcal{G}[k+1]$-conditional probability density, $q^i[k+1](x[k+1])$:

$$\bar{E}[\mathbf{e}_i' \phi[k+1] \bar{\Lambda}[k+1] f(x[k+1]) \,|\, \mathcal{G}[k+1]] = \int_{\Omega} f(z) q^i[k+1](z) dz. \tag{9.15}$$

If the $\{q^i[k+1]; i \in \mathbf{S}\}$ were known, the joint density $\times$ mass function of the zygostate would be

$$p^i[k+1](z) = \frac{q^i[k+1](z)}{\sum_j \int_{\Omega} q^j[k+1](u) du}, \tag{9.16}$$

from which $\hat{x}[k+1]$ and $\hat{\phi}[k+1]$ along with higher moments can be calculated.

To find $\{q^i[k+1](z); i \in \mathbf{S}\}$, expand (9.15) by replacing $\phi[k+1]$, $\bar{\Lambda}[k+1]$, and $x[k+1]$ with their values. This yields

$$\bar{E}[\mathbf{e}_i' \phi[k+1] \bar{\Lambda}[k+1] f(x[k+1]) \,|\, \mathcal{G}[k+1]]$$
$$= \bar{E} \Bigg[\mathbf{e}_i' (\Pi \phi[k] + m[k+1]) \bar{\Lambda}[k] z[k+1]' \mathbf{P} \mathbf{e}_i |F|^{-1}$$
$$\times \Phi(F^{-1}(y[k+1] - H_i(x[k+1] + \chi \mathbf{e}_i))) \frac{S}{\Phi(y[k+1])}$$
$$\times f\left(\sum_j (A_j x[k] + C_j w[k+1]) \mathbf{c}_j' \phi[k] - \chi(\mathbf{c}_i - \phi[k]) \right) \Bigg| \mathcal{G}[k+1] \Bigg]. \tag{9.17}$$

It is true that $\sum_j \phi[k]'\mathbf{e}_j \equiv 1$. Substituting this into (9.17), we obtain

$$\begin{aligned}
&\bar{E}\left[\mathbf{e}_i'\phi[k+1]\bar{\Lambda}[k+1]f(x[k+1])\,|\,\mathcal{G}[k+1]\right] \\
&= \bar{E}\Bigg[\sum_j (\phi[k]'\mathbf{e}_j)(\mathbf{e}_i'\Pi\mathbf{e}_j)\bar{\Lambda}[k]z[k+1]'\mathbf{P}\mathbf{e}_i|F|^{-1} \\
&\quad\times \Phi(F^{-1}(y[k+1]-H_i(x[k+1]+\boldsymbol{\chi}\mathbf{e}_i))) \\
&\quad\times \frac{S}{\Phi(y[k+1])} f(A_jx[k]+C_jw[k+1]-\boldsymbol{\chi}(\mathbf{e}_i-\mathbf{e}_j))\,\Bigg|\,\mathcal{G}[k+1]\Bigg].
\end{aligned}$$

Under $\bar{\mathcal{P}}$, $\{\mathcal{G}[k]\}$ is uninformative with respect to the zygostate. Since an unnormalized distribution is sought, factors common to all regimes can and will be ignored in what follows:

$$\begin{aligned}
&\bar{E}[\mathbf{e}_i'\phi[k+1]\bar{\Lambda}[k+1]f(x[k+1])\,|\,\mathcal{G}[k+1]] \\
&= \int_\Omega \sum_j \Pi_{ij}z[k+1]'\mathbf{P}_{.i}\Phi(F^{-1}(y[k+1] \\
&\quad - H_i(A_j\zeta + C_jw - \boldsymbol{\chi}(\mathbf{e}_i-\mathbf{e}_j)+\boldsymbol{\chi}\mathbf{e}_i))) \\
&\quad \times f(A_j\zeta + C_jw - \boldsymbol{\chi}(\mathbf{e}_i-\mathbf{e}_j))q^j[k](\zeta)\Phi(w)\,d\zeta\,dw.
\end{aligned}$$

To simplify this, make the change of variable $z = A_j\zeta + C_jw - \chi_i + \chi_j$. Then $d\zeta\,dw = |C_j|^{-1}d\zeta dz$ and

$$\begin{aligned}
&\int_\Omega \sum_j \Pi_{ij}z[k+1]'\mathbf{P}_{.i}|C_j|^{-1}\Phi(F^{-1}(y[k+1]-H_i(z+\chi_i))) \\
&\quad \times f(z)q^j[k](\zeta)\Phi(C_j^{-1}(z-A_j\zeta+\chi_i-\chi_j))d\zeta\,dz \\
&= \int_\Omega f(z)q^i[k+1](z)\,dz.
\end{aligned}$$

Hence

COM-estimator

$$\begin{aligned}
q^i[k+1](z) = &\sum_j \Pi_{ij}z[k+1]'\mathbf{P}_{.i}|C_j|^{-1}\Phi(F^{-1}(y[k+1] \\
&-H_i(z+\chi_i)))\int_\Omega \Phi(C_j^{-1}(z-A_j\zeta+\chi_i-\chi_j)) \\
&\times q^j[k](\zeta)\,d\zeta. \qquad (9.18)
\end{aligned}$$

Equation (9.18) is the recurrence formula that delineates the COM-estimator. Variants on (9.18) have been derived for kindred plant representations:

- If $\chi = 0$, see [EE99, Equation (4)].
- If $\chi = 0$ and the noise in $\{z[k]\}$ is Gaussian, see [EDS96, Equation (8)].
- If $\chi = 0$, the noise in $\{z[k]\}$ is Gaussian, and $g[k] = y[k] + z[k]$, see [EvdH99, Equation (10)].

Unfortunately, (9.18) is not an algorithm of the form we seek – although recursive, it is infinite dimensional.

To approximate (9.18) with a finite-dimensional recurrence, suppose $q[k]$ is an unnormalized Gaussian sum. The number of terms in the sum is limited only by the computational complexity permitted in the application. We will assume that N terms will suffice in every regime:

$$q^j[k](\zeta) = \sum_{l=1}^{N} \alpha_l^j[k] |D_l^j[k]|^{\frac{1}{2}} \exp - \frac{1}{2}(\zeta - m_l^j[k])' D_l^j[k](\zeta - m_l^j[k]). \tag{9.19}$$

In (9.19), the unnormalized $\mathcal{G}[k]$-condition probability of the event

$$\{x[k] \in [\zeta, \zeta + d\zeta], \phi[k] = \mathbf{e}_j\}$$

is given under $\mathcal{P}$ by a sum of N Gaussian pattern functions,

$$\{\mathbf{N}_\zeta(m_l^j[k], P_l^j[k]); l \in \mathbf{N}\}$$

with means $m_l^j[k]$ and positive covariances $P_l^j[k] = D_l^j[k]^{-1}$. In this approximation, $m_l^j[k]$ translates the lth pattern function, $D_l^j[k]$ adjusts its shape, and N circumscribes the span of the sum. The pattern functions are weighted by $\{\alpha_l^j[k]; l \in \mathbf{N}\}$. All of the coefficients are $\mathcal{G}[k]$-adapted. There are NS elements in $q[k]$ though many could be zero.

It is shown in Appendix 2 that there is a recurrence formula for the coefficients, $\{\alpha_l^j[k], m_l^j[k], P_l^j[k]; j \in \mathbf{S}, l \in \mathbf{N}\}$. The recurrence can be most concisely stated in the mixed covariance-information form used to delineate the IMM:

base-state recurrence

Extrapolation:

$$m_l^i[k+1]^- = A_i m_l^i[k]; i \in \mathbf{S}, l \in \mathbf{N}; \tag{9.20}$$

$$P_l^i[k+1]^- = A_i P_l^i[k] A_i' + R_\chi(i); i \in \mathbf{S}, l \in \mathbf{N}. \tag{9.21}$$

Update:

$$d^i[k+1](l;j)^+ = d^j_l[k+1]^- + H_i' D_x (y[k+1] - H_i \chi_i) - D^j_l[k+1]^- (\chi_i - \chi_j);\ i, j \in \mathbf{S}, l \in \mathbf{N}; \quad (9.22)$$

$$D^i[k+1](l;j)^+ = D^j_l[k+1]^- + H_i' D_x H_i;\ i, j \in \mathbf{S}, l \in \mathbf{N}, \quad (9.23)$$

where

$$P^i[k+1](l;j)^+ = (D^i[k+1](l;j)^+)^{-1}$$

and

$$m^i[k+1](l;j)^+ = P^i[k+1](l;j)^+ d^i[k+1](l;j)^+.$$

As in the path-length-one MM approaches, there are S parallel Kalman filter extrapolations. But in the COM-estimator, each filter extrapolates N initial conditions. The PL1-MM filters extrapolate only one.

The update equations are more complex than those appearing in the MM filters. At an update, the possibility of a regime transition cannot be ignored: In the non-communicating PL1-MM filter, modal transitions are superfluous, and in the IMM, the transitions enter into the mixing step. In either multiple model algorithm (after adjusting for the set point influence on $\{y[k]\}$), the update equation would read

$$\Delta d^i[k+1]^+ = H_i' D_x y[k+1]; \qquad \Delta D^i[k+1]^+ = H_i' D_x H_i;\ i \in \mathbf{S}.$$

The update in the information matrix of the COM-estimator is close to that of the Kalman filter. There is a different indexing, but the increment in the information matrix is the same $(H_i' D_x H_i)$ from every (l, j).

The update of the conditional mean, $m^i[k+1](l;j)^+$, involves more bookkeeping. As with the information matrix, modal mixing in the COM-estimator takes place even in the absence of set point discontinuities: Even if $\chi = 0$, $d^i[k+1](l;j)^+ = d^j_l[k+1]^- + H_i' D_x y[k+1]$. For mode $\mathbf{e}_i$, the array of means, $\{m^i[k+1](l;j)^+\}$, has NS columns and each is identified with a particular modal transition. If $\mathbf{e}_i \mapsto \mathbf{e}_i$, $\Delta d^i_l[k+1]^+ = H_i' D_x (y[k+1] - H_i \chi_i)$, a simple translation. If $\{\phi[k]\}$ makes an $\mathbf{e}_j \mapsto \mathbf{e}_i$ transition, the base-state observation is referenced to the current modal-state $(y[k+1] - H_i \chi_i)$, and a base-state discontinuity, $(-\chi_i + \chi_j)$, must be included in the mean.

The weightings in $d^i[k+1](l;j)^+$ balance measurement fidelity and the shape of the pattern functions. The size of jump

$$d^i[k+1](l;j)^+ - d^j_l[k+1]^-$$

in the $(l;j)$-th subfilter will be increased if the measurement noise is small and if

$y[k+1]$ differs from $H_i\chi_i$. Similarly, the set point offset $(\chi_j - \chi_i)$ is important when the probability of the lth pattern function is concentrated in a small region (i.e., when $D_l^j[k+1]^-$ is big). Suppose the modal measurement is quite good and an $\mathbf{e}_j \mapsto \mathbf{e}_i$ transition occurs at time $t = (k+1)T$. In the absence of the base-state measurement, the extrapolation in the base-state mean should be

$$m^i[k+1](l; j)^+ \approx m_l^j[k+1]^- + \chi_j - \chi_i;$$

the mean translates by the amount of the base-state discontinuity. The size of the difference, $m_l^j[k+1]^+ - (m_l^j[k+1]^- + \chi_j - \chi_i)$, is a coarse measure of the influence of $y[k+1]$ since this difference is that between the corrected and uncorrected extrapolation.

Modal mixing is isolated in the IMM but is distributed in the COM-estimator. It appears in the filter update in (9.22) and (9.23) and also in the weighting coefficients, $\{\alpha^i[k+1](l; j)^+; i, j \in \mathbf{S}, l \in \mathbf{N}\}$. The equation is

modal mixing

$$\begin{aligned}\alpha^i[k+1](l; j)^+ = &\alpha_l^j[k]\Pi_{ij} z[k+1]'\mathbf{P}_{.i} \big| D_l^j[k](D_l^j[k] \\ &+ A_j' D_\chi(j) A_j)^{-1} P^i[k+1](l; j)\big|^{\frac{1}{2}} \\ &\times \exp -\frac{1}{2}\{m^i[k+1](l; j)' D^i[k+1](l; j) \\ &\times m^i[k+1](l; j) - (m_l^j[k+1]^- \\ &+ \chi_j - \chi_i)' D_l^j[k+1]^- (m_l^j[k+1]^- + \chi_j - \chi_i) \\ &+ (y[k+1] - H_i\chi_i)' D_x (y[k+1] - H_i\chi_i)\}. \quad (9.24)\end{aligned}$$

The weighting coefficient $\alpha^i[k+1](l; j)^+$ is proportional to its predecessor $(\alpha_l^j[k])$, proportional to the probability of an $\mathbf{e}_j \mapsto \mathbf{e}_i$ transition (Π_{ij}), and proportional to the probability that the modal observation was generated from $\phi[k+1] = \mathbf{e}_i$ $(z[k+1]'\mathbf{P}_{.i})$. Similar factors appear in the IMM and in the image-enhanced IMM [EE99, LDB98].

In the IMM, the update of the modal probabilities depends exclusively upon the behavior of the subfilter residuals, $r^j[k+1] = y[k+1] - H\hat{x}^j[k+1]^-$. The smaller the residuals in a particular subfilter are, weighted by the associated information matrix, the more likely the associated mode is thought to be. The unnormalized modal update has the form

$$q[k] = \hat{\phi}[k]^- * [|D^i[k+1]^-| \exp\left\{-\frac{1}{2} r^i[k+1]' D^i[k+1]^- r^i[k+1]\right\}. \quad (9.25)$$

The coefficients in (9.24) are responsive, not to the residuals themselves, but instead to the effect these residuals have on the update. The weight $\alpha^i[k+1](l;j)^+$ is generated by a continuing exponential product of the quadratic differences of the base-state mean before and after a base-state measurement (a difference of quadratic forms, not a quadratic form of the differences). With coefficients thus determined, the unnormalized $\mathcal{G}[k+1]$-conditional distribution of the zygostate can be written immediately:

$$q^i[k+1]^+ = \sum_{j,l} \alpha^i[k+1](l;j)^+ \mathbf{N}(m^i[k+1](l;j)^+, P^i[k+1](l;j)^+). \tag{9.26}$$

From (9.26), the various moments of interest can be computed, that is,

$$\hat{\phi}[k+1] = \left[\sum_{j,l} \alpha^i[k+1]^+(l;j) \Big/ \sum_{i,j,l} \alpha^i[k+1]^+(l;j) \right] \tag{9.27}$$

and

$$\hat{x}[k+1] = \sum_{i,j,l} \alpha^i[k+1]^+(l;j) m^i[k+1](l;j)^+ \Big/ \sum_{i,j,l} \alpha^i[k+1]^+(l;j). \tag{9.28}$$

Unfortunately, there are not N terms in (9.26) but NS. To satisfy the complexity constraint, pruning and/or merging must be used to reduce the number of terms. Computational experience with (9.26) is wanting. For now we will simply keep the largest term from each modal transition hypothesis. For all $i, j \in \mathbf{S}$ let $l(i, j)^*$ satisfy:

$$\alpha^i[k+1](l(i,j)^*; j)^+ \geq \alpha^i[k+1](l;j)^+.$$

(This is the maximizing index and will be assumed to be unique.) The next iteration of the COM-estimator begins with a set of S terms ($N = S$):

updated distribution

$$q^i[k+1] = \sum_j \alpha^i[k+1](l(i,j)^*; j)^+ \times \mathbf{N}(m^i[k+1](l(i,j)^*; j)^+, P^i[k+1](l(i,j)^*; j)^+). \tag{9.29}$$

The COM-estimator is inherently more complex than the path-length-one MM filter or the IMM. The underlying Gaussian sum approximation to the

$\mathcal{G}[k]$-distribution of the zygostate requires S times as many terms as do these MM filters. The use of pruning instead of merging can make the estimates of the base-states less robust than those found in the IMM. However, the approximation preserves the identity of the modal state more clearly. In applications in which the regime probabilities are important, the COM-estimator is a useful alternative to other MM filters.

9.2 Gaussian Minimum Shift Keying

To illustrate the utility of the COM approach, consider an unconventional application in which we seek to determine the modal-state instead of the base-state. The block diagram of a mobile communication link is shown in Figure 9.1. A source generates a primitive data sequence, $\{\iota[k]\}$, that is symmetric, iid, and bipolar: $\iota[k] \in \{-1, 1\}$. There is a companion time-continuous process $\{\iota_t\}$, which is constant on intervals, $t \in [kT, (k+1)T)$, with $\iota_{kT} = \iota[k]$. The binary process is recast by a transmit filter to create the baseband information signal. This signal is modulated, transmitted, and then demodulated to yield the baseband signal at the receiver. This latter is sampled to generate an observation sequence $\{y[k]\}$ from which $\{\iota[k]\}$ is reconstructed.

An advantage accruing to a properly chosen transmit filter is that the radio frequency (RF) spectrum required for transmission is significantly reduced. Denote the impulse response of the transmit filter by $\{g_t\}$. The baseband signal at the transmitter, $\{f_t\}$, is the convolution of $\{g_t\}$ and $\{\iota_t\}$. A popular (noncausal) transmit filter used in Gaussian minimum shift keying (GMSK) mobile applications is selected from the parametric family

$$g_t = (\alpha/\sqrt{\pi})\epsilon^{-(\alpha t)^2};\ t \in (-\infty, \infty).$$

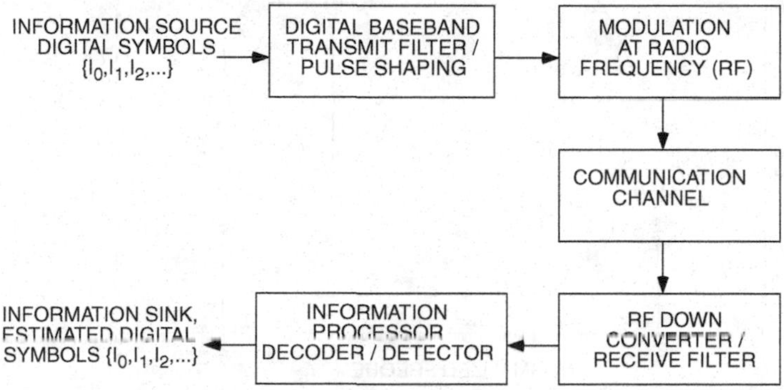

Figure 9.1. A wireless communication link.

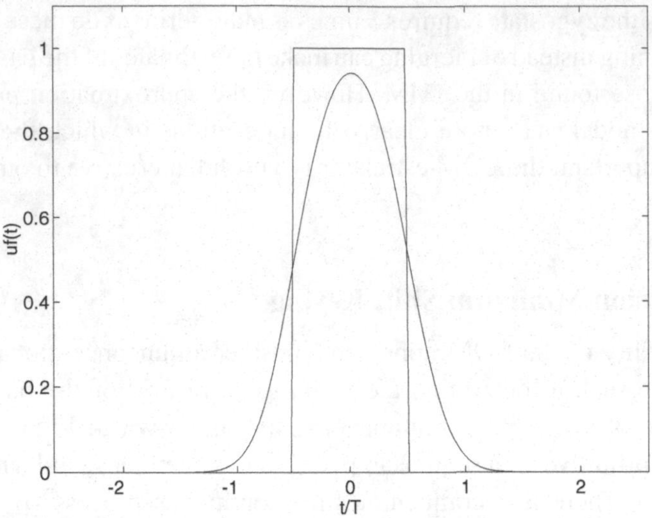

Figure 9.2. Impulse response of the transmit filter for $\beta = 0.5$.

The index for the transmit filter is more often given in terms of β, the bandwidth of the filter relative to the bit rate: $\alpha = \pi(2\ln 2)\beta T$. Figure 9.2 shows a centered unit pulse of length T along with the output of the transmit filter, $\{I_\beta(t);\ \beta = 0.5\}$. (The unit pulse corresponds to the output of a transmit filter with $\beta = \infty$.) The (normalized) power spectral density of the baseband GMSK($\beta = 0.5$) signal is contrasted in Figure 9.3 with direct transmission (GMSK($\beta = \infty$,) also called

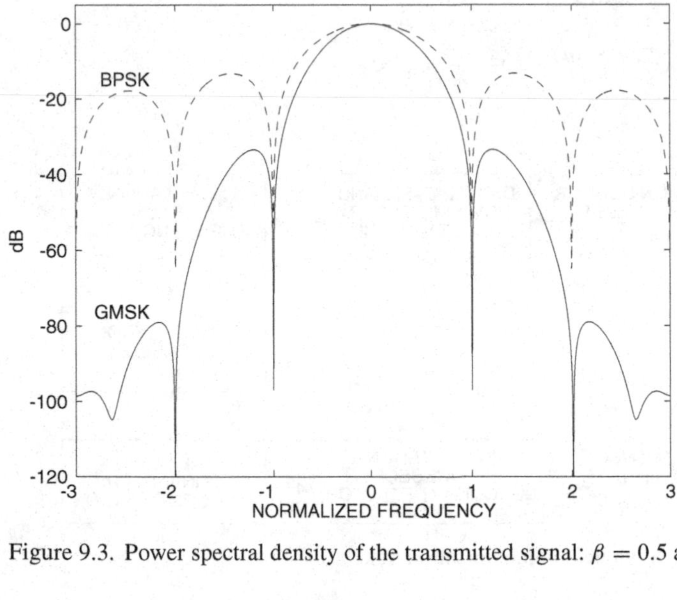

Figure 9.3. Power spectral density of the transmitted signal: $\beta = 0.5$ and $\beta = \infty$.

binary phase shift keying (BPSK)). The reduction in spectral bandwidth is apparent [Pro95, MH81].

In the basic GMSK link, the baseband signal is given by

$$f_t = \sum_{i=-\infty}^{\infty} \iota_i I_\beta(t - iT). \tag{9.30}$$

At any time t, the baseband signal is given as an infinite and noncausal weighted sum of the data symbols. In the temporal bin identified with the ith symbol, f_t is composed of a term related to the current symbol, $\iota_i I_\beta(t - iT)$, along with a contribution from all of the other symbols, $\sum_{j \neq i} \iota_j I_\beta(t - jT)$. This out-of-bin aggregate, called *intersymbol interference* (ISI), contaminates the baseband signal and makes interpreting it more difficult. Fortunately, as Figure 9.2 shows clearly, when $\beta = 0.5$, the influence of the interfering symbols for $|i - j| \geq 1$ is so small that they can be safely ignored. So for $t \in [kT, (k+1)T)$,

$$f_t \approx \sum_{i=k-1}^{k+1} \iota_i I_\beta(t - (k - i)T). \tag{9.31}$$

Although the out-of-bin symbols are viewed as interference in elementary analyses, they can be interpreted as part of the signal if the modal state space is properly constructed. Order the combinations of $(\iota_{t+T}, \iota_t, \iota_{t-T})$ in the natural way beginning with $(-1, -1, -1)$, and let $\phi_t = \mathbf{e}_i$ if $(\iota_{t+T}, \iota_t, \iota_{t-T})$ takes on its ith value. Equation (9.31) can be written

$$f_t \approx H_t \phi_t. \tag{9.32}$$

In this construction, the modal-state is eight dimensional for $\beta = 0.5$ even though the symbol dimension is two. Nevertheless, defining ϕ_t in this way converts the energy in the out-of-bin symbols into part of the observation.

Figure 9.4 displays the set of eight possible baseband signals over two symbol periods as the modal-state varies over its range; Figure 9.4 is called the *eye chart*. Suppose the effect of the channel is to add white Gaussian noise and the baseband signal at the receiver is sampled at the center of the temporal bin once every period. The observation is

$$y[k] = f[k] + n[k]. \tag{9.33}$$

There are various ways in which the original data sequence could be reconstructed at the receiver. The simplest is if $y[k]$ is positive, $\iota[k] = 1$ is declared with $\iota[k] = -1$ otherwise. This detector uses the maximum value of $\{f_t\}$, and the untoward effects of additive channel noise should be reduced. Unfortunately, this detector ignores ISI and it is sensitive to the timing of the sampler.

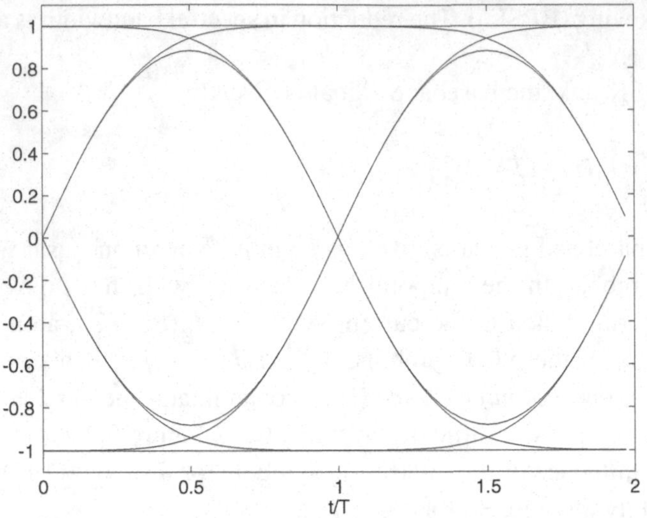

Figure 9.4. The eye chart showing ISI for $\beta = 0.5$.

In narrowband mobile communication systems, the transmission protocol is more complex. Instead of $\{f_t\}$, a signal equivalent to the integral of $\{f_t\}$ is transmitted. Integration increases the smoothness of the baseband process but creates more temporal overlap. In continuous phase modulation (CPM) systems, the data sequence is expressed within a phase process $\{\theta_t\}$ given by (e.g., [Kor90, Equation (47)])

$$\theta_t = 2\pi \sum_{i=-\infty}^{\infty} h_i \iota_i p(t - iT), \tag{9.34}$$

where θ_t is to be interpreted modulo 2π, and h_i is the modulation index for the ith symbol (h_i is assumed identically 1/2 henceforth). For GMSK, the phase-response-function, $\{p_t\}$, is a scaled integral of $I_\beta(t)$ (see [YMF88, Equation (7)]). The variation in $p(t)$ is spread over the temporal bins that support $I_\beta(t)$, and the scaling is such that $p(-\infty) = 0$ and $p(\infty) = 1/2$. Again, conserving bandwidth with small values of β leads to significant intersymbol interference.

Mobile channels are subject to deep fades and sudden phase changes. Conventional methods of channel equalization have not proven to be adequate for signal reconstruction. Although the support for the variation of $p(t)$ extends across multiple temporal bins, for a specified β there is a finite – and usually small – integer L such that $p(\tau) \approx 1/2$ for $\tau \geq LT$ and $p(\tau) \approx 0$ for $\tau \leq -LT$. Neglecting small terms, this permits us to write θ_t as (see [YMF88, Table 1] or [SW84])

$$\theta_t = \frac{1}{2}\pi \sum_{i=-\infty}^{k-L-1} \iota_i + \pi \sum_{i=-L}^{L} \iota_{k+i} p(t - (k+i)T) \tag{9.35}$$

for times in the kth temporal bin. Note that there are $2L$ interfering symbols.

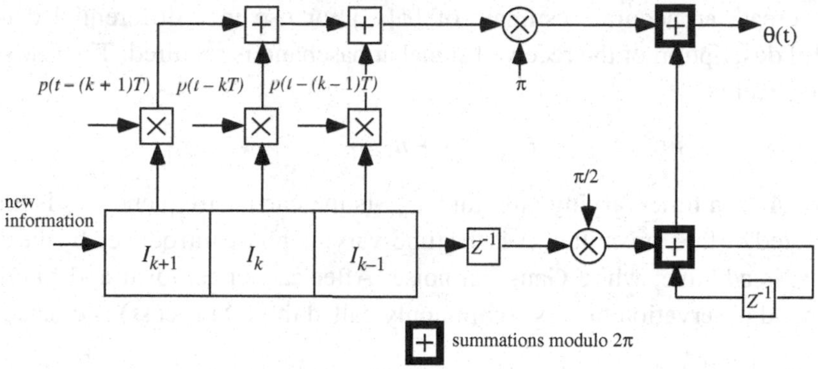

Figure 9.5. Partial response CPM with memory.

Equation (9.35) can be displayed pictorially as in Figure 9.5 (shown for $L = 1$, which will suffice for $\beta = 0.5$). The first term in (9.35) is the memory-state shown to the right of the figure as a delay and an accumulator. The status of this term can be represented by a unit vector ρ_t in $\mathbb{R}^4$ (e.g., if $\frac{1}{2}\pi \sum_{i=-\infty}^{k-L-1} \iota_i = 0$, then $\rho_t = e_1$ and so on). The second term in (9.35) is displayed as the shift register in the figure and its status can be represented by a 2^{2L+1}-dimension unit vector, ζ_t; for example, if arrayed in increasing order, when all relevant $\{\iota[k]\}$ are -1,

$$\pi h \sum_{i=-L}^{L} \iota_{k+i}\, p(t-(k+i)T) = -\pi h \sum_{i=-L}^{L} p(t-(k+i)T)$$

and $\zeta_t = \mathbf{e}_1$. The modal-state at any time is simply $\phi_t = \rho_t \otimes \zeta_t$. The unit vectors in the modal-state space are also called the phase-states.

As constructed, the phase-state process is Markovian and can be particularized using conventional procedures: $\phi[k] = \Pi\phi[k-1] + m[k]$, where Π is the transition matrix and $\{m[k]\}$ is a martingale difference. For future reference, observe that only two elements of $\Pi_{i.}$ are nonzero for each i – any phase-state has precisely two progenitors. Note also that the state space of the phase process is of higher dimension than is the symbol space to account for the presence of ISI and memory: For $L = 1$ ($\beta = 0.5$), the dimension of the state space of the phase process is thirty-two whereas that of the symbol process is two.

Detection is easy if a good approximation to $\{\phi[k]\}$ exists. In the CPM system the information is conveyed in the integral (or sum) of the symbol process. Because of its form, $\Delta\rho[k] = \pm\pi/2$, depending on whether ι_{k-L-1} is ± 1. Hence, faithful identification of the data symbol is achieved by differential detection:

$$\hat{\iota}_{k-L-1} = \mathrm{sgn}(\Delta\rho[k])$$

[SW83]. There is an intrinsic delay of $(L+1)T$ using differential detection, but this is of no consequence.

To create an accurate estimate of $\{\phi[k]\}$ for use in a differential detector, a careful description of the received signal at baseband is required. The raw signal at the receiver is

$$s_t = A_t \sin(2\pi f_c t + \theta_t + \varphi_t) + n_t, \tag{9.36}$$

where A_t is a time-varying amplitude, f_c is the carrier frequency, θ_t is the phase generated by the transmitter, φ_t is the time-varying phase introduced by the channel, and n_t is additive, white Gaussian noise. After carrier removal and filtering, the baseband observation process (commonly called the IQ process) is created:

$$y_t = \begin{bmatrix} I_t \\ Q_t \end{bmatrix} = K_t \begin{bmatrix} \cos(\theta_t)\cos(\varphi_t) - \sin(\theta_t)\sin(\varphi_t) \\ \cos(\theta_t)\sin(\varphi_t) + \sin(\theta_t)\cos(\varphi_t) \end{bmatrix} + \begin{bmatrix} n_1 \\ n_2 \end{bmatrix}, \tag{9.37}$$

where $\{K_t, \varphi_t\}$ is the channel gain-phase process, and $\{n_1\}$ and $\{n_2\}$ are noise processes.

The channel gain-phase vector delineates the influence of the transmission link, and we will refer to it as the channel state. The channel state has continuous, though volatile, sample paths and will be identified with the plant state of the hybrid model. The phase-state sequence has a finite state space and will be identified with the modal-state of the hybrid model. In a twist on the hybrid estimation problem, primary interest lies in estimating the modal-state instead of the plant state.

Despite the name, the plant state dynamics do not lend themselves to modeling as in (1.1). Researchers have found it easier to describe the channel characteristics in an altered state space. Let $\{X_t\}$ be a two-dimensional, stationary, Gaussian random process with independent and identically distributed components. The power spectral density (PSD) common to both components of $\{X_t\}$ has the peculiar form given in [ACW73, Equation (1)]:

$$\Phi_X(f) = \begin{cases} \dfrac{C}{[1-(f/f_D)^2]^{\frac{1}{2}}} & \text{if } |f| \leq f_D, \\ 0 & \text{otherwise,} \end{cases} \tag{9.38}$$

where C is a scaling coefficient and f_D is the Doppler shift corresponding to the vehicle speed and carrier wavelength. In these terms, $\{y_t\}$ can be expressed as

$$y_t = \begin{bmatrix} I_t \\ Q_t \end{bmatrix} = \begin{bmatrix} \cos(\theta_t) & -\sin(\theta_t) \\ \cos(\theta_t) & \sin(\theta_t) \end{bmatrix} \begin{bmatrix} X_1 \\ X_2 \end{bmatrix} + \begin{bmatrix} n_1 \\ n_2 \end{bmatrix}. \tag{9.39}$$

The conventional approach to system modeling would translate the PSD in (9.38) into an LGM model with the state variables of the shaping filter acting as the base-state. The form of the PSD precludes doing this. Although the PSD of $\{X_t\}$ can be duplicated to any degree of accuracy using an LGM model, the more accurate the approximation (the higher the order), the more complex the resulting detection algorithm will be. Let us render $\{X_t\}$ in terms of a pair of uncoupled, continuous shaping filters; both are lightly damped ($\xi = 0.05$), second-order transfer functions with resonant frequency f_D. The state (the base-state), x_t, of the composite channel model is a 4-vector with component ordering: X_1 and its rate; respectively, X_2 and its rate. The time-discrete model for the base-state process is found by sampling the two uncoupled time-continuous filters:

$$x[k+1] = Ax[k] + w[k+1], \tag{9.40}$$

where $\{w[k]\}$ is a Gaussian white noise sequence with covariance $R_\chi > 0$.

The baseband observation at time kT can be written concisely as

$$y[k] = \sum_i H_i \mathbf{e}_i' \phi[k] x[k] + n[k], \tag{9.41}$$

where H_i is found by integrating (9.39) and (9.41), and $\{n[k]\}$ is a white-noise Gaussian sequence with covariance $R_x > 0$. There is no separate modal measurement in this application: $\mathcal{Y}[k] \equiv \mathcal{G}[k]$.

The COM-algorithm first generates $\{\hat{\phi}[k]\}$ and then uses a one-bit differential detector to create an estimate of $\{\iota_{k-L-1}\}$. Begin with $q[k]$, a Gaussian sum approximation to the unnormalized $\mathcal{Y}[k]$-density of the zygostate. Pruning retains one term from each permissible modal transition. The COM-estimator is initialized with the family

$$\{\alpha_l^j[k], m_l^j[k], D_l^j[k];\ j, l \in \mathbf{S}\},$$

but because of the nature of Π, all but two of the $\alpha^i[k+1](j;l)$ vanish for every $i \in \mathbf{S}$; thus the actual dimension of the algorithm is $2S$, not S^2.

The COM-algorithm is simplified considerably when $\chi = 0$:

base-state recurrence

Extrapolation:

$$m_l^i[k+1]^- = Am_l^i[k];\ i, l \in \mathbf{S}, \tag{9.42}$$

$$P_l^i[k+1]^- = AP_l^i[k]A' + R_\chi;\ i, l \in \mathbf{S}. \tag{9.43}$$

Update:

$$\Delta d_l^i[k+1]^+ = H_i' D_x y[k+1];\ i, l \in \mathbf{S}, \tag{9.44}$$

$$\Delta D_l^i[k+1]^+ = H_i' D_x H_i;\ i, l \in \mathbf{S}. \tag{9.45}$$

The weighting coefficients can be simplified too. There are no modal observations and factors common across modes can be neglected. For $i, j, l \in \mathbf{S}$:

$$\alpha^i[k+1](l;j)^+ = \alpha_l^j[k]\Pi_{ij}\big|D_l^j[k](D_l^j[k] + A'D_\chi A)^{-1}P^i[k+1](l;j)\big|^{\frac{1}{2}} \times \exp -\frac{1}{2}\Delta\big(m_l^i[k+1]'D_l^i[k+1]m_l^i[k+1]\big).$$

From $\hat{\phi}[k]$, $\hat{\rho}[k]$ can be produced by direct enumeration. Signal detection is accomplished using a differential rule:

Let $\iota = \arg\max(\hat{\rho}[k])_i$.

$$\text{If} \quad |\hat{\rho}_\iota[k] - \hat{\rho}_{\iota-1}[k]| \le |\hat{\rho}_\iota[k] - \hat{\rho}_{\iota+1}[k]|$$

$$\text{then} \quad \hat{\imath}_{k-L-1} = 1,$$

$$\text{otherwise} \quad \hat{\imath}_{k-L-1} = -1.$$

Although not required for detection, the $\mathcal{Y}[k]$-conditional mean of $x[k]$ can be found by summing on i and suitably normalizing the result. From $\hat{x}[k]$ estimates of the channel coefficients can be produced by inverting the map $(K_t, \varphi_t) \Rightarrow X_t$. Because of their formal construction, these estimates of channel gain and phase should not be confused with the associated $\mathcal{Y}_t$-conditional means. Recall that $\hat{X}_1[k] = \mathbf{e}_1'\hat{x}[k]$ and $\hat{X}_2[k] = \mathbf{e}_3'\hat{x}[k]$. Plausible estimates of channel gain and phase are

channel amplitude:

$$\hat{K}[k] = (\hat{x}[k]'(\mathbf{E}_1 + \mathbf{E}_3)\hat{x}[k])^{\frac{1}{2}}, \tag{9.46}$$

channel phase:

$$\hat{\varphi}[k] = \tan^{-1}(\mathbf{e}_3'\hat{x}[k]/\mathbf{e}_1'\hat{x}[k]). \tag{9.47}$$

The COM-estimator does provide the information required for adaptive channel estimation [Ses94], but it differs from "the most common approach (to blind equalization which filters) the output by an estimate of the inverse channel followed by some decision device" [GW95].

Pruning is required because, for every hypothesis $\mathbf{e}_i$ in the state space of $\phi[k]$, there are $2S$ terms in the Gaussian sum; for example, in $\alpha^i[k+1]^+(l;j)$, (j,l)

Table 9.1. *The mobile environment.*

Carrier frequency	836 MHz
Symbol rate	8 kHz
Doppler shift	38.7 Hz
Vehicle speed	50 km/h

runs over 2S. The pruning procedure selects the subhypothesis with the largest weight from each of the possible progenitors. This is a very simple rule and can be made more sophisticated if the sample rate is sufficiently slow. The COM-estimator avoids the geometric growth in dynamic hypotheses that bedevils other approaches involving blind equalization (see [GW95], [LM90, p. 778], and [Ilt90, p. 199]).

9.3 An Example

To be more concrete, consider the mobile environment described in Table 9.1 [Ken96]. Figures 9.6 and 9.7 show a sample path of the channel gain and phase over a range of 5,000 symbol times. On this interval, the channel gain decreases by more than 20 dB with a similar variation in the phase.

Let us contrast the performance of three algorithms:

BPSK: When the transmit filter has a unity transfer function, $\Delta\theta_t$ is 0 or π according as ι is ± 1 (*differentially encoded, binary-phase-shift-keying*). There is no ISI because the influence of the kth symbol is confined to the

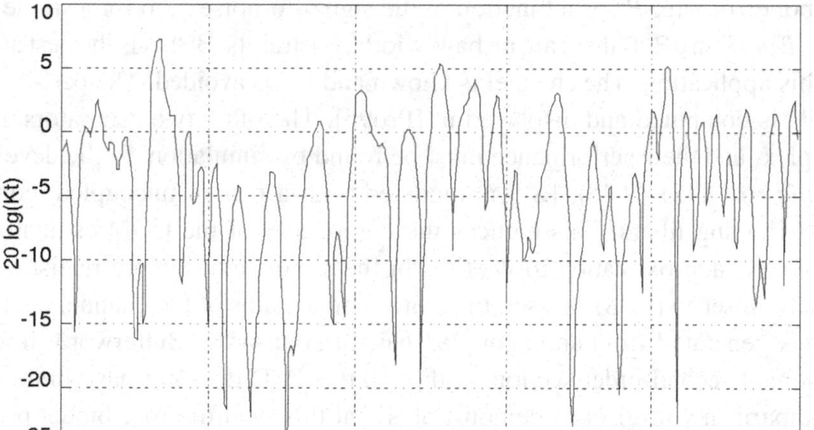

Figure 9.6. Channel gain; fading channel $v = 50$ km/hr, $f_c = 836$ MHz.

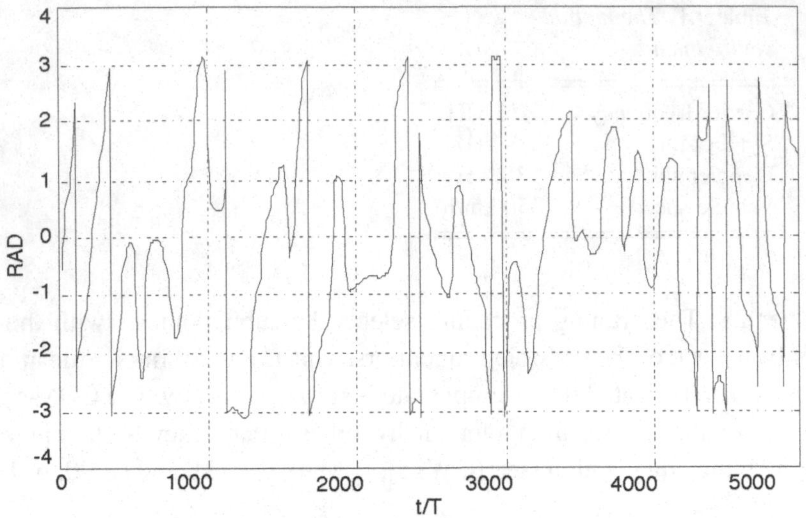

Figure 9.7. Channel phase; fading channel $v = 50$ km/hr, $f_c = 836$ MHz.

kth temporal bin. There is considerable spectral spreading as indicated in Figure 9.3. The signal will be coherently detected after passing over a fading channel with perfect knowledge of channel phase.

GMSK: This is the basic GMSK link with a measured channel state ($\hat{x}_t = x_t$) and $\beta = 0.5$ [KSE94]. The algorithm is linear and of low dimension.

COM: This is the COM-estimator given in (9.42)–(9.45) for the fading channel and $\beta = 0.5$.

The performance of a detection algorithm is commonly phrased in terms of the symbol error rate, P_e, as a function of the signal-to-noise ratio for a single symbol time, E_b/N_o in dB. If there are no bandwidth constraints, BPSK is the best algorithm for this application. The channel is known and ISI is avoided. The performance of BPSK is computed and displayed in [Pro95]. The other two estimators are more complex, and their performance must be found by simulation. In the development leading to (9.42)–(9.45), $\{X_t\}$ is modeled with a pair of uncoupled, continuous LGM shaping filters. This reduces the complexity of the COM-estimator, but it is a coarse approximation to $\Phi_X(f)$. In the simulation, a more realistic channel model (closer to (9.38)) is used to create sample paths of the channel coefficients: $\{X_t\}$ is generated from an uncoupled pair of sixth-order Butterworth filters with a resonant second-order section. Although this PSD does not have finite support, a comparison with $\Phi_X(f)$ demonstrates that this twelfth-order model provides a reasonable channel representation.

Consider the performance of the decoding algorithms in the mobile environment. GMSK was derived assuming the channel state is measured and uses $\beta = 0.5$ to

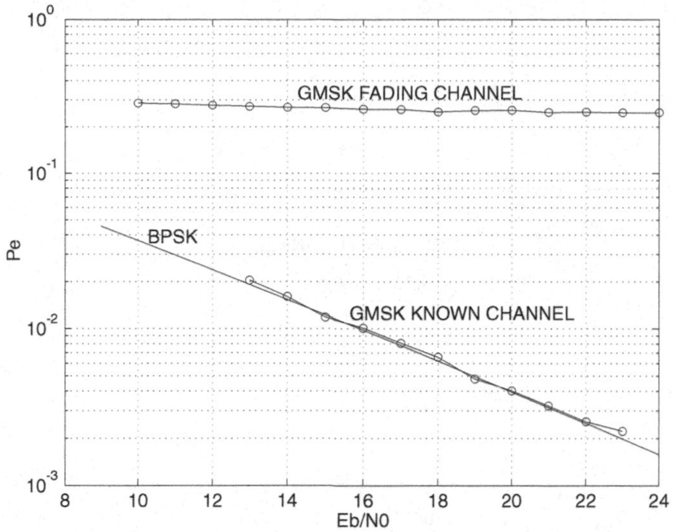

Figure 9.8. BPSK and GMSK in the fading channel.

reduce the transmission bandwidth. The lower two curves in Figure 9.8 show the bit error rate as a function of channel SNR (E_b/N_o) for the mobile channel. The performance of GMSK is essentially indistinguishable from that of BPSK despite the considerable intersymbol interference in GMSK and the spectral spreading in BPSK. This shows that when the channel state is known, the performance degradation is minimal even when the bandwidth is severely constrained.

When GMSK is used in a channel with random state, its error rate degrades to an unacceptable degree. The top curve in Figure 9.8, labeled "fading channel," shows what happens when GMSK measures the actual channel fade but wrongly sets the channel phase to zero. The channel phase is volatile (see Figure 9.7), and when GMSK ignores the changing phase it samples the eye curve at a less advantageous point. Although $P_e < 0.5$, it does not decrease significantly with SNR.

COM integrates the channel dynamics into its description of the operating environment, though not very meticulously since it uses an uncoupled pair of second-order filters for the channel processes. Figure 9.9 contrasts the performance of COM with BPSK in the severe fading created by a mobile platform. The COM-estimator knows neither the channel gain nor the phase. (This is an advantage given to BPSK.) Nor does it use the extended transmission band required by BPSK. Nevertheless, COM is seen to have a performance comparable to BPSK with a much lower sensitivity to channel fades than found in GMSK. (P_e degrades by a factor of less than two for $E_b/N_o = 15$.)

Although detection using COM requires no explicit channel identification, it is interesting to see how well it accomplishes this ancillary task. This is done in [Ken96]

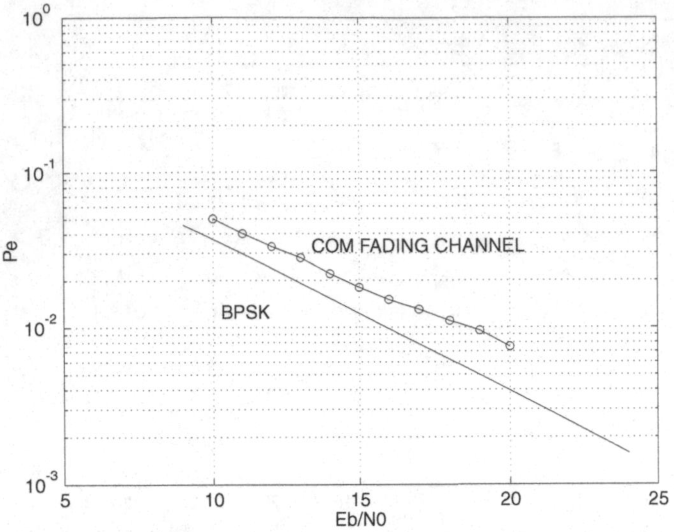

Figure 9.9. BPSK and COM in the fading channel.

using the estimate of channel gain as computed in (9.46) and (9.47). Despite the fact that COM uses only a simplified model in its development, the channel estimates are surprisingly good. At a low SNR ($E_b/N_0 = 10$), the amplitude estimate lags the true amplitude and tends to overshoot to some degree. The phase estimate has intervals of sizable error. Except at a discrete set of transition times, these errors are essentially of magnitude $k\pi$ and are of little consequence in differential detection.

In this mobile channel, COM provides a good estimate of the signal symbol sequence. Even though the channel state model is very coarse, the estimator is able to track the true channel state closely. Although not discussed here, the algorithm is simple enough that multiple samples/symbol could be processed. Other studies (not shown) indicate that the performance of the COM-algorithm does not significantly degrade when the spectral specifications on the channel are made tighter (e.g., $\beta = 0.25$).

Appendix 1

PME Derivation Details

A1.1 Introduction

In this appendix, the PME algorithm will be developed. Let $(\Omega, \mathcal{F}, \mathcal{P})$ be a probability space and let $\{\mathcal{F}_t\}$ be a right-continuous filtration that generates $\mathcal{F}$: $\sigma\{\vee_t \mathcal{F}_t\} = \mathcal{F}$. There are several exogenous processes defined on this space, all right-continuous and $\mathcal{F}_t$-adapted. The comprehensive state is expressible in terms of these primary processes, and it has a decomposition into the base-state, x_t, and the modal-state, ϕ_t. The *zygostate* of the hybrid system, (x_t, ϕ_t), is associated with a similarly partitioned observation, $\mathrm{vec}(y_t, z_t) = g_t$.

There are several subfiltrations on $\{\mathcal{F}_t\}$ that are important in what follows. The observation $\{g_t\}$ generates $\{\mathcal{G}_t\}$. With perfect modal-state measurement (i.e., $g_t = \mathrm{vec}(y_t, \phi_t)$), we indicate the g_t-generated filtration by $\{\mathcal{G}_t^{\phi}\}$. With no modal measurements (i.e., $g_t = y_t$), we would have $\{\mathcal{G}_t\} = \{\mathcal{Y}_t\}$.

Associated with an observation process, there is an innovations process. Of most interest here is the $\mathcal{G}_t$-innovations process, and this can be partitioned compatibly with the observation: $\nu_t = \mathrm{vec}(\nu_x, \nu_{\phi})$. It will be assumed that $\{\nu_t\}$ also generates $\{\mathcal{G}_t\}$. We seek a causal map from $\{\nu_t\}$ to $\{\hat{x}_t\}$ and $\{\hat{\phi}_t\}$. This is a difficult construction because of the nonlinear and discontinuous system dynamics.

The approximation used in developing the PME derives from the premise that the proximate source of information about ϕ_t is conveyed by $\{z_t\}$: If $\{\zeta_t\}$ is a process for which the $\mathcal{G}_t^{\phi}$-expectation is independent of $\{y_t\}$, the contribution of $d\nu_x$ to $d\hat{\zeta}_t$ is negligible as compared to that of $d\nu_{\phi}$. This does not mean, however, that the PME ignores the information content in $\{y_t\}$: We are not equating the $\mathcal{G}_t$-expectation of ζ_t with the $\mathcal{Z}_t$-expectation. Rather, we are saying that the coincident contribution of the base-state measurement to $d\hat{\zeta}_t$ is comparatively small. This hypothesis is not always appropriate, but when it is, an implementable algorithm can be developed.

The average rate of modal measurements is λ/s. These observations are distributed across the S modal categories. The $\mathcal{F}_t$-vector observation rate is $\boldsymbol{\lambda}_t = \lambda \mathbf{P} \phi_t$

(or $\boldsymbol{\lambda}_t = h'\phi_t$ where $h = \lambda\mathbf{P}'$), the ith component of which is the probability that the observation is classified in the ith modal bin. To develop the PME note that the observations can be written in two ways:

$$dy_t = H(x_t + \chi\phi)\,dt + dn_x \quad \text{and} \quad dz_t = h'\phi\,dt + dn_\phi, \tag{A1.1}$$

where the observation noise, $\{\mathrm{vec}(n_x, n_\phi)\}$, is an $\mathcal{F}_t$-martingale. Alternatively,

$$dy_t = H(\hat{x}_t + \chi\hat{\phi}_t)\,dt + d\nu_x \quad \text{and} \quad dz_t = h'\hat{\phi}_t\,dt + d\nu_\phi, \tag{A1.2}$$

where the innovations process, $\{\mathrm{vec}(\nu_x, \nu_\phi)\}$, is a $\mathcal{G}_t$-martingale.

The processes $\{\nu_x\}$ and $\{n_x\}$ are continuous, and their predictable quadratic variation is simply written

$$d\langle n_x, n_x; \mathcal{F}_t\rangle_t = d\langle \nu_x, \nu_x; \mathcal{G}_t\rangle_t = R_x\,dt. \tag{A1.3}$$

The optional quadratic variation of both processes is also $R_x t$, and it will be supposed that the system is such that $R_x > 0$.

The modal noise process, $\{n_\phi\}$, is discontinuous, and its predictable quadratic variation is random:

$$\begin{aligned} d\langle n_\phi, n_\phi; \mathcal{F}_t\rangle_t &= E[dz\,dz' \,|\, \mathcal{F}_t] = \mathrm{diag}(\boldsymbol{\lambda}_t)\,dt, \\ d\langle n_\phi, n_\phi; \mathcal{G}_t\rangle_t &= \mathrm{diag}(\hat{\boldsymbol{\lambda}}_t)\,dt = R_\phi\,dt, \end{aligned} \tag{A1.4}$$

where (A1.4) is taken to be the definition of R_ϕ. The system will be assumed to be such that each component of $\boldsymbol{\lambda}_t$ is positive, and consequently, $R_\phi > 0$. The optional quadratic variation requires the additional accumulation of the jumps $\{\Delta n_\phi \Delta n_\phi'\}$.

The modal observations enter the PME in a peculiar manner. Let $\{\vartheta_t\}$ be a process that is constant between modal observations and with increments $\Delta\vartheta_t = h(\hat{\boldsymbol{\lambda}}_t^{-1} * \Delta z_t)$, where $\hat{\boldsymbol{\lambda}}_t^{-1}$ is understood componentwise.

In what follows, a special notation is useful to make the PME more intuitive. Recall that x_t (respectively ϕ_t) is the state with $\mathcal{G}_t$-mean and error given by $\hat{x}_t$ and $\tilde{x}_t$ (respectively $\hat{\phi}_t$ and $\tilde{\phi}_t$). The second moments of these variables are $R_{x\phi}(t) = E[x_t\phi_t' \,|\, \mathcal{G}_t]$ and $P_{x\phi}(t) = E[\tilde{x}_t\phi_t' \,|\, \mathcal{G}_t]$, with similar definitions for $P_{xx}(t)$, $R_{xx}(t)$, etc. Let us extend this notational convention to third moments as follows. Consider the two vectors x_t, ϕ_t and the scalar ϕ_i. Let us display third moments as

$$\begin{aligned} R_{x\phi\phi_i}(t) &= E[x_t\phi_t'\phi_i \,|\, \mathcal{G}_t], \\ P_{x\phi\phi_i}(t) &= E[\tilde{x}_t\tilde{\phi}_t'\tilde{\phi}_i \,|\, \mathcal{G}_t], \end{aligned}$$

with similar definitions for $P_{xx\phi_i}(t)$, $R_{xx\phi_i}(t)$, etc.

This notation can be extended to compound moments; for example,

$$P_{(xx\phi_i)\phi_m} = E[\widetilde{(x_t x_t' \phi_i)}\tilde{\phi}_m \,|\, \mathcal{G}_t],$$
$$P_{(x\phi_i)x\phi_m} = E[\widetilde{(x_t\phi_i)}\tilde{x}_t'\tilde{\phi}_m \,|\, \mathcal{G}_t].$$

The parentheses in the subscript act as delimiters in the calculation. One fourth moment appears in what follows:

$$P_{xx\phi_i\phi_m} = E[\tilde{x}_t\tilde{x}_t'\tilde{\phi}_r\tilde{\phi}_m \,|\, \mathcal{G}_t].$$

A1.2 Modal Estimation

To begin, consider the modal-state estimation algorithm:

modal estimation

Between modal measurements:

$$d\hat{\phi}_t = Q'\hat{\phi}_t\, dt. \tag{A1.5}$$

At a modal measurement:

$$\hat{\phi}_t^+ = \hat{\phi}_t^- * \Delta\vartheta_t. \tag{A1.6}$$

A1.2.1 Discussion

The modal dynamics are given in (1.12):

$$d\phi = Q'\phi\, dt + dm_t,$$

where $\{m_t\}$ is a purely discontinuous $\mathcal{F}_t$-martingale. Decomposing the semimartingale $\{\hat{\phi}_t\}$ (see [Ell82, Theorem 18.11]), we have

$$d\hat{\phi}_t = E[d\phi_t \,|\, \mathcal{G}_t] + \gamma_t\, d\nu_t,$$

where γ_t is a $\mathcal{G}_t$-predictable matrix process. Since $\{\hat{\phi}_t\}$ is trivially adapted to $\{\mathcal{G}_t^\phi\}$ (is a ϕ-dominated moment), $\{\nu_x\}$ can be neglected and the estimate can be written

$$d\hat{\phi} = E[d\phi_t \,|\, \mathcal{G}_t] + \gamma_{\phi\phi}\, d\nu_\phi.$$

For notational convenience, let $Q'\phi_t = F_\phi$. Then $E[d\phi_t \,|\, \mathcal{G}_t] = \hat{F}_\phi\, dt$. So we have

$$d\hat{\phi}_t = \hat{F}_\phi\, dt + \gamma_{\phi\phi}\, d\nu_\phi, \tag{A1.7}$$

where $d\nu_\phi = h'\tilde{\phi}\, dt + dn_\phi$. To find $\gamma_{\phi\phi}$ explicitly, the formalism used successfully by Elliott [Ell82] will be employed. First note that

$$d(\phi z') = d\phi\, z' + \phi\, dz' + d\phi\, dz'. \tag{A1.8}$$

It will be assumed throughout that changes in $\{\phi_t\}$ are not simultaneous with receipt of an observation so that $d\phi_t\, dz_t' = 0$. But

$$\begin{aligned} d\phi\, z' &= (Q'\phi\, dt + dm)z' \\ &= F_\phi z'\, dt + d\mu, \end{aligned} \tag{A1.9}$$

where, in the development that follows, the process $\{\mu_t\}$ will represent the matrix $\mathcal{F}_t$-martingale appropriate to the equation in which it appears. The second term in (A1.8) can be written

$$\phi\, dz' = \phi\phi' h\, dt + \phi\, dn_\phi'.$$

Combining these equations, it follows that

$$d(\phi z)' = (\phi\phi' h + F_\phi z')\, dt + d\mu.$$

Taking the $\mathcal{G}_t$-expectation of this expression, we obtain

$$E[d(\phi z') \,|\, \mathcal{G}_t]/dt = R_{\phi\phi} h + \hat{F}_\phi z'. \tag{A1.10}$$

We can express $E[d(\phi z') \,|\, \mathcal{G}_t]$ in another way. Using (A1.7), we have

$$d\hat{\phi}\, dz' = (\gamma_{\phi\phi}\, d\nu_\phi)\, dn_\phi'.$$

So

$$d\hat{\phi}\, dz' = \gamma_{\phi\phi}\, dn_\phi\, dn_\phi'.$$

It is a direct calculation to show that

$$\begin{aligned} d\hat{\phi}\, z' &= \hat{F}_\phi z'\, dt + d\mu, \\ \hat{\phi}\, dz' &= \hat{\phi}\phi' h\, dt + d\mu. \end{aligned}$$

Collecting the terms, we get

$$d(\hat{\phi} z') = (\gamma_{\phi\phi}\, dn_\phi\, dn_\phi' + \hat{\phi}\phi' h + \hat{F}_\phi z')\, dt + d\mu,$$

and taking the $\mathcal{G}_t$-expectation of this leads to

$$E[d(\hat{\phi} z') \,|\, \mathcal{G}_t]/dt = \gamma_{\phi\phi} R_\phi + \hat{\phi}\hat{\phi}' h + \hat{F}_\phi z'. \tag{A1.11}$$

The predictable compensators, (A1.10) and (A1.11), must be equal:

$$\gamma_{\phi\phi} R_\phi + \hat{\phi}\hat{\phi}' h + \hat{F}_\phi z' = (P_{\phi\phi} + \hat{\phi}\hat{\phi}')h + \hat{F}_\phi z'.$$

From this it follows that

$$\gamma_{\phi\phi} = P_{\phi\phi} h R_{\phi}^{-1}. \tag{A1.12}$$

Substituting this into (A1.7) yields

$$d\hat{\phi}_t = Q'\hat{\phi}_t\,dt + P_{\phi\phi} h R_{\phi}^{-1}\,d\nu_{\phi}, \tag{A1.13}$$

which can be simplified by observing that

$$R_{\phi}^{-1} h'\hat{\phi}_t = \mathrm{diag}(\hat{\boldsymbol{\lambda}}_t^{-1})\hat{\boldsymbol{\lambda}}_t = \mathbf{1}.$$

So

$$hR_{\phi}^{-1}h'\hat{\phi} = \lambda\mathbf{P}'\mathbf{1} = \lambda\mathbf{1}.$$

Also, $P_{\phi_i\phi}\mathbf{1} = 0$ for every $i \in \boldsymbol{S}$. Hence, $\gamma_{\phi\phi}\hat{\boldsymbol{\lambda}}_t = 0$. So $d\hat{\phi}_t = Q'\hat{\phi}_t\,dt$ if $dz = 0$, and $\Delta\hat{\phi}_t = P_{\phi\phi}\Delta\vartheta_t$ if $dz \neq 0$. Continuing, we have

$$\Delta\hat{\phi}_t = (\mathrm{diag}(\hat{\phi}_t) - \hat{\phi}_t\hat{\phi}_t')\Delta\vartheta_t. \tag{A1.14}$$

However,

$$\hat{\phi}_t'\Delta\vartheta_t = \hat{\phi}_t' h R_{\phi}^{-1}\Delta z_t = (R_{\phi}^{-1}h'\hat{\phi}_t)'\Delta z_t = \mathbf{1}'\Delta z_t = 1.$$

So

$$\hat{\phi}_t^{+} - \hat{\phi}_t^{-} = \mathrm{diag}(\hat{\phi}_t^{-})\Delta\vartheta_t - \hat{\phi}_t^{-}.$$

This can be written

$$\hat{\phi}_t^{+} = \hat{\phi}_t^{-} * \Delta\vartheta_t.$$

Note that the PME modal estimate is $\mathcal{Z}_t$-adapted.

A1.3 Base-State Estimation

The dynamic equation of the base-state estimate is:

base-state estimation

Between modal measurements:

$$d\hat{x}_t = \left(\sum_i A_i R_{x\phi_i} + B_i\left(u_t\hat{\phi}_i - \upsilon P_{\phi\phi_i}\right) + \rho'\hat{\phi}_t\right)dt + P_{x\chi}H'R_x^{-1}d\nu_x. \tag{A1.15}$$

At a modal measurement:

$$\Delta\hat{x}_t = P_{x\phi}\,\Delta\vartheta_t. \tag{A1.16}$$

A1.3.1 Discussion

The base-state dynamics are given in (A1.15):

$$dx_t = \sum_i \left((A_i x_t + B_i(u_t - \upsilon\tilde{\phi}_t))\phi_i + \rho'\phi_t\right) dt$$
$$+ \sum_i C_i\phi_i \, dw_t + \sum_{i,l}(M(i,l)x_t + \theta(i,l))\phi_i \, dm_l.$$

To determine the equation of evolution of the $\mathcal{G}_t$-conditional mean of x_t, both components of the innovations process must be exploited:

$$d\nu_t = \text{vec}(d\nu_\phi, d\nu_x).$$

Let

$$\sum_i \left(A_i x_t + B_i(u_t - \upsilon\tilde{\phi}_t)\right)\phi_i + \rho'\phi_t = F_x.$$

Direct calculation indicates that

$$\hat{F}_x = \sum_i \left(A_i R_{x\phi_i} + B_i(u_t\hat{\phi}_i - \upsilon P_{\phi\phi_i})\right) + \rho'\hat{\phi}_t. \tag{A1.17}$$

But $E[dx_t \,|\, \mathcal{G}_t] = \hat{F}_x \, dt$ and

$$d\hat{x}_t = \hat{F}_x \, dt + \gamma_\phi \, d\nu_\phi + \gamma_x \, d\nu_x. \tag{A1.18}$$

The filter will be complete when the $\{\gamma_\phi, \gamma_x\}$ process is determined.

The observation can be reformed as an array, $g_t = (z_t, y_t)$, from which $d(x_t g_t') = dx\, g' + x\, dg' + dx\, dg'$. Since $d\phi_t\, dz_t' = 0$ and $dw_t\, dn_x' = 0$, it follows that $dx\, dg' = 0$. However,

$$dx\, g' = (F_x z'\, dt, F_x y'\, dt) + d\mu,$$

where $\{\mu_t\}$ is again a matrix $\mathcal{F}_t$-martingale. Also,

$$x\, dg' = (x\phi' h\, dt + x\, dn_\phi', x\chi' H'\, dt + x\, dn_x').$$

Collecting terms, we obtain

$$\frac{d}{dt}(xg') = (x\phi' h + F_x z', x\chi' H' + F_x y') + \frac{d\mu}{dt}.$$

Taking the $\mathcal{G}_t$-expectation of this gives

$$\frac{E[d(x_t g_t') \,|\, \mathcal{G}_t]}{dt} = (R_{x\phi} h + \hat{F}_x z', R_{x\chi} H' + \hat{F}_x y'). \tag{A1.19}$$

Let us again write $E[d(x_t g_t') \mid \mathcal{G}_t]$ using (A1.18). Look first at the martingale product term, that is,

$$d\hat{x}\, dg' = (\gamma_\phi\, d\nu_\phi + \gamma_x\, d\nu_x)(dn_\phi', dn_x').$$

The components of the innovations process can be decomposed into

$$d\nu_x = H\tilde{\chi}_t\, dt + dn_x.$$

But $\{n_x\}$ is continuous, and $\{n_\phi\}$ is purely discontinuous. It follows, therefore, that

$$d\hat{x}\, dg' = (\gamma_\phi\, dn_\phi\, dn_\phi', \gamma_x R_x\, dt).$$

It is a direct calculation to show that

$$d\hat{x}\, g' = (\hat{F}_x z'\, dt, \hat{F}_x y'\, dt) + d\mu,$$

$$\hat{x}\, dg' = (\hat{x}\phi' h\, dt, \hat{x}\chi' H\, dt) + d\mu.$$

Collecting terms, we get

$$d(\hat{x}g') = \left(\gamma_\phi\, dn_\phi\, dn_\phi' + \hat{x}\phi' h + \hat{F}_x z', \quad \hat{F}_x y' + \hat{x}\chi' H + \gamma_x R_x\right) dt + d\mu. \tag{A1.20}$$

Taking the $\mathcal{G}_t$-expectation of (A1.20) yields

$$\frac{E[d(\hat{x}_t g') \mid \mathcal{G}_t]}{dt} = (\gamma_\phi R_\phi + \hat{x}\hat{\phi}' h + \hat{F}_x z', \hat{F}_x y' + \hat{x}\hat{\phi}' H + \gamma_x R_x). \tag{A1.21}$$

The predictable compensators, (A1.18) and (A1.21), must be equal. Therefore,

$$\gamma_\phi R_\phi + \hat{x}\hat{\phi}' h + \hat{F}_x z' = (P_{x\phi} + \hat{x}\hat{\phi}')h + \hat{F}_x z'. \tag{A1.22}$$

From this it follows that

$$\gamma_\phi = P_{x\phi} h R_\phi^{-1}, \tag{A1.23}$$

with a similar calculation showing that

$$\gamma_x = P_{x\chi} H' R_x^{-1}. \tag{A1.24}$$

But $\gamma_\phi h' \hat{\phi} = 0$. Simplifying yields (A1.15)–(A1.16).

A1.4 Some Mixed Moments

The implementation of the filter requires that the gains $\gamma_x = P_{x\chi} H' R_x^{-1}$ and $\gamma_\phi = P_{x\phi} h R_\phi^{-1}$ be evaluated. This in turn requires the dynamic equations for three canonical covariance matrices,

$$P_{xx}(t) \triangleq E[\tilde{x}_t\tilde{x}_t' \,|\, \mathcal{G}_t],$$

$$P_{x\phi}(t) \triangleq E[\tilde{x}_t\tilde{\phi}_t' \,|\, \mathcal{G}_t],$$

and

$$P_{xx\phi_m}(t) \triangleq E[\tilde{x}_t\tilde{x}_t'\tilde{\phi}_m \,|\, \mathcal{G}_t],$$

to be deduced. The dynamic equations involve a variety of higher moments (e.g., $P_{x\phi\phi_m}(t) = E[\tilde{x}_t\tilde{\phi}_t'\tilde{\phi}_m \mid \mathcal{G}_t]$, $P_{xx\phi_m\phi_r}(t) = E[\tilde{x}_t\tilde{\phi}_t'\tilde{\phi}_r\tilde{\phi}_m \mid \mathcal{G}_t]$, etc.), but these sequent moments can be expressed as algebraic functions of the canonical moments. For use in the later development, the required identities will now be displayed.

A1.4.1 The Mixed Third Moment $P_{x\phi\phi_m}$

The relevant equation is

$$P_{x\phi\phi_m} = P_{x\phi_m}(\mathbf{e}_m - \hat{\phi})' - \hat{\phi}_m P_{x\phi}.$$

Clearly,

$$(P_{x\phi\phi_m})_{i,j} = E[\tilde{x}_i\tilde{\phi}_j\tilde{\phi}_m \,|\, \mathcal{G}_t] = E[\tilde{x}_i(\phi_j - \hat{\phi}_j)(\phi_m - \hat{\phi}_m) \,|\, \mathcal{G}_t].$$

If $j = m$, this becomes

$$\begin{aligned}(P_{x\phi\phi_m})_{i,j} &= E\left[\tilde{x}_i(\phi_m - 2\hat{\phi}_m\phi_m + \hat{\phi}_m^2) \,\middle|\, \mathcal{G}_t\right] \\ &= (P_{x\phi})_{i,m}(1 - 2\hat{\phi}_m),\end{aligned}$$

while if $j \neq m$

$$\begin{aligned}(P_{x\phi\phi_m})_{i,j} &= E[\tilde{x}_i(-\hat{\phi}_m\phi_j - \hat{\phi}_j\phi_m + \hat{\phi}_j\hat{\phi}_m) \,|\, \mathcal{G}_t] \\ &= -(P_{x\phi})_{i,j}\hat{\phi}_m - (P_{x\phi})_{i,m}\hat{\phi}_j.\end{aligned}$$

A1.4.2 The Mixed Fourth Moment $P_{xx\phi_m\phi_r}$

Here we have

$$P_{xx\phi_m\phi_r} = \delta_{r,m}(P_{xx\phi_m} + P_{xx}\hat{\phi}_m) - \hat{\phi}_m P_{xx\phi_r} - \hat{\phi}_r P_{xx\phi_m} - P_{xx}\hat{\phi}_r\hat{\phi}_m.$$

Clearly,

$$(P_{xx\phi_m\phi_r})_{i,j} = E[\tilde{x}_i\tilde{x}_j\tilde{\phi}_r\tilde{\phi}_m \,|\, \mathcal{G}_t] = E[\tilde{x}_i\tilde{x}_j(\phi_r - \tilde{\phi}_r)(\phi_m - \tilde{\phi}_m) \,|\, \mathcal{G}_t].$$

Table A1.1. *PME moment identities.*

$$
\begin{aligned}
P_{x\phi\phi_m} &= P_{x\phi_m}(\mathbf{e}_m - \hat{\phi})' - \hat{\phi}_m P_{x\phi}\\
P_{\phi\phi\phi_m} &= ((\mathbf{e}_m - \hat{\phi})(\mathbf{e}_m - \hat{\phi})' - P_{\phi\phi})\hat{\phi}_m\\
P_{xx\phi_m\phi_r} &= \delta_{r,m}(P_{xx\phi_m} + P_{xx}\hat{\phi}_m) - \hat{\phi}_m P_{xx\phi_r} - \hat{\phi}_r P_{xx\phi_m} - P_{xx}\hat{\phi}_r\hat{\phi}_m\\
P_{(x\phi_i)x} &= P_{xx\phi_i} + P_{xx}\hat{\phi}_i + \hat{x}P_{\phi_i x}\\
P_{(x\phi_i)\phi} &= P_{x\phi\phi_i} + P_{x\phi}\hat{\phi}_i + \hat{x}P_{\phi_i\phi}\\
R_{xx\phi_i} &= P_{xx\phi_i} + P_{xx}\hat{\phi}_i + P_{x\phi_i}\hat{x}' + \hat{x}P'_{x\phi_i} + \hat{x}\hat{x}'\hat{\phi}_i\\
P_{(x\phi_i)x\phi_m} &= P_{xx\phi_i\phi_m} + \hat{\phi}_i P_{xx\phi_m} + \hat{x}P_{\phi_i x\phi_m} - P_{x\phi_i}P_{\phi_m x}\\
P_{(xx\phi_i)\phi_m} &= (\delta_{i,m} - \hat{\phi}_m)R_{xx\phi_i}
\end{aligned}
$$

If $m = r$, this becomes

$$
\begin{aligned}
(P_{xx\phi_m\phi_r})_{i,j} &= E\left[\tilde{x}_i\tilde{x}_j(\phi_m - 2\hat{\phi}_m\phi_m + \hat{\phi}_m^2) \mid \mathcal{G}_t\right]\\
&= (P_{xx\phi_m} + \hat{\phi}_m P_{xx})_{i,j}(1 - 2\hat{\phi}_m) + \hat{\phi}_m^2(P_{xx})_{i,j},
\end{aligned}
$$

while if $m \neq r$

$$
\begin{aligned}
(P_{xx\phi_m\phi_r})_{i,j} &= E[\tilde{x}_i\tilde{x}_j(-\hat{\phi}_m\phi_r - \hat{\phi}_r\phi_m + \hat{\phi}_r\hat{\phi}_m) \mid \mathcal{G}_t]\\
&= -\hat{\phi}_r(P_{xx\phi_m} + \hat{\phi}_m P_{xx})_{i,j} - \hat{\phi}_m(P_{xx\phi_r} + \hat{\phi}_r P_{xx})_{i,j}\\
&\quad + (P_{xx})_{i,j}\hat{\phi}_r\hat{\phi}_m.
\end{aligned}
$$

Continuing with this direct evaluation, we can produce the list of identities shown in Table A1.1.

A1.5 Error Dynamics

Before computing the three canonical higher moments, the $\mathcal{F}_t$-dynamics of $\{\tilde{x}_t\tilde{x}_t'\}$, $\{\tilde{x}_t\tilde{\phi}_t'\}$, and $\{\tilde{x}_t\tilde{x}_t'\tilde{\phi}_m\}$ are required. These and the underlying error dynamics of the $\{\tilde{x}_t\}$ and $\{\tilde{\phi}_t\}$ processes are tabulated in Tables A1.2 and A1.3.

Table A1.2. *Error process dynamics.*

$\{\tilde{x}_t\}$	$d\tilde{x}_t = \sum_i (A_i(\widetilde{x_t\phi_i}) + B_i(\tilde{s}_i u_t - v(\widetilde{\hat{\phi}\phi_i})) + (\rho' - \gamma_\phi h')\tilde{\phi}_t - \gamma_x H\tilde{\chi}_t)\,dt - \gamma_x\,dn_x - \gamma_\phi\,dn_\phi + \sum_i C_i\phi_i\,dw_t + \sum_{i,l}(M(i,l)x_t + \theta(i,l))\phi_i\,dm_l$
$\{\tilde{\phi}_t\}$	$d\tilde{\phi}_t = (Q' - \gamma_{\phi\phi}h')\tilde{\phi}_t\,dt - \gamma_{\phi\phi}\,dn_\phi + dm_t$

Table A1.3. *Mixed error process dynamics.*

$(\tilde{x}_t\tilde{x}_t')$	$d(\tilde{x}_t\tilde{x}_t') = \Big(\Big(\sum_i (\boldsymbol{A}_i(\widetilde{x_t\phi_i}) + B_i(\tilde{\phi}_i u_t - \boldsymbol{\upsilon}(\widetilde{\tilde{\phi}\phi_i}))) + (\boldsymbol{\rho}' - \gamma_\phi h')\tilde{\phi}_t - \gamma_x H\tilde{\chi}_t\Big) dt + \sum_i C_i\phi_i\, dw_t + \sum_{i,l} (M(i,l)x_t + \theta(i,l))\,\phi_i\, dm_l - \gamma_x\, dn_x - \gamma_\phi\, dn_\phi\Big)\tilde{x}_t' + \tilde{x}_t(\cdot)' + \Big(\gamma_x R_x \gamma_x' + \sum_i R_\chi(i)\phi_i\Big) dt + \gamma_\phi\, dn_\phi\, dn_\phi' \gamma_\phi' + \sum_{i,l,r}(M(i,l)x_t + \theta(i,l))(M(i,r)x_t + \theta(i,r))'\phi_i\, dm_l\, dm_r$
$(\tilde{x}_t\tilde{\phi}_t')$	$d(\tilde{x}_t\tilde{\phi}_t') = \Big(\Big(\sum_i (\boldsymbol{A}_i(\widetilde{x_t\phi_i}) + B_i(\tilde{\phi}_i u_t - \boldsymbol{\upsilon}(\widetilde{\tilde{\phi}\phi_i}))) + (\boldsymbol{\rho}' - \gamma_\phi h')\tilde{\phi}_t - \gamma_x H\tilde{\chi}_t\Big) dt - \gamma_x\, dn_x - \gamma_\phi\, dn_\phi + \sum_i C_i\phi_i\, dw_t + \sum_{i,l}(M(i,l)x_t + \theta(i,l))\phi_i\, dm_l\Big)\tilde{\phi}_t' + \tilde{x}_t((Q' - \gamma_{\phi\phi}h')\tilde{\phi}_t\, dt - \gamma_{\phi\phi}\, dn_\phi + dm)' + \gamma_\phi\, dn_\phi\, dn_\phi' \gamma_{\phi\phi}' + \sum_{i,l}(M(i,l)x_t + \theta(i,l))\phi_i\, dm_l\, dm'$
$(\tilde{x}_t\tilde{x}_t'\tilde{\phi}_m)$	$d(\tilde{x}_t\tilde{x}_t'\tilde{\phi}_m) = \Big[\Big(\Big(\sum_i (\boldsymbol{A}_i(\widetilde{x_t\phi_i}) + B_i(\tilde{\phi}_i u_t - \boldsymbol{\upsilon}(\widetilde{\tilde{\phi}\phi_i}))) + (\boldsymbol{\rho}' - \gamma_\phi h')\tilde{\phi}_t - \gamma_x H\tilde{\chi}_t\Big) dt - \gamma_x\, dn_x - \gamma_\phi\, dn_\phi + \sum_i C_i\phi_i\, dw_t + \sum_{i,l}(M(i,l)x_t + \theta(i,l))\phi_i\, dm_l\Big)\tilde{x}_t' + \tilde{x}_t(\cdot)' + \Big(\gamma_x R_x\gamma_x' + \sum_i R_\chi(i)\phi_i\Big) dt + \sum_{i,l,r}(M(i,l)x_t + \theta(i,l))(M(i,r)x_t + \theta(i,r))'\phi_i\, dm_l\, dm_r + \gamma_\phi\, dn_\phi\, dn_\phi'\gamma_\phi'\Big]\tilde{\phi}_m + \tilde{x}_t\tilde{x}_t'((Q' - \gamma_{\phi\phi}h')\tilde{\phi}_t\, dt - \gamma_{\phi\phi}\, dn_\phi + dm)_m - (\gamma_\phi\, dn_\phi\tilde{x}_t' + (\gamma_\phi\, dn_\phi\tilde{x}_t')' - \gamma_\phi\, dn_\phi\, dn_\phi'\gamma_\phi)(-\gamma_{\phi\phi}\, dn_\phi)_m + \Big(\Big(\sum_{i,l}(M(i,l)x_t + \theta(i,l))\phi_i\, dm_l\tilde{x}_t'\Big) + (\cdot)' + \sum_{i,l,r}(M(i,l)x_t + \theta(i,l))(M(i,r)x_t + \theta(i,r))'\phi_i\, dm_l\, dm_r\Big) dm_m$

A1.5.1 Discussion

It follows directly from Tables A1.2 and A1.3 that

$$d\tilde{x}_t = \left(\sum_i (A_i(\widetilde{x_t\phi_i}) + B_i(\tilde{\phi}_i u_t - v(\tilde{\phi}\phi_i - P_{\phi\phi_i}))) + \rho'\tilde{\phi}_t\right) dt$$

$$- \gamma_x\, d\nu_x - \gamma_\phi\, d\nu_\phi + \sum_i C_i\phi_i\, dw_t + \sum_{i,l}(M(i,l)x_t$$

$$+ \theta(i,l))\phi_i\, dm_l.$$

The innovations terms can be expanded in terms of $\mathcal{F}_t$-martingales to yield

$$d\tilde{x}_t = \left(\sum_i (A_i(x_t\phi_i - R_{x\phi_i}) + B_i(\tilde{\phi}_i u_t - v(\widetilde{\tilde{\phi}\phi_i}))) + \rho'\tilde{\phi}_t\right) dt$$

$$- \gamma_x\, dn_x - \gamma_\phi\, dn_\phi + \sum_i C_i\phi_i\, dw_t + \sum_{i,l}(M(i,l)x_t$$

$$+ \theta(i,l))\phi_i\, dm_l - \gamma_x H\tilde{\phi}_t\, dt - \gamma_\phi h'\tilde{\phi}_t\, dt.$$

To find the dynamic equation for $\{\tilde{x}_t\tilde{x}_t'\}$, observe that

$$(d\tilde{x}_t)\tilde{x}_t' = \left(\sum_i \left(A_i(\widetilde{x_t\phi_i}) + B_i(\tilde{\phi}_i u_t - v(\widetilde{\tilde{\phi}\phi_i}))\right.\right.$$

$$\left. + (\rho' Q' - \gamma_\phi h')\tilde{\phi}_t - \gamma_x H\tilde{\chi}_t\right) dt - \gamma_x\, dn_x - \gamma_\phi\, dn_\phi$$

$$\left. + \sum_i C_i\phi_i\, dw_t + \sum_{i,l}(M(i,l)x_t + \theta(i,l))\phi_i\, dm_l\right)\tilde{x}_t'.$$

Also,

$$(d\tilde{x}_t)d\tilde{x}_t' = \left(-\gamma_x\, dn_x - \gamma_\phi\, dn_\phi + \sum_i C_i\phi_i\, dw_t + \rho'\, dm_t\right)(\cdot)'.$$

Neglecting orthogonal products, we obtain

$$(d\tilde{x}_t)d\tilde{x}_t' = \left(\gamma_x R_x\gamma_x' + \sum_i R_\chi(i)\phi_i\right) dt + \sum_{i,l,r}(M(i,l)x_t$$

$$+ \theta(i,l))(M(i,r)x_t + \theta(i,r))'\phi_i\, dm_l\, dm_r + \gamma_\phi\, dn_\phi\, dn_\phi'\gamma_\phi',$$

and similarly,

$$d\tilde{\phi}_t = (Q' - \gamma_{\phi\phi}h')\tilde{\phi}_t\, dt - \gamma_{\phi\phi}\, dn_\phi + dm_t.$$

To find the dynamic equation for $\{\tilde{x}_t \tilde{\phi}'_t\}$, observe that

$$(d\tilde{x}_t)\tilde{\phi}'_t = \Bigg(\Bigg(\sum_i (A_i(\widetilde{x_t\phi_i}) + B_i(\tilde{\phi}_i u_t - v(\widetilde{\tilde{\phi}\phi_i})))$$

$$+(\rho' - \gamma_\phi h')\tilde{\phi}_t - \gamma_x H \tilde{\chi}_t\Bigg) dt - \gamma_x\, dn_x - \gamma_\phi\, dn_\phi$$

$$+\sum_i C_i\phi_i\, dw_t + \sum_{i,l}(M(i,l)x_t + \theta(i,l))\phi_i\, dm_l\Bigg)\tilde{\phi}'_t.$$

Also,

$$\tilde{x}_t(d\tilde{\phi}_t)' = \tilde{x}_t((Q' - \gamma_{\phi\phi}h')\tilde{\phi}_t\, dt - \gamma_{\phi\phi}\, dn_\phi + dm)'.$$

Finally,

$$(d\tilde{x}_t)d\tilde{\phi}'_t = \Bigg(-\gamma_x\, dn_x - \gamma_\phi\, dn_\phi + \sum_i C_i\phi_i\, dw_t$$

$$+\sum_{i,l}(M(i,l)x_t + \theta(i,l))\phi_i\, dm_l\Bigg)(-\gamma_{\phi\phi}\, dn_\phi + dm)'$$

$$= \gamma_\phi\, dn_\phi\, dn'_\phi \gamma'_{\phi\phi} + \sum_{i,l}(M(i,l)x_t + \theta(i,l))\phi_i\, dm_l\, dm'.$$

Direct substitution gives the dynamic equation for $d(\tilde{x}_t\tilde{\phi}'_t)$. The development of the equation for $d(\tilde{x}_t\tilde{x}'_t\tilde{\phi}_m)$ follows the same pattern.

Before computing the base-state error covariance matrix, some moment identities related to the discontinuous martingales are useful. These simplify some of the special versions of the PME.

A1.5.2 The Predictable Quadratic Variation of $\{m_t\}$

We have

$$\begin{aligned} E[\,dm\, dm' \,|\, \mathcal{F}_t] &= d\langle m, m; \mathcal{F}_t\rangle_t \\ &= [\mathrm{diag}(Q'\phi_t) - (\mathrm{diag}\phi_t)Q - Q'(\mathrm{diag}\phi_t)]\, dt \\ &= \boldsymbol{V}(\phi_t)\, dt, \end{aligned}$$

where $\boldsymbol{V}(\phi_t) = \sum_i \boldsymbol{V}(\mathbf{e}_i)\phi_i$ [EAM95, Appendix B]. The $\mathcal{G}_t$-predictable quadratic variation of $\{m_t\}$ is the expected value of $d\langle m, m; \mathcal{F}_t\rangle_t$:

$$d\langle m, m; \mathcal{G}_t\rangle_t = \sum_i \boldsymbol{V}(\mathbf{e}_i)\hat{\phi}_i\, dt = \boldsymbol{V}(\hat{\phi}_t)\, dt.$$

A1.5.3 The Predictable Cubic Variation of $\{m_t\}$

Let

$$E[\,dm\,dm'\,dm_k\,|\,\phi_t = \mathbf{e_r}] = \boldsymbol{U}_k(\mathbf{e_r})\,dt.$$

Then direct evaluation yields

$$\boldsymbol{U}_k(\mathbf{e}_r) = (\mathbf{e_r} - \mathbf{e}_k)\,Q_{rk}(\mathbf{e_r} - \mathbf{e}_k)' + [\mathbf{e_r} \otimes (\mathbf{e}'_\mathbf{r}Q) \\ + (\mathbf{e_r} \otimes (\mathbf{e}'_\mathbf{r}Q))' - \operatorname{diag}(\mathbf{e}'_\mathbf{r}Q)]\,\delta_{r,k}.$$

Hence,

$$E[\,dm\,dm'\,dm_k\,|\,\mathcal{F}_t] = \sum_r \boldsymbol{U}_k(\phi_r)\,dt,$$

and

$$E[\,dm\,dm'\,dm_k\,|\,\mathcal{G}_t] = \sum_r \boldsymbol{U}_k(\mathbf{e_r})\hat{\phi}_r\,dt = \boldsymbol{U}_k(\hat{\phi}_t)\,dt.$$

A1.5.4 The Predictable Cubic Variation of $\{n_t\}$

In contrast to $\{m_t\}$, only one component of $\{n_\phi\}$ changes at a time. Suppose that $\Delta z_t = \mathbf{e}_i$. Then $dn_\phi\,dn'_\phi(dn_i) = \mathbf{E_i}$. The probability of this event occurring if $\phi_{t-} = \mathbf{e}_k$ is $\lambda\mathbf{P}_{i,k}\,dt$. So

$$E[dn_\phi\,dn'_\phi(\,dn_i)\,|\,\phi = \mathbf{e}_k] = h_{ki}\mathbf{E_i}\,dt.$$

This can be written as

$$E[dn_\phi\,dn'_\phi(dn_i)\,|\,\mathcal{F}_t] = \boldsymbol{\lambda}_i\mathbf{E_i}\,dt,$$

$$E[dn_\phi\,dn'_\phi(dn_i)\,|\,\mathcal{G}_t] = \hat{\boldsymbol{\lambda}}_i\mathbf{E_i}\,dt.$$

A1.5.5 Moment Identities

Since $\sum_{i\in S}\hat{\phi}_i \equiv 1$, $\sum_{i\in S}\tilde{\phi}_i \equiv 0$. Therefore,

$$\sum_i P_{xx\phi_i} = 0,$$

$$P_{x\phi\phi_m}\mathbf{1} = 0,$$

and

$$\sum_r P_{xx\phi_m\phi_r} = \sum_m P_{xx\phi_m\phi_r} = 0.$$

A1.6 Base-State Covariance

The dynamic equation of the base-state error covariance is:

base-state covariance

Between modal measurements:

$$\frac{d}{dt}P_{xx}=\left(\sum_i (A_i(P_{xx\phi_i}+P_{xx}\hat{\phi}_i+\hat{x}_t P_{\phi_i x})+B_i(u_t P_{\phi i x}-\upsilon(P_{\phi x\phi_i} -P_{\phi x}\hat{\phi}_i)))+\rho' P_{\phi x}\right)+(\cdot)'-\gamma_x R_x\gamma_x'+\sum_i R_\chi(i)\hat{\phi}_i +\sum_{i,l} Q_{il}(M(i,l)R_{xx\phi_i}M(i,l)'+\theta(i,l)\hat{\phi}_i\theta(i,l)') +\sum_{i,l} Q_{il}(M(i,l)R_{x\phi_i}\theta(i,l)'+\theta(i,l)R_{\phi_i x}M(i,l)'). \quad (A1.25)$$

At a modal measurement:

$$\Delta P_{xx}=-\Delta\hat{x}\Delta\hat{x}'+\sum_k P_{xx\phi_k}\Delta\vartheta_k. \quad (A1.26)$$

Discussion

To compute the error covariance, note first that the dynamic equation of the outer product of the base-state error is given above:

$$\begin{aligned} d(\tilde{x}_t\tilde{x}_t') &= \left(\left(\sum_i (A_i(\widetilde{x_t\phi_i})+B_i(\tilde{\phi}_i u_t-\upsilon(\widetilde{\phi\phi_i})))+(\rho'-\gamma_\phi h')\tilde{\phi}_t -\gamma_x H\tilde{\chi}_t\right)dt-\gamma_x\,dn_x-\gamma_\phi\,dn_\phi+\sum_i C_i\phi_i\,dw_t \right.\\ &\quad \left. +\sum_{i,l}(M(i,l)x_t+\theta(i,l))\phi_i\,dm_l\right)\tilde{x}_t'+(\cdot)'+\gamma_\phi\,dn_\phi\,dn_\phi'\gamma_\phi' \\ &\quad +\gamma_x R_x\gamma_x'\,dt+\sum_i R_\chi(i)\phi_i\,dt+\sum_{i,l,r}(M(i,l)x_t \\ &\quad +\theta(i,l))(M(i,r)x_t+\theta(i,r))'\phi_i\,dm_l\,dm_r \\ &= F_{xx}\,dt+d\mu_t, \end{aligned} \quad (A1.27)$$

where F_{xx} is the $\mathcal{F}_t$-compensator of $\{\tilde{x}_t\tilde{x}_t'\}$, and $\{\mu_t\}$ is a matrix $\mathcal{F}_t$-martingale. Direct calculation indicates that

$$\hat{F}_{xx} = \Bigg(\sum_i (A_i(P_{xx\phi_i} + P_{xx}\hat{\phi}_i + \hat{x}_t P_{\phi_i x}) + B_i(u_t P_{\phi_i x} - v(P_{\phi x \phi_i} - P_{\phi x}\hat{\phi}_i))) + (\rho' - \gamma_\phi h')P_{\phi x} - \gamma_x R_x \gamma_x' \Bigg) + (\cdot)' + \gamma_x R_x \gamma_x' + \sum_i R_\chi(i)\hat{\phi}_i + \sum_{i,l} Q_{il}(M(i,l)R_{xx\phi_i}M(i,l)' + \theta(i,l)\hat{\phi}_i\theta(i,l)') + \sum_{i,l} Q_{il}(M(i,l)R_{x\phi_i}\theta(i,l)' + \theta(i,l)R_{\phi_i x}M(i,l)') + \gamma_\phi R_\phi \gamma_\phi'. \tag{A1.28}$$

It is known that $\{P_{xx}\}$ is a ϕ-dominated moment of $\{y_t\}$. Consequently, only $d\nu_\phi$ need be considered in the computation of dP_{xx}:

$$dP_{xx} = \hat{F}_{xx}\,dt + \sum_k \gamma_{xx}(k)d(\nu_\phi)_k, \tag{A1.29}$$

where $\{\gamma_{xx}(k); k \in S\}$ is a set of $\mathcal{G}_t$-predictable gain matrices. Since $\{\nu_\phi\}$ is the only part of the innovations process that enters into this calculation, $(\nu_\phi)_k$ will be written ν_k. As we have done before, we note that

$$d(\tilde{x}_t\tilde{x}_t'z_t') = d(\tilde{x}_t\tilde{x}_t')z' + \tilde{x}_t\tilde{x}_t\,dz' + d(\tilde{x}_t\tilde{x}_t')\,dz'.$$

First,

$$d(\tilde{x}_t\tilde{x}_t')\,dz_i = \Bigg(\Bigg(- \gamma_x\,dn_x - \gamma_\phi\,dn_\phi + \sum_{i,l}(M(i,l)x_t + \theta(i,l))\phi_i\,dm_l)\tilde{x}_t' + \sum_i C_i\phi_i\,dw_t\Bigg) + (\cdot)' + \gamma_\phi\,dn_\phi\,dn_\phi'\gamma_\phi'\Bigg)\,dn_i + \sum_{i,l}(M(i,l)x_t + \theta(i,l))(M(i,l)x_t + \theta(i,l))'\phi_i\,dm_l,$$

where only the martingale products have been retained, and $(n_\phi)_i$ has been written n_i. We need to consider terms like

$$\gamma_\phi\,dn_\phi\tilde{x}_t'\,dn_i = \gamma_\phi\mathbf{e}_i\,dn_i\tilde{x}_t'.$$

So

$$E[\gamma_\phi\,dn_\phi\tilde{x}_t'\,dn_i \mid \mathcal{F}_t] = \gamma_\phi\mathbf{e}_i\sum_k \tilde{x}_t'h_{ik}\phi_k.$$

Continuing, we have

$$E[\gamma_\phi \, dn_\phi \tilde{x}_t' \, dn_i \,|\, \mathcal{G}_t] = \gamma_\phi \mathbf{e}_i \sum_k h_{ik} P_{\phi_k x}$$
$$= \gamma_\phi \mathbf{e}_i (h' P_{\phi x})_i.$$

Further, $dm \; dn_i = 0$. Completing the expectation yields

$$E[d(\tilde{x}_t \tilde{x}_t') \, dz_i \,|\, \mathcal{G}_t] = \hat{\boldsymbol{\lambda}}_i \gamma_\phi \mathbf{e}_i \gamma_\phi' - \gamma_\phi \mathbf{e}_i h' P_{\phi_i x} - (\gamma_\phi \mathbf{e}_i h' P_{\phi_i x})'$$
$$= -\hat{\boldsymbol{\lambda}}_i \gamma_\phi \mathbf{e}_i \gamma_\phi'. \tag{A1.30}$$

From (A1.27) we have

$$d(\tilde{x}_t \tilde{x}_t') z_i = F_{xx} z_i \, dt + d\mu.$$

Taking the expectation of this, we find

$$E[d(\tilde{x}_t \tilde{x}_t') z_i \,|\, \mathcal{G}_t]/dt = \hat{F}_{xx} z_i.$$

Also,

$$(\tilde{x}_t \tilde{x}_t') \, dz_i = (\tilde{x}_t \tilde{x}_t')((\phi' h)_i \, dt + dn_i),$$

and taking the expectation of this gives

$$E[(\tilde{x}_t \tilde{x}_t') \, dz_i \,|\, \mathcal{G}_t]/dt = \sum_k (P_{xx\phi_k} + P_{xx} \hat{\phi}_k) h_{ki}. \tag{A1.31}$$

Combining terms yields

$$E[d(\tilde{x}_t \tilde{x}_t' z_i) \,|\, \mathcal{G}_t]/dt = -\hat{\boldsymbol{\lambda}}_i \gamma_\phi \mathbf{e}_i \gamma_\phi' + \hat{F}_{xx} z_i + \sum_k (P_{xx\phi_k} + P_{xx} \hat{\phi}_k) h_{ki}. \tag{A1.32}$$

To develop a comparable representation for $\{P_{xx}\}$ recall that

$$dP_{xx} = \hat{F}_{xx} \, dt + \sum_k \gamma_{xx}(k)(h' \tilde{\phi}_t \, dt + dn_\phi)_k.$$

Since $dn_k \, dn_i = \delta_{k,i}$, it follows that

$$dP_{xx} \, dz_i = \sum_k \gamma_{xx}(k) \, dn_k \, dn_i$$
$$= \gamma_{xx}(i) \, dn_i.$$

Taking the expectation leads to

$$E[(dP_{xx}) \, dz_i \,|\, \mathcal{G}_t]/dt = \gamma_{xx}(i) \hat{\boldsymbol{\lambda}}_i.$$

Further, $E[dP_{xx}\, z_i \,|\, \mathcal{G}_t]/dt = \hat{F}_{xx} z_i$ and $E[P_{xx}\, dz_i \,|\, \mathcal{G}_t]/dt = P_{xx}\hat{\boldsymbol{\lambda}}_i$. Combining terms gives

$$E[dP_{xx}\, z_i \,|\, \mathcal{G}_t]/dt = (\gamma_{xx}(i) + P_{xx})\hat{\boldsymbol{\lambda}}_i + \hat{F}_{xx} z_i. \tag{A1.33}$$

Equating (A1.32) and (A1.33) yields

$$\begin{aligned}(\gamma_{xx}(i) &+ P_{xx})\hat{\boldsymbol{\lambda}}_i + \hat{F}_{xx} z_i \\ &= -\hat{\boldsymbol{\lambda}}_i \gamma_\phi \mathbf{e}_i \gamma_\phi' + \hat{F}_{xx} z_i + \sum_k (P_{xx\phi_k} + P_{xx}\hat{\phi}_k) h_{ki}.\end{aligned}$$

Solving for $\gamma_{xx}(i)$, it follows that

$$\gamma_{xx}(i) = -\gamma_\phi \mathbf{e}_i \gamma_\phi' + \hat{\boldsymbol{\lambda}}_i^{-1} \sum_k P_{xx\phi_k} h_{ki}. \tag{A1.34}$$

The innovations dependent term in (A1.29) can now be written as

$$\sum_i \gamma_{xx}(i)\, d\nu_i = -\sum_i \gamma_\phi \mathbf{e}_i \gamma_\phi\, d\nu_i + \sum_k P_{xx\phi_k}(h R_\phi^{-1}\, d\nu_\phi)_k. \tag{A1.35}$$

Substituting (A1.28) and (A1.35) into (A1.29), we have

$$\begin{aligned} dP_{xx} = \Bigg(\Bigg(&\sum_i (\mathbf{A}_i(P_{xx\phi_i} + P_{xx}\hat{\phi}_i + \hat{x}_t P_{\phi_i x}) + B_i(u_t P_{\phi i x} \\ &-\boldsymbol{v}(P_{\phi x \phi_i} - P_{\phi x}\hat{\phi}_i))) + (\rho' - \gamma_\phi h')P_{\phi x} - \gamma_x R_x \gamma_x'\Bigg) + (\cdot)' \\ &+\gamma_x R_x \gamma_x' + \sum_i R_\chi(i)\hat{\phi}_i + \gamma_\phi R_\phi \gamma_\phi'\Bigg)\, dt - \sum_i \gamma_\phi \mathbf{e}_i \gamma_\phi\, d\nu_i \\ &+\sum_{i,l} Q_{il}(M(i,l) R_{xx\phi_i} M(i,l)' + \theta(i,l)\hat{\phi}_i \theta(i,l)')\, dt \\ &+\sum_{i,l} Q_{il}(M(i,l) R_{x\phi_i} \theta(i,l)' + \theta(i,l) R_{\phi_i x} M(i,l)')\, dt \\ &+\sum_k P_{xx\phi_k}(h R_\phi^{-1}\, d\nu_\phi)_k. \end{aligned}$$

There are ways in which this can be simplified. Between jumps in $\{z_t\}$, $h R_\phi^{-1}\, d\nu_\phi = -\lambda\mathbf{1}$. So,

$$\sum_k P_{xx\phi k}(h R_\phi^{-1}\, d\nu_\phi)_k = 0.$$

Also, $\sum_i \gamma_\phi \mathbf{e}_i \gamma_\phi \hat{\boldsymbol{\lambda}}_i = \gamma_\phi h' P_{\phi x}$. When there is no observation, this can be written as

$$\frac{d}{dt} P_{xx} = \left(\sum_i (\boldsymbol{A}_i (P_{xx\phi_i} + P_{xx}\hat{\phi}_i + \hat{x}_t P_{\phi_i x}) + B_i (u_t P_{\phi_i x} - \boldsymbol{v}(P_{\phi x \phi_i} - P_{\phi x}\hat{\phi}_i))) + (\rho' Q' - \gamma_\phi h') P_{\phi x} - \gamma_x R_x \gamma_x' \right) + (\cdot)' + \gamma_x R_x \gamma_x' + 2\gamma_\phi R_\phi \gamma_\phi' + \sum_i R_\chi(i)\hat{\phi}_i + \sum_{i,l} Q_{il}(M(i,l) R_{xx\phi_i} M(i,l)' + \theta(i,l)\hat{\phi}_i \theta(i,l)') + \sum_{i,l} Q_{il}(M(i,l) R_{x\phi_i} \theta(i,l)' + \theta(i,l) R_{\phi_i x} M(i,l)').$$

When there is an observation

$$\Delta P_{xx} = -\gamma_\phi \mathrm{diag}(dz)\gamma_\phi' + \sum_k P_{xx\phi_k}(k)(h R_\phi^{-1} dz)_k,$$

or

$$\Delta P_{xx} = -\Delta\hat{x}\Delta\hat{x}' + \sum_k P_{xx\phi_k}\Delta\vartheta_k.$$

Simplifying yields the result.

A1.7 Base-State Modal-State Cross Covariance

base-state, modal-state cross covariance

Between modal measurements:

$$\frac{d}{dt} P_{x\phi} = \sum_i (\boldsymbol{A}_i (P_{x\phi\phi_i} + P_{x\phi}\hat{\phi}_i + \hat{x}_t P_{\phi_i \phi}) + B_i (u_t P_{\phi_i \phi} - \boldsymbol{v}(P_{\phi\phi\phi_i} - P_{\phi\phi}\hat{\phi}_i))) + \rho' P_{\phi\phi} - \gamma_x H P_{\chi\phi} + P_{x\phi} Q + \sum_{i,l} Q_{il}(M(i,l) R_{x\phi_i} + \theta(i,l)\hat{\phi}_i)\mathbf{e}_l'.$$

At a modal measurement:

$$\Delta P_{x\phi} = -\Delta\hat{x}\Delta\hat{\phi}' + \sum_k P_{x\phi\phi_k}\Delta\vartheta_k. \tag{A1.36}$$

A1.7.1 Discussion

To compute the error cross covariance, note first that the dynamic equation of the error outer product is presented in Table A1.3:

$$d(\tilde{x}_t\tilde{\phi}_t') = \Bigg(\bigg(\sum_i (A_i(\widetilde{x_t\phi_i}) + B_i(\tilde{\phi}_i u_t - v(\widetilde{\phi\phi_i})))$$
$$+ (\rho' - \gamma_\phi h')\tilde{\phi}_t - \gamma_x H\tilde{\chi}_t\bigg) dt - \gamma_x\, dn_x - \gamma_\phi\, dn_\phi$$
$$+ \sum_i C_i\phi_i\, dw_t + \sum_{i,l}(M(i,l)x_t + \theta(i,l))\phi_i\, dm_l\Bigg)\tilde{\phi}_t'$$
$$+ \tilde{x}_t((Q' - \gamma_{\phi\phi}h')\tilde{\phi}_t\, dt - \gamma_{\phi\phi}\, dn_\phi + dm)' + \gamma_\phi\, dn_\phi\, dn_\phi'\gamma_{\phi\phi}'$$
$$+ \sum_{i,l}(M(i,l)x_t + \theta(i,l))\phi_i\, dm_l\, dm,$$

so that

$$d(\tilde{x}_t\tilde{\phi}_t') = F_{x\phi}\, dt + d\mu_t, \tag{A1.37}$$

where $F_{x\phi}$ is the $\mathcal{F}_t$-compensator of $\{\tilde{x}_t\tilde{\phi}_t\}$. Direct calculation indicates that

$$\hat{F}_{x\phi} = \sum_i (A_i(P_{x\phi\phi_i} + P_{x\phi}\hat{\phi}_i + \hat{x}_t P_{\phi_i\phi}) + B_i(u_t P_{\phi_i\phi}$$
$$-v(P_{\phi\phi\phi_i} - P_{\phi\phi}\hat{\phi}_i))) + \rho' P_{\phi\phi} - \gamma_x H P_{\chi\phi} + P_{x\phi}Q$$
$$+ \sum_{i,l} Q_{il}(M(i,l)R_{x\phi_i} + \theta(i,l)\hat{\phi}_i)\mathbf{e}_l' - \gamma_\phi h' P_{\phi\phi}. \tag{A1.38}$$

But $\{P_{x\phi}\}$ is a ϕ-dominated moment (trivially so). Consequently,

$$dP_{x\phi} = \hat{F}_{x\phi}\, dt + \sum_k \gamma_{x\phi}(k)\, dv_k, \tag{A1.39}$$

where $\{\gamma_{x\phi}(k);\, k \in \mathbf{S}\}$ is a set of $\mathcal{G}_t$-predictable gain matrices. Hence

$$d(\tilde{x}_t\tilde{\phi}_t')\, dz_i = (\gamma_\phi\, dn_\phi\, dn_\phi'\gamma_{\phi\phi}' - \gamma_\phi\, dn_\phi\tilde{\phi}' + \tilde{x}_t(-\gamma_{\phi\phi}\, dn_\phi)')\, dn_i.$$

Therefore,

$$E[d(\tilde{x}_t\tilde{\phi}_t')\, dz_i \,|\, \mathcal{G}_t]/dt = -\hat{\boldsymbol{\lambda}}_i\gamma_\phi\mathbf{E}_{\mathbf{i}}\gamma_{\phi\phi}', \tag{A1.40}$$

and from (A1.40) we have

$$E[d(\tilde{x}_t\tilde{\phi}_t')z_i \,|\, \mathcal{G}_t]/dt = \hat{F}_{x\phi}z_i. \tag{A1.41}$$

Also,

$$(\tilde{x}_t\tilde{\phi}_t')\,dz_i = (\tilde{x}_t\tilde{\phi}_t')((\phi_t{}'h)_i\,dt + dn_i)$$

with expectation

$$E[(\tilde{x}_t\tilde{\phi}_t')\,dz_i \,|\, \mathcal{G}_t]/dt = \sum_k (P_{x\phi\phi_k} + P_{x\phi}\hat{\phi}_k)h_{ki}. \tag{A1.42}$$

Now combine (A1.40)–(A1.42) to get

$$E[d(\tilde{x}_t\tilde{\phi}_t'z_i) \,|\, \mathcal{G}_t]/dt = -\hat{\boldsymbol{\lambda}}_i\gamma_\phi\mathbf{E_i}\gamma_{\phi\phi}' + \hat{F}_{x\phi}z_i + \sum_k (P_{x\phi\phi_k} + P_{x\phi}\hat{\phi}_k)h_{ki}dt. \tag{A1.43}$$

To develop a comparable equation, (A1.39) can be expanded to obtain a representation for $\{P_{x\phi}\}$:

$$dP_{x\phi} = \hat{F}_{x\phi}\,dt + \sum_k \gamma_{x\phi}(k)(h'\tilde{\phi}_t\,dt + dn_\phi)_k.$$

Thus,

$$\begin{aligned} dP_{x\phi}\,dz_i &= \left(\sum_k \gamma_{x\phi}(k)\,dn_k\right) dn_i \\ &= \gamma_{x\phi}(i)\,dn_i. \end{aligned}$$

Taking the expectation of this yields

$$E[dP_{x\phi}\,dz_i \,|\, \mathcal{G}_t]/dt = \gamma_{x\phi}(i)\hat{\boldsymbol{\lambda}}_i.$$

It follows directly that

$$E[(dP_{x\phi})z_i \,|\, \mathcal{G}_t]/dt = \hat{F}_{x\phi}z_i,$$

and

$$E[(P_{x\phi})\,dz_i \,|\, \mathcal{G}_t]/dt = P_{x\phi}\hat{\boldsymbol{\lambda}}_i.$$

Combining these, we obtain

$$E[(d(P_{x\phi}z_i) \,|\, \mathcal{G}_t]/dt = (\gamma_{x\phi}(i) + P_{x\phi})\hat{\boldsymbol{\lambda}}_i + \hat{F}_{x\phi}z_i. \tag{A1.44}$$

Equating (A1.43) and (A1.44) gives

$$(\gamma_{x\phi}(i) + P_{x\phi})\hat{\boldsymbol{\lambda}}_i + \hat{F}_{x\phi} z_i$$
$$= -\hat{\boldsymbol{\lambda}}_i \gamma_\phi \mathbf{e}_i \gamma'_{\phi\phi} + \hat{F}_{x\phi} z_i + \sum_k (P_{x\phi\phi_k} + P_{x\phi}\hat{\phi}_k) h_{ki}. \qquad \text{(A1.45)}$$

Thus,

$$\gamma_{x\phi}(i) = -\gamma_\phi \mathbf{E_i} \gamma'_{\phi\phi} + \hat{\boldsymbol{\lambda}}_i^{-1} \sum_k P_{xx\phi_k} h_{ki}. \qquad \text{(A1.46)}$$

The innovations dependent term in (A1.39) can now be written as

$$\sum_i \gamma_{x\phi}(i)\, d\nu_i = -\sum_i \gamma_\phi \mathbf{E_i} \gamma_{\phi\phi}\, d\nu_i + \sum_k P_{x\phi\phi_k} (h R_\phi^{-1}\, d\nu_\phi)_k. \qquad \text{(A1.47)}$$

Substituting (A1.38) and (A1.47) into (A1.39) we have

$$\begin{aligned} dP_{x\phi} = \Bigg(& \sum_i (\mathbf{A}_i (P_{x\phi\phi_i} + P_{x\phi}\hat{\phi}_i + \hat{x}_t P_{\phi_i\phi}) + B_i (u_t P_{\phi_i\phi} \\ & - \upsilon(P_{\phi\phi\phi_i} - P_{\phi\phi}\hat{\phi}_i)) + \rho' P_{\phi\phi} - \gamma_x H P_{\chi\phi} + P_{x\phi} Q \\ & + \sum_{i,l} Q_{il} (M(i,l) R_{x\phi_i} + \theta(i,l)\hat{\phi}_i)\mathbf{e}_l' - \gamma_\phi h' P_{\phi\phi} \Bigg) dt \\ & - \sum_i \gamma_\phi \mathbf{E_i} \gamma_{\phi\phi}\, d\nu_i + \sum_k P_{x\phi\phi_k} (h R_\phi^{-1}\, d\nu_\phi)_k. \end{aligned}$$

There are ways in which this can be simplified. Between jumps in $\{z_t\}$, $\sum_k P_{x\phi\phi_k}$ $(hR_\phi^{-1}\, d\nu_\phi)_k = 0$. Also,

$$\sum_i \gamma_\phi \mathbf{e}_i \gamma_{\phi\phi} \hat{\boldsymbol{\lambda}}_i = \gamma_\phi h' P_{\phi\phi}.$$

When there is no observation, this can be written as

$$\begin{aligned} \frac{dP_{x\phi}}{dx} = & \sum_i (\mathbf{A}_i (P_{x\phi\phi_i} + P_{x\phi}\hat{\phi}_i + \hat{x}_t P_{\phi_i\phi}) + B_i (u_t P_{\phi_i\phi} \\ & -\upsilon(P_{\phi\phi\phi_i} - P_{\phi\phi}\hat{\phi}_i))) + \rho' P_{\phi\phi} - \gamma_x H P_{\chi\phi} + P_{x\phi} Q \\ & + \sum_{i,l} Q_{il} (M(i,l) R_{x\phi_i} + \theta(i,l)\hat{\phi}_i)\mathbf{e}_l'. \end{aligned}$$

When there is an observation,

$$\Delta P_{x\phi} = -\gamma_\phi \mathrm{diag}(dz)\gamma'_{\phi\phi} + \sum_k P_{x\phi\phi_k} (h R_\phi^{-1} dz)_k,$$

or

$$\Delta P_{xx} = -\Delta\hat{x}\Delta\hat{\phi}' + \sum_k P_{x\phi\phi_k}\Delta\vartheta_k.$$

A1.8 A Mixed Third Central Moment

To complete the equation for the second moment matrices, the mixed third central moments, $\{P_{xx\phi_i}, i \in S\}$, must be evaluated. This requires even more detailed analysis than that leading to the second moments. However, the underlying procedure is the same.

a mixed third central moment

Between modal measurements:

$$\begin{aligned}
\frac{d}{dt}P_{xx\phi_m} = \Bigg(& \sum_i (\mathbf{A}_i (P_{xx\phi_i\phi_m} + \hat{\phi}_i P_{xx\phi_m} + \hat{x} P_{\phi_i x\phi_m} - P_{x\phi_i} P_{\phi_m x}) \\
& + B_i (u_t P_{\phi_i x\phi_m} - \boldsymbol{v}((\mathbf{e}_i - \hat{\phi}) P_{\phi_i x} (\delta_{i,m} - \hat{\phi}_m \\
& - P_{\phi\phi_i} P_{x\phi_m})))) + \boldsymbol{\rho}' Q' P_{\phi x\phi_m}) - \gamma_x H P_{\chi x\phi_m} \Bigg) + (\cdot)' \\
& + \sum_j P_{xx\phi_j} Q_{jm} + R_\chi(i) P_{\phi_i\phi_m} + (\boldsymbol{\rho}' P_{\phi_x\phi_m} + P_{x\phi\phi_m}\boldsymbol{\rho}) \\
& + \sum_{i,l} Q_{il} (M(i,l)(P_{(xx\phi_i)\phi_m} + \delta_{l,m} R_{xx\phi_i}) M(i,l)' \\
& + M(i,l)((P_{(x\phi_m)\phi})_{\cdot i} + \delta_{l,m} R_{x\phi_i})\theta(i,l)' \\
& + \theta(i,l)((P'_{(x\phi_m)\phi})_{i\cdot} + \delta_{l,m} R_{\phi_i x}) M(i,l)' \\
& + \theta(i,l)(P_{\phi_i\phi_m} + \delta_{l,m}\hat{\phi}_i)\theta(i,l)') \\
& + \sum_i Q_{im} (M(i,m) P_{(x\phi_i)x} + \theta(i,m) R_{\phi_i x} \\
& + P_{(x\phi_i)x} M(i,m)' + R_{x\phi_i}\theta(i,m)'). \qquad \text{(A1.48)}
\end{aligned}$$

At a modal measurement:

$$\begin{aligned}
\Delta P_{xx\phi_m} = & -\Delta\hat{\phi}_m \Delta\hat{x}\Delta\hat{x}' - \Delta\hat{\phi}_m P_{xx}{}^+ - \Delta\hat{x} P^+_{\phi_m x} \\
& - P^+_{x\phi_m}\Delta\hat{x}' + \sum_i P_{xx\phi_m\phi_i}\Delta\vartheta_i. \qquad \text{(A1.49)}
\end{aligned}$$

A1.8.1 Discussion

The procedure used earlier will be followed step-by-step:

$$
\begin{aligned}
d(\tilde{x}_t\tilde{x}_t'\tilde{\phi}_m) = {} & \Bigg(\Bigg(\Bigg(\sum_i (A_i(\widetilde{x_t\phi_i}) + B_i(\tilde{\phi}_i u_t - v(\widetilde{\phi\phi_i})))\\
& + (\rho' - \gamma_\phi h')\tilde{\phi}_t - \gamma_x H\tilde{\chi}_t\Bigg) dt - \gamma_x\, dn_x - \gamma_\phi\, dn_\phi\\
& + \sum_i C_i\phi_i\, dw_t + \sum_{i,l}(M(i,l)x_t + \theta(i,l))\phi_i\, dm_l\Bigg)\tilde{x}_t'\\
& + (\cdot)' + \left(\gamma_x R_x\gamma_x' + \sum_i R_\chi(i)\phi_i\right) dt + \sum_{i,l,r}(M(i,l)x_t\\
& + \theta(i,l))(M(i,r)x_t + \theta(i,r))'\phi_i\, dm_l\, dm_r\\
& + \gamma_\phi\, dn_\phi\, dn_\phi'\gamma_\phi'\Bigg)\tilde{\phi}_m + (\tilde{x}_t\tilde{x}_t')((Q' - \gamma_{\phi\phi}h')\tilde{\phi}_t\, dt\\
& - \gamma_{\phi\phi}\, dn_\phi + dm)_m + (-\gamma_\phi\, dn_\phi\tilde{x}_t' + (-\gamma_\phi\, dn_\phi\tilde{x}_t')'\\
& + \gamma_\phi\, dn_\phi\, dn_\phi'\gamma_\phi)(-\gamma_{\phi\phi}\, dn_\phi)_m + \Bigg(\sum_{i,l}(M(i,l)x_t\\
& + \theta(i,l))\phi_i\, dm_l\tilde{x}_t'\Bigg) + (\cdot)' + \sum_{i,l,r}((M(i,l)x_t\\
& + \theta(i,l))(M(i,r)x_t + \theta(i,r))'\phi_i\, dm_l\, dm_r)\, dm_m,
\end{aligned}
$$

or

$$
d(\tilde{x}_t\tilde{x}_t'\tilde{\phi}_m) = F_{xx\phi_m}\, dt + d\mu_t, \tag{A1.50}
$$

where $F_{xx\phi_m}$ is the $\mathcal{F}_t$-compensator of $\{\tilde{x}_t\tilde{x}_t'\tilde{\phi}_m\}$. Direct calculation indicates that

$$
\begin{aligned}
\hat{F}_{xx\phi_m} = {} & \Bigg(\sum_i (A_i(P_{xx\phi_i\phi_m} + \hat{\phi}_i P_{xx\phi_m} + \hat{x}P_{\phi_i x\phi_m} - P_{x\phi_i}P_{\phi_m x})\\
& + B_i(u_t P_{\phi_i x\phi_m} - v((\mathbf{e}_i - \hat{\phi})P_{\phi_i x}(\delta_{i,m} - \hat{\phi}_m - P_{\phi\phi_i}P_{x\phi_m}))))\\
& + (\rho' - \gamma_\phi h')P_{\phi x\phi_m} - \gamma_x H P_{\chi x\phi_m}\Bigg) + (\cdot)' + \gamma_\phi\sum_k \mathbf{e}_k h_{ik} P_{\phi_i\phi_m}\gamma_\phi'\\
& + \sum_j P_{xx\phi_j}(Q' - \gamma_{\phi\phi}h')_{jm} + \gamma_\phi\sum_j \mathbf{e}_j(h'P_{\phi x})_{j\cdot}(\gamma_{\phi\phi})_{mj}
\end{aligned}
$$

$$+\left(\gamma_\phi \sum_j \mathbf{e}_j (h' P_{\phi x})_{j\cdot} (\gamma_{\phi\phi})_{mj}\right)' - \gamma_\phi \sum_j \mathbf{E_j} \hat{\boldsymbol{\lambda}}_j (\gamma_{\phi\phi})_{mj} \gamma_\phi'$$

$$+ \sum_i R_x(i) P_{\phi_i \phi_m} + \sum_{i,l} Q_{il} \big(M(i,l) P_{(xx\phi_i)\phi_m} M(i,l)'$$

$$+ M(i,l) \big(P_{(x\phi_m)\phi}\big)_{\cdot i} \theta(i,l)' + \theta(i,l) \big(P'_{(x\phi_m)\phi}\big)_{\cdot i} M(i,l)'$$

$$+ \theta(i,l) P_{\phi_i \phi_m} \theta(i,l)'\big) + \sum_i Q_{im} \big(M(i,m) R_{xx\phi_i} M(i,m)'$$

$$+ M(i,m) R_{x\phi_i} \theta(i,m)' + \theta(i,m) R_{\phi_i x} M(i,m)'$$

$$+ \theta(i,m) \hat{\phi}_i \theta(i,m)'\big) + \sum_i Q_{im} \big(M(i,m) P_{(x\phi_i)x}$$

$$+ \theta(i,m) R_{\phi_i x} + P_{x(x\phi_i)} M(i,m)' + R_{x\phi_i} \theta(i,m)'\big).$$

This can be written in a simpler form:

$$\hat{F}_{xx\phi_m} = \left(\sum_i \big(\boldsymbol{A}_i \big(P_{xx\phi_i\phi_m} + \hat{\phi}_i P_{xx\phi_m} + \hat{x} P_{\phi_i x\phi_m} - P_{x\phi_i} P_{\phi_m x}\big) \right.$$

$$+ B_i \big(u_t P_{\phi_i x\phi_m} - \boldsymbol{v}((\mathbf{e}_i - \hat{\phi}) P_{\phi_i x} (\delta_{i,m} - \hat{\phi}_m - P_{\phi\phi_i} P_{x\phi_m}))\big)\big)$$

$$\left. + (\boldsymbol{\rho}' - \gamma_\phi h') P_{\phi x\phi_m} - \gamma_x H P_{\chi x\phi_m} \right) + (\cdot)'$$

$$+ \sum_j P_{xx\phi_j} (Q' - \gamma_{\phi\phi} h')_{jm} + R_\chi(i) P_{\phi_i \phi_m}$$

$$+ 2\gamma_\phi \mathrm{diag}(\hat{\boldsymbol{\lambda}} * (\gamma_{\phi\phi})_{m\cdot}) \gamma_\phi' + \sum_{i,l} Q_{il} \big(M(i,l) \big(P_{(xx\phi_i)\phi_m}$$

$$+ \delta_{l,m} R_{xx\phi_i}\big) M(i,l)' + M(i,l) \big(\big(P_{(x\phi_m)\phi}\big)_{\cdot i} + \delta_{l,m} R_{x\phi_i} \big) \theta(i,l)'$$

$$+ \theta(i,l) \big(\big(P'_{(x\phi_m)\phi}\big)_{i\cdot} + \delta_{l,m} R_{\phi_i x} \big) M(i,l)'$$

$$+ \theta(i,l) \big(P_{\phi_i\phi_m} + \delta_{l,m} \hat{\phi}_i \big) \theta(i,l)'\big) + \sum_i Q_{im} \big(M(i,m) P_{(x\phi_i)x}$$

$$+ \theta(i,m) R_{\phi_i x} + P'_{(x\phi_i)x} M(i,m)' + R_{x\phi_i} \theta(i,m)'\big).$$

But $\{P_{xx\phi_m}\}$ is a ϕ-dominated moment (again trivially so). Consequently,

$$dP_{xx\phi_m} = \hat{F}_{xx\phi_m}\, dt + \sum_k \gamma_{xx\phi_m}(k)\, d\nu_k, \tag{A1.51}$$

where $\{\gamma_{xx\phi_m}(k); k \in \mathbf{S}\}$ are the set of $\mathcal{G}_t$-predictable gain matrices. But

$$d(\tilde{x}_t\tilde{x}_t'\tilde{\phi}_m)\,dz_i = ((-\gamma_\phi\,dn_\phi\tilde{x}_t' - \tilde{x}_t\,dn_\phi'\gamma_\phi' + \gamma_\phi\,dn_\phi\,dn_\phi'\gamma_\phi')$$
$$\times(\tilde{\phi}_m - \gamma_{\phi\phi}\,dn_\phi)_m - (\tilde{x}_t\tilde{x}_t')(\gamma_{\phi\phi}\,dn_\phi)_m)\,dn_k.$$

Taking the expectation and simplifying, we get

$$E[d(\tilde{x}_t\tilde{x}_t'\tilde{\phi}_m)\,dz_i\,|\,\mathcal{G}_t]/dt$$
$$= -\gamma_\phi\mathbf{e}_i\sum_\alpha h_{\alpha i}(P_{\phi_m x\phi_\alpha} + \hat{\phi}_\alpha P_{\phi_m x} - h'P_{\phi_i x}(\gamma_{\phi\phi})_{mi}) + (\cdot)'$$
$$-\sum_\alpha h_{\alpha i}(P_{xx\phi_\alpha} + \hat{\phi}_\alpha P_{xx})(\gamma_{\phi\phi})_{mi}.$$

Also,

$$E[d(\tilde{x}_t\tilde{x}_t'\tilde{\phi}_m)z_i\,|\,\mathcal{G}_t]/dt = \hat{F}_{xx\phi_m}z_i$$

and

$$E[(\tilde{x}_t\tilde{x}_t'\tilde{\phi}_m)dz_i\,|\,\mathcal{G}_t]/dt = \hat{\boldsymbol{\lambda}}_i P_{xx\phi_m} + \sum_j P_{xx\phi_m\phi_j}h_{jk}.$$

Combining these yields

$$E[d(\tilde{x}_t\tilde{x}_t'\tilde{\phi}_m z_i)\,|\,\mathcal{G}_t]/dt$$
$$= -\gamma_\phi\mathbf{e}_i\sum_\alpha h_{\alpha i}(P_{\phi_m x\phi_\alpha} + \hat{\phi}_\alpha P_{\phi_m x} - h'P_{\phi_i x}(\gamma_{\phi\phi})_{mi}$$
$$+ (\cdot)' - \sum_\alpha h_{\alpha,i}(P_{xx\phi_\alpha} + \hat{\phi}_\alpha P_{xx})(\gamma_{\phi\phi})_{mi}$$
$$+ \hat{F}_{xx\phi}z_i + \hat{\boldsymbol{\lambda}}_i P_{xx\phi_m} + \sum_j P_{xx\phi_m\phi_j}h_{jk}. \tag{A1.52}$$

To develop an equation comparable to (A1.52), (A1.51) can be expanded to obtain a representation for $\{P_{xx\phi_m}\}$: $(dP_{xx\phi_m})\,dz_i = \gamma_{xx\phi_m}(i)\,dn_i$. Taking the expectation of this we get

$$E\big[(dP_{xx\phi_m})dz_i\,\big|\,\mathcal{G}_t\big]/dt = \gamma_{xx\phi_m}(i)\hat{\boldsymbol{\lambda}}_i.$$

From (A1.51), it follows directly that $E[(dP_{xx\phi_m})z_i\,|\,\mathcal{G}_t]/dt = \hat{F}_{xx\phi}z_i$. Further,

$$E\big[P_{xx\phi_m}dz_i\,\big|\,\mathcal{G}_t\big]/dt = P_{xx\phi_m}\hat{\boldsymbol{\lambda}}_i.$$

Combining these expressions gives

$$E\big[(d(P_{xx\phi_m}z_i)\,\big|\,\mathcal{G}_t\big]/dt = (\gamma_{xx\phi_m}(i) + P_{xx\phi_m})\hat{\boldsymbol{\lambda}}_i + \hat{F}_{xx\phi_m}z_i. \tag{A1.53}$$

Equating (A1.52) and (A1.53), we get

$$
\begin{aligned}
&(\gamma_{xx\phi_m}(i) + P_{xx\phi_m})\hat{\boldsymbol{\lambda}}_i + \hat{F}_{xx\phi} z_i \\
&\quad = -\gamma_\phi \mathbf{e}_i \sum_\alpha h_{\alpha i}(P_{\phi_m x \phi_\alpha} + \hat{\phi}_\alpha P_{\phi_m x}) - h' P_{\phi_i x}(\gamma_{\phi\phi})_{mi} + (\cdot)' \\
&\quad - \sum_\alpha h_{\alpha i}(P_{xx\phi\alpha} + \hat{\phi}_\alpha P_{xx})(\gamma_{\phi\phi})_{mi} + \hat{F}_{xx\phi_m} z_i \\
&\quad + \hat{\boldsymbol{\lambda}}_i P_{xx\phi_m} + \sum_j P_{xx\phi_m\phi_j} h_{jk},
\end{aligned}
$$

or

$$
\begin{aligned}
\gamma_{xx\phi_m}(i) = \hat{\boldsymbol{\lambda}}_i^{-1} \Bigg(&-\gamma_\phi \mathbf{e}_i \sum_\alpha h_{\alpha i}(P_{\phi_m x \phi_\alpha} + \hat{\phi}_\alpha P_{\phi_m x}) - h' P_{\phi_i x}(\gamma_{\phi\phi})_{mi} \\
&+ (\cdot)' - \sum_\alpha h_{\alpha i}(P_{xx\phi_\alpha} + \hat{\phi}_\alpha P_{xx})(\gamma_{\phi\phi})_{mi} + \sum_j P_{xx\phi_m\phi_j} h_{ji} \Bigg).
\end{aligned} \tag{A1.54}
$$

Equations (A1.51) and (A1.54) yield the sought after result. We can simplify this. Since, $dv_i = dz_i - \hat{\boldsymbol{\lambda}}_i\, dt$, we can look at the innovations dependent terms when there is no observation:

$$
\begin{aligned}
\sum_i \gamma_{xx\phi_m}(i)\hat{\boldsymbol{\lambda}}_i = \sum_i \Bigg(&-\gamma_\phi \mathbf{e}_i \sum_\alpha h_{\alpha i}(P_{\phi_m x \phi_\alpha} + \hat{\phi}_\alpha P_{\phi_m x}) \\
&- h' P_{\phi_i x}(\gamma_{\phi\phi})_{mi} \Bigg) + (\cdot)' \\
&- \sum_\alpha h_{\alpha i}(P_{xx\phi_\alpha} + \hat{\phi}_\alpha P_{xx})(\gamma_{\phi\phi})_{mi} + \sum_j P_{xx\phi_m\phi_j} h_{ji} \\
= \Big(&-\gamma_\phi h' P_{\phi x \phi_m} - \gamma_\phi \hat{\boldsymbol{\lambda}} P_{\phi_m x} + \gamma_\phi \mathrm{diag}(\hat{\boldsymbol{\lambda}} * (\gamma_{\phi\phi})_{m.})\gamma_\phi' \Big) \\
&+ (\cdot)' - \sum_\alpha (\gamma_{\phi\phi} h')_{m\alpha} P_{xx\phi_\alpha} + \sum_i \hat{\boldsymbol{\lambda}}_i (\gamma_{\phi\phi})_{mi} P_{xx} \\
&+ \sum_j P_{xx\phi_m\phi_j} \sum_i h_{ji}.
\end{aligned}
$$

But

$$
\sum_i \hat{\boldsymbol{\lambda}}_i (\gamma_{\phi\phi})_{m,i} P_{xx} + \sum_j P_{xx\phi_m\phi_j} \sum_i h_{ji} = 0,
$$

$$
\gamma_\phi \hat{\boldsymbol{\lambda}} P_{\phi_m x} = 0,
$$

and

$$
\begin{aligned}
&\hat{F}_{xx\phi_m} - \sum_i \gamma_{xx\phi_m}(i)\hat{\boldsymbol{\lambda}}_i \\
&\quad = \left(\sum_i (\boldsymbol{A}_i (P_{xx\phi_i\phi_m} + \hat{\phi}_i P_{xx\phi_m} + \hat{\phi} P_{\phi_i x\phi_m} - P_{x\phi_i} P_{\phi_m x}) \right. \\
&\qquad + B_i (u_t P_{\phi_i x\phi_m} - \boldsymbol{\upsilon}((\mathbf{e}_i - \hat{\phi}) P_{\phi_i x} (\delta_{i,m} - \hat{\phi}_m - P_{\phi\phi_i} P_{x\phi_m})))) \\
&\qquad \left. + \rho' Q' P_{\phi x\phi_m} - \gamma_x H P_{\chi x\phi_m} \right) + (\cdot)' + \sum_j P_{xx\phi_j} Q'_{mj} \\
&\qquad + R_\chi(i) P_{\phi_i\phi_m} + (\rho' P_{\phi x\phi_m} + P_{x\phi\phi_m}\rho) + \sum_{i,l} Q_{il}(M(i,l) \\
&\qquad \times (P_{(xx\phi_i)\phi_m} + \delta_{l,m} R_{xx\phi_i}) M(i,l)' + M(i,l)((P_{(x\phi_m)\phi})_{.i} \\
&\qquad + \delta_{l,m} R_{x\phi_i})\theta(i,l)' + \theta(i,l)((P'_{(x\phi_m)\phi})_{i.} + \delta_{l,m} R_{\phi_i x}) M(i,l)' \\
&\qquad + \theta(i,l)(P_{\phi_i\phi_m} + \delta_{l,m}\hat{\phi}_i)\theta(i,l)') + \sum_i Q_{im}(M(i,m) P_{(x\phi_i)x} \\
&\qquad + \theta(i,m) R_{\phi_i x} + P_{(x\phi_i)x} M(i,m)' + R_{x\phi_i}\theta(i,m)').
\end{aligned}
$$

Equation (A1.48) then follows. When there is an observation

$$
\begin{aligned}
\Delta P_{xx\phi_m} &= -P_{x\phi}\Delta\vartheta\Delta\vartheta'\mathbf{e}_m P_{\phi_m x} - (P_{x\phi}\Delta\vartheta\Delta\vartheta'\mathbf{e}_m P_{\phi_m x})' + P_{xx\phi_m} \\
&\qquad \times(\Delta\vartheta_m - 1) + \hat{\phi}_m \Delta\vartheta_m \left(2P_{x\phi}\Delta\vartheta\Delta\vartheta' P_{\phi x} - \sum_i P_{xx\phi_i}\Delta\vartheta_i \right).
\end{aligned}
$$

The update equation can be simplified by observing the following:

$$
\begin{aligned}
&\hat{\phi}_m \Delta\vartheta_m \left(2P_{x\phi}\Delta\vartheta\Delta\vartheta' P_{\phi x} - \sum_i P_{xx\phi_i}\Delta\vartheta_i \right) \\
&\quad = \hat{\phi}_m^+ (\Delta\hat{x}\Delta\hat{x}' - \Delta P_{xx}) \sum_i P_{xx\phi_m\phi_i}\Delta\vartheta_i P_{xx\phi_m}(\Delta\vartheta_m - 1) \\
&\qquad + \hat{\phi}_m^+ P_{xx} - \hat{\phi}_m \left(P_{xx} + \sum_i P_{xx\phi_i}\Delta\vartheta_i \right).
\end{aligned}
$$

Or

$$P_{xx\phi_m}(\Delta\vartheta_m - 1) = \sum_i P_{xx\phi_m\phi_i}\Delta\vartheta_i - \Delta\hat{\phi}_m P_{xx} + \hat{\phi}_m(\Delta P_{xx} + \Delta\hat{x}\Delta\hat{x}').$$

Additionally,

$$\Delta\hat{x}\Delta\vartheta'\mathbf{e}_m P_{\phi_m x} = \Delta\hat{x} P^+_{\phi_m x} + \hat{\phi}^+_m \Delta\hat{x}\Delta\hat{x}'.$$

Combining these expressions gives

$$\begin{aligned}\Delta P_{xx\phi_m} &= -\Delta\hat{\phi}_m \Delta\hat{x}\Delta\hat{x}' - \Delta\hat{\phi}_m P^+_{xx} - \Delta\hat{x} P^+_{\phi_m x} \\ &\quad - P^+_{x\phi_m}\Delta\hat{x}' + \sum_i P_{xx\phi_m\phi_i}\Delta\vartheta_i.\end{aligned}$$

When the base-state observations are discrete, ($y[k] = H\chi[k] + n[k]$; $E[n[k]\, n[k]'] = R_x > 0$), the PME must be modified. As in the Kalman filter, between observations the $\mathcal{G}_t$-moments are extrapolated without dependence on the observation ($R_x = \infty$). At a measurement there is a correction. The predictable quadratic variation of the innovations process, $\Delta\nu_x[k] = y[k] - \hat{y}[k]$, differs from that found in the continuous case: $E[(d\nu_x)\, d\nu_x' \mid \mathcal{G}^\phi_t]/dt = R_x$ for continuous observations;

$$E\left[\Delta\nu[k]\Delta\nu[k]' \middle| \mathcal{G}^\phi_t\right] = H P_{\chi\chi} H' + R_x$$

for discrete observations. In the discrete PME , the innovations gain are modified:

$$\gamma_x = P_{\chi x} H' \left(E\left[(d\nu_x)\, d\nu_x' \middle| \mathcal{G}^\phi_t\right]/dt\right)^{-1}$$

for continuous observations, and

$$\gamma_x = P_{\chi x} H' \left(E\left[\Delta\nu_k \Delta\nu_k' \middle| \mathcal{G}^\phi_t\right]\right)^{-1}$$

for discrete observations. This leads to the update equations for the Kalman filter:

$$\Delta\hat{x} = \gamma_x \Delta\nu_x;\ \Delta P_{xx} = -\gamma_x (H P_{\chi\chi} H' + R_x)\gamma_x'$$

with $\gamma_x = P_{\chi x} H'(H P_{\chi\chi} H' + R_x)^{-1}$. To implement the PME with discrete measurements, the same replacements will be made.

We can combine these results into a table showing the PME algorithm.

the PME: time-continuous state; time-discrete measurements

Between observations:

$$\frac{d}{dt}\hat{\phi}_t = Q'\hat{\phi}_t,$$

$$\frac{d}{dt}\hat{x}_t = \sum_i M_1(i) + \rho'\hat{\phi}_t,$$

$$\frac{d}{dt}P_{x\phi} = \sum_i (M_2(i) + N_2(i)) + P_{x\phi}Q + \rho' P_{\phi\phi},$$

$$\frac{d}{dt}P_{xx} = \sum_i (M_3(i) + M_3(i)' + N_3(i) + R_\chi(i)\hat{\phi}_i) + \rho' P_{\phi x} + P_{x\phi}\rho,$$

$$\begin{aligned}\frac{d}{dt}P_{xx\phi_m} = \sum_i \big(& M_4(i,m) + M_4(i,m)' + R_\chi(i)P_{\phi_i\phi_m} \\ &+ P_{xx\phi_i}Q_{im} + N_4(i)\big) + \rho' P_{\phi x\phi_m} + P_{x\phi\phi_m}\rho.\end{aligned}$$

At a modal observation:

$$\hat{\phi}^+ = \hat{\phi}^- * \Delta\vartheta,$$

$$\Delta\hat{x} = P_{x\phi}\Delta\vartheta,$$

$$\Delta P_{x\phi} = -\Delta\hat{x}\Delta\hat{\phi}' + \sum_k P_{x\phi\phi_k}\Delta\vartheta_k,$$

$$\Delta P_{xx} = -\Delta\hat{x}\Delta\hat{x}' + \sum_k P_{xx\phi_k}\Delta\vartheta_k,$$

$$\begin{aligned}\Delta P_{xx\phi_m} = &-\Delta\hat{\phi}_m\Delta\hat{x}\Delta\hat{x}' - \Delta\hat{\phi}_m P^+_{xx} - \Delta\hat{x}P^+_{\phi_m x} \\ &- P^+_{x\phi_m}\Delta\hat{x}' + \sum_k P_{xx\phi_m\phi_k}\Delta\vartheta_k.\end{aligned}$$

At a base-state observation:

$$\Delta\hat{x} = \gamma_x\Delta\upsilon_x,$$

$$\Delta P_{x\phi} = -\gamma_x H P_{\chi\phi},$$

$$\Delta P_{xx} = -\gamma_x(H P_{\chi\chi}H' + R_x)\gamma_x',$$

$$\Delta P_{xx\phi_m} = -\gamma_x H P_{\chi_x\phi_m} - P_{x\chi\phi_m}H'\gamma_x'.$$

The coefficients are:

coefficient identities for the PME with time-discrete observations

$$P_{\chi x} = P_{xx} + P_{\phi x} \text{(and similarly for } R_{\chi\phi}\text{, etc., respectively, } P_{\chi\phi}\text{, etc.),}$$

$$\gamma_x = P_{x\chi} H'(H P_{\chi\chi} H' + R_x)^{-1},$$

$$\boldsymbol{V}(\mathbf{e}_i) = \operatorname{diag}(\mathbf{e}_i' Q) - \mathbf{e}_i \otimes (\mathbf{e}_i' Q) - (\mathbf{e}_i \otimes (\mathbf{e}_i' Q))',$$

$$\boldsymbol{U}_k(\mathbf{e}_r) = (\mathbf{e}_r - \mathbf{e}_k) Q_{rk} (\mathbf{e}_r - \mathbf{e}_k)' + [\mathbf{e}_\mathbf{r} \otimes (\mathbf{e}_\mathbf{r}' Q) + (\mathbf{e}_\mathbf{r} \otimes (\mathbf{e}_\mathbf{r}' Q))' - \operatorname{diag}(\mathbf{e}_\mathbf{r}' Q)] \delta_{r,k},$$

$$M_1(i) = (\boldsymbol{A}_i \hat{x}_t + B_i u_t) \hat{\phi}_i + \boldsymbol{A}_i P_{x\phi_i} - B_i \boldsymbol{v} P_{\phi\phi_i},$$

$$M_2(i) = (\boldsymbol{A}_i \hat{x}_t + B_i u_t) P_{\phi_i \phi} + \boldsymbol{A}_i (P_{x\phi\phi_i} + P_{x\phi} \hat{\phi}_i) - B_i \boldsymbol{v} (P_{\phi\phi\phi_i} - P_{\phi\phi} \hat{\phi}_i),$$

$$N_2(i) = \sum_l Q_{il} (M(i,l) R_{x\phi_i} + \theta(i,l) \hat{x}_i) \mathbf{e}_l',$$

$$M_3(i) = (\boldsymbol{A}_i \hat{x}_t + B_i u_t) P_{\phi_i x} + \boldsymbol{A}_i (P_{xx\phi_i} + P_{xx} \hat{\phi}_i) - B_i \boldsymbol{v} (P_{\phi x \phi_i} - P_{\phi x} \hat{\phi}_i),$$

$$N_3(i) = \sum_l Q_{il} (M(i,l) R_{xx\phi_i} M(i,l)' + M(i,l) R_{x\phi i} \theta(i,l)' + \theta(i,l) R_{\phi_i x} M(i,l)' + \theta(i,l) \hat{\phi}_i \theta(i,l)'),$$

$$M_4(i,m) = (\boldsymbol{A}_i \hat{x}_t + B_i u_t) P_{\phi_i x \phi_m} + \boldsymbol{A}_i (P_{xx\phi_i\phi_m} + P_{xx\phi_m} \hat{\phi}_i - P_{x\phi_i} P_{\phi_m x}) - B_i \boldsymbol{v} ((\mathbf{e}_i - \hat{\phi}) P_{\phi_i x} (\delta_{i,m} - \hat{\phi}_m - P_{\phi\phi_i} P_{x\phi_m})),$$

$$N_4(i,m) = \sum_l Q_{il} (M(i,l) (P_{(xx\phi_i)\phi_m} + \delta_{l,m} R_{xx\phi_i}) M(i,l)' + M(i,l) ((P_{(x\phi_m)\phi})_{.i} + \delta_{l,m} R_{x\phi_i}) \theta(i,l)' + \theta(i,l) ((P_{(x\phi_m)\phi})_{i.} + \delta_{l,m} R_{\phi_i x}) M(i,l)' + \theta(i,l) (P_{\phi_i\phi_m} + \delta_{l,m} \hat{\phi}_i) \theta(i,l)'). + Q_{im} (M(i,m) P_{(x\phi_i)x} + \theta(i,m) R_{\phi_i x} + P_{(x\phi_i)x} M(i,m)' + R_{x\phi_i} \theta(i,m)').$$

Appendix 2

COM Derivation Details

At time $t = kT$, $[q^j[k](\zeta)]$ is a Gaussian sum with N terms in each component:

$$q^j[k](\zeta) = \sum_{l=1}^{N} \alpha_l^j[k] |D_l^j[k]|^{\frac{1}{2}} \exp -\frac{1}{2}\{(\zeta - m_l^j[k])' D_l^j[k](\zeta - m_l^j[k])\}.$$

The recurrence relation for $q^i[k+1]$ is given in (9.18):

$$q^i[k+1](z) = \sum_j \Pi_{ij} z[k+1]' \mathbf{P}_{.i} |C_j|^{-1} \Phi(F^{-1}(y[k+1] - H_i(z + \chi_i)))$$
$$\times \int_\Omega \Phi(C_j^{-1}(z - A_j\zeta + \chi_i - \chi_j)) q^j[k](\zeta)\, d\zeta.$$

Expand the product of Gaussian pattern functions in (9.18). In what follows, factors common across all modal hypotheses will be ignored without comment. We get

$$\Phi(F^{-1}(y[k+1] - H_i(z + \chi_i)))\Phi(C_j^{-1}(z - A_j\zeta + \chi_i - \chi_j)) q[k]^j(\zeta)$$
$$= \sum_{l=1}^{N} \alpha_l^j[k] |D_l^j[k]|^{\frac{1}{2}} \exp -\frac{1}{2} J_1(l),$$

where

$$J_1(l) = (\zeta - m_l^j[k])' D_l^j[k](\zeta - m_l^j[k])$$
$$+ (C_j^{-1}(z - A_j\zeta + \chi_i - \chi_j))'(C_j^{-1}(z - A_j\zeta + \chi_i - \chi_j))$$
$$+ (F^{-1}(y[k+1] - H_i(z + \chi_i)))'(F^{-1}(y[k+1] - H_i(z + \chi_i))).$$

The argument of the exponential is a quadratic form. Define a new set of coefficient matrices:

$$A_l^j[k] = D_l^j[k] + A_j' D_\chi(j) A_j,$$

$$B^j[k](l;i) = D_l^j[k]\,m_l^j[k] + A_j' D_\chi(j)(z + \chi_i - \chi_j),$$

$$C^j[k](l;i) = m_l^j[k]' D_l^j[k] m_l^j[k] + (z + \chi_i - \chi_j)' D_\chi(j)(z + \chi_i - \chi_j)$$
$$+ (y[k+1] - H_i(z + \chi_i))' D_x (y[k+1] - H_i(z + \chi_i)).$$

The exponential factor can be written

$$\exp -\frac{1}{2} J_1(l) = \left|A_l^j[k]\right|^{-\frac{1}{2}} \exp -\frac{1}{2}\{C^j[k](l;i) - B^j[k](l;i)'$$
$$\times A_l^j[k]^{-1} B^j[k](l;i)\}\mathbf{N}_\zeta(A_l^j[k]^{-1} B^j[k](l;i), A_l^j[k]^{-1}).$$

Next complete the squares in z to get

$$B^j[k](l;i) = H^j z + E^i[k](l;j),$$

$$H^j = A_j' D_\chi(j),$$

$$E^j[k](l;i) = D_l^j[k] m_l^j[k] + A_j' D_\chi(j)(\chi_i - \chi_j)$$

and

$$C^j[k](l;i) = z' F^j[k](l;i) z - 2z' K^j[k](l;i) + G^j[k](l;i),$$

$$F^j[k](l;i) = D_\chi(j) + H_i' D_x H_i,$$

$$K^j[k](l;i) = H_i' D_x (y[k+1] - H_i \chi_i) - D_\chi(j)(\chi_i - \chi_j),$$

$$G^j[k](l;i) = m_l^j[k]' D_l^j[k] m_l^j[k] + (\chi_i - \chi_j)' D_\chi(j)(\chi_i - \chi_j)$$
$$+ (y[k+1] - H_i \chi_i)' D_x (y[k+1] - H_i \chi_i).$$

Completing the squares and combining gives

$$\exp -\frac{1}{2} J_1(l) = \left|A_l^j[k]\right|^{-\frac{1}{2}} \left|F^j[k](l;i) - (H^j)' A_l^j[k]^{-1} H^j\right|^{-\frac{1}{2}}$$
$$\times \exp -\frac{1}{2}\{G^j[k](l;i) - E^j[k](l;i)' A_l^j[k]^{-1} E^j[k](l;i)$$
$$- J_2(l)' J_3(l)^{-1} J_2(l)\}\mathbf{N}_\zeta(A_l^j[k]^{-1} B^j[k](l;i), A_l^j[k]^{-1})$$
$$\times \mathbf{N}_z(J_3(l)^{-1} J_2(l), J_3(l)^{-1}),$$

where

$$J_2(l) = (H^j)' A_l^j[k]^{-1} E^j[k](l;i) + K^j[k](l;i),$$

$$J_3(l) = F^j[k](l;i) - (H^j)' A_l^j[k]^{-1} H^j.$$

Now integrate to obtain

$$\int_{\Omega} \Phi(F^{-1}(y[k+1] - H_i(z + \chi_i)))\Phi(C_j^{-1}(z - A_j\zeta + \chi_i - \chi_j))q[k]^j(\zeta)\, d\zeta$$

$$= \sum_{l=1}^{N} \alpha_l^j[k]|D_l^j[k]|^{\frac{1}{2}}|A_l^j[k]|^{-\frac{1}{2}}|F^j[k](l;i) - (H^j)'A_l^j[k]^{-1}H^j|^{-\frac{1}{2}}$$

$$\times \exp -\frac{1}{2}\{G^j[k](l;i) - E^j[k](l;i)'A_l^j[k]^{-1}E^j[k](l;i)$$

$$-J_2(l)'J_3(l)^{-1}J_2(l)\}\mathbf{N}_z(J_3(l)^{-1}J_2(l), J_3(l)^{-1}).$$

Then $q^i[k+1]$ is a Gaussian sum with

$$D^i[k+1](l;j) = F^j[k](l;i) - (H^j)'A_l^j[k]^{-1}H^j,$$

$$m^i[k+1](l;j) = P^i[k+1](l;j)((H^j)'A_l^j[k]^{-1}E^j[k](l;i) + K^j[k](l;i)),$$

$$\alpha^i[k+1](l;j) = \Pi_{ij}z[k+1]'\mathbf{P}_{.i}|C_j|^{-1}\alpha_l^j[k]|D_l^j[k]|^{\frac{1}{2}}|A_l^j[k]|^{-\frac{1}{2}}$$

$$\times|D^i[k+1](l;j)|^{-\frac{1}{2}} \exp -\frac{1}{2}\{G^j[k](l;i)$$

$$-E^j[k](l;i)'A_l^j[k]^{-1}E^j[k](l;i) - m^i[k+1](l;j)'$$

$$\times D^i[k+1](l;j)m^i[k+1](l;j)\}.$$

The expressions for the coefficients can be simplified. Note first that

$$D^i[k+1](l;j) = D_\chi(j) + H_i'D_xH_i - D_\chi(j)A_j(D_l^j[k]$$

$$+A_j'D_\chi(j)A_j)^{-1}A_j'D_\chi(j)$$

$$= H_i'D_xH_i + (R_\chi(j) + A_jP_l^j[k]A_j')^{-1}.$$

This can be written in more conventional form as

$$P_l^j[k+1]^- = A_jP_l^j[k]A_j' + R_\chi(j)$$

followed by

$$D^i[k+1](l;j) = D_l^j[k+1]^- + H_i'D_xH_i.$$

With $d^i[k](l;j) - D^i[k](l;j)m^i[k](l;j)$, we have

$$d^i[k+1](l;j) = (H^j)'A_l^j[k]^{-1}E^j[k](l;i) + K^j[k](l;i)$$

$$= D_\chi(j)A_j(D_l^j[k] + A'_j D_\chi(j)A_j)^{-1}(D_l^j[k]m_l^j[k]$$
$$+A'_j D_\chi(j)(\chi_i - \chi_j))$$
$$+H'_i D_x(y[k+1] - H_i\chi_i) - D_\chi(j)(\chi_i - \chi_j)$$
$$= D_\chi(j)A_j(D_l^j[k] + A'_j D_\chi(j)A_j)^{-1}d_l^j[k]$$
$$+(D_\chi(j)A_j(D_l^j[k] + A'_j D_\chi(j)A_j)^{-1}A'_j D_\chi(j)$$
$$-D_\chi(j))(\chi_i - \chi_j) + H'_i D_x(y[k+1] - H_i\chi_i).$$

Write this as

$$d_l^j[k+1]^- = D_\chi(j)A_j(D_l^j[k] + A'_j D_\chi(j)A_j)^{-1}d_l^j[k],$$
$$d^i[k+1](l;j) = d_l^j[k+1]^- + H'_i D_x(y[k+1] - H_i\chi_i)$$
$$-D_l^j[k+1]^-(\chi_i - \chi_j).$$

Note that

$$(D_l^j[k] + A'_j D_\chi(j)A_j)^{-1}$$
$$= A_j^{-T}(I - (A_j^{-1}R_\chi(j)A_j^{-T} + P_l^j[k])^{-1}A_j^{-1}R_\chi(j)A_j^{-T}).$$

So

$$P_l^j[k+1]^- D_\chi(j)A(D_l^j[k] + A'_j D_\chi(j)A_j)^{-1}$$
$$= A_j(A_j^{-1}R_\chi(j)A_j^{-T} + P_l^j[k])(I - (A_j^{-1}R_\chi(j)A_j^{-T}$$
$$+P_l^j[k])^{-1}R_\chi(j)A_j^{-T}$$
$$= A_j(A_j^{-1}R_\chi(j)A_j^{-T} + P_l^j[k]) - A_jA_j^{-1}R_\chi(j)A_j^{-T}$$
$$= A_jP_l^j[k].$$

Hence

$$P_l^j[k+1]^- d_l^j[k+1]^- = A_jP_l^j[k]d_l^j[k]$$
$$= A_jm_l^j[k].$$

This can be written

$$m_l^j[k+1]^- = A_jm_l^j[k].$$

The expression for the weighting coefficients is complicated in appearance. The exponential argument in $\alpha^i[k+1](l;j)$ can be rewritten as

$$\begin{aligned} &G^j[k](l;i) - E^j[k](l;i)' A_l^j[k]^{-1} E^j[k](l;i) - J_2(l)' J_3(l)^{-1} J_2(l) \\ &\quad = d_l^j[k]' P_l^j[k] d_l^j[k] + (\chi_i - \chi_j)' D_\chi(j)(\chi_i - \chi_j) \\ &\qquad + (y[k+1] - H_i \chi_i)' D_x (y[k+1] - H_i \chi_i) \\ &\qquad - (d_l^j[k] + A_j' D_\chi(j)(\chi_i - \chi_j))' A_l^j[k]^{-1} (d_l^j[k] + A_j' D_\chi(j)(\chi_i - \chi_j)) \\ &\qquad - d^i[k+1](l;j)' P^i[k+1](l;j) d^i[k+1](l;j). \end{aligned}$$

But

$$\begin{aligned} &d_l^j[k]' P_l^j[k] d_l^j[k] - d_l^j[k]' A_l^j[k]^{-1} d_l^j[k] \\ &\quad = d_l^j[k]' (P_l^j[k] - A_l^j[k]^{-1}) d_l^j[k] \\ &\quad = d_l^j[k]' P_l^j[k] A_j' P_l^j[k+1]^- A_j P_l^j[k] d_l^j[k] \\ &\quad = m_l^j[k]' A_j' D_l^j[k+1]^- A_j m_l^j[k]. \end{aligned}$$

Also,

$$\begin{aligned} &(\chi_i - \chi_j)' D_\chi(j)(\chi_i - \chi_j) - (A_j' D_\chi(j)(\chi_i - \chi_j))' A_l^j[k]^{-1} (A_j' D_\chi(j)(\chi_i - \chi_j)) \\ &\quad = (\chi_i - \chi_j)' D_l^j[k+1]^- (\chi_i - \chi_j). \end{aligned}$$

Finally,

$$\begin{aligned} &-2 d_l^j[k]' A_l^j[k]^{-1} A_j' D_\chi(j)(\chi_i - \chi_j) = 2(d_l^j[k+1]^-)'(\chi_i - \chi_j) \\ &\quad = -2(m_l^j[k+1]^-)' D_l^j[k+1]^- (\chi_i - \chi_j). \end{aligned}$$

Therefore, with some cancellation of terms, we obtain

$$\begin{aligned} &G^j[k](l;i) - E^j[k](l;i)' A_l^j[k]^{-1} E^j[k](l;i) - J_2(l)' J_3(l)^{-1} J_2(l) \\ &\quad = m^i[k+1](l;j)' D^i[k+1](l;j) m^i[k+1](l;j) \\ &\qquad - (m_l^j[k+1]^- + \chi_j - \chi_i)' D_l^j[k+1]^- (m_l^j[k+1]^- + \chi_j - \chi_i) \\ &\qquad + (y[k+1] - H_i \chi_i)' D_x (y[k+1] - H_i \chi_i). \end{aligned}$$

Also,

$$\begin{aligned} &|D_l^j[k]|^{\frac{1}{2}} |A_l^j[k]|^{-\frac{1}{2}} |J_3(l)|^{-\frac{1}{2}} \\ &\quad = |D_l^j[k]|^{\frac{1}{2}} |D_l^j[k] + A_j' D_\chi(j) A_j|^{-\frac{1}{2}} |P^i[k+1](l;j)|^{\frac{1}{2}}. \end{aligned}$$

The result thus follows.

BIBLIOGRAPHY

[AB91] L. Adelman and T. Bresnick. Examining the effect of information sequence on expert judgement. *Proceedings of the Symposium on Command and Control Research Command and Control Research*, pp. 306–317, June 1991.

[ACW73] G. A. Arredondo, W. H. Chriss, and E. H. Walker. A multipath fading simulator for mobile radio. *IEEE Trans. on Vehicular Technology*, 22(4):241–247, Nov. 1973.

[AIK91] A. Averbuch, S. Itzikowitz, and T. Kapon. Radar target tracking Viterbi versus IMM. *IEEE Trans. on Aerospace and Electronic Systems*, 27:550–563, May 1991.

[AKG86] D. Andrisani, F. P. Kuhl, and D. Gleason. A nonlinear tracker using attitude measurements. *IEEE Trans. on Aerospace and Electronic Systems*, 22:533–539, Sept. 1986.

[AKKS91] D. Andrisani, E. T. Kim, F. P. Kuhl, and J. Schierman. A nonlinear helicopter tracker using attitude measurements. *IEEE Trans. on Aerospace and Electronic Systems*, 27:328–333, Jan. 1991.

[AM79] B. D. O Anderson and J. B. Moore. *Optimal Filtering*. Prentice-Hall, 1979.

[AM90] B. D. O. Anderson and J. B. Moore. *Optimal Control: Linear Quadratic Methods*. Prentice-Hall, 1990.

[AS87] D. Andrisani and J. Schierman. Tracking maneuvering helicopters using attitude and rotor angle measurements. *Proceedings of the Asilomar Conf. on Circuits, Systems, and Computers*, Nov. 1987.

[Ath87] M. Athans. Command and control (C2) theory: A challenge to control science. *IEEE Trans. Automatic Control*, 32:286–293, April 1987.

[AW95] K. J. Astrom and B. Wittenmark. *Adaptive Control*. Addison-Wesley, 2nd ed., 1995.

[BB96] E. K. Boukas and K. Benjelloun. Robust control for linear systems with Markovian jumping parameters. *Proc. of the IFAC Conference*, pp. 433–438, June 1996.

[BBS88] H. A. P. Blom and Y. Bar-Shalom. The interacting multiple model algorithm for systems with Markov switching coefficients. *IEEE Trans. Automatic Control*, 3(8):780–783, Aug. 1988.

[Bek83] E. Bekir. Adaptive Kalman filter for tracking maneuvering targets. *Journal of Guidance Control and Dynamics*, 6(5):414–416, Sept. 1983.

[Ber83] R. F. Berg. Estimation and prediction for maneuvering target trajectories. *IEEE Trans. Automatic Control*, 28:294–304, March 1983.

[Bha86] B. Bhanu. Automatic target recognition: State of the art survey. *IEEE Trans. on Aerospace and Electronic Systems*, 22:364–379, July 1986.

[Bjo82] T. Bjork. Finite optimal filters for a class of nonlinear diffusions with jumping parameters. *Stochastics*, 6:121–138, 1982.

[Bog87] P. L. Bogler. Tracking a maneuvering target using input estimation. *IEEE Trans. on Aerospace and Electronic Systems*, 23:298–301, May 1987.

[Boy96] J. E. Boyd. *Nonlinear Filtering for Multimodal Systems with Mode-Dependent Observations*. PhD dissertation, University of California, San Diego, June 1996.

[BPL82] H. H. Burk, T. R. Perkins, and J.F Lethrum. Maneuvering vehicle path simulator. Tech Report AMSAA TR 331, U.S. Army Motevial Systems Analysis Activity, 1982.

[Brè72] P. M. Brémaud. *A Martingale Approach to Point Processes*. PhD thesis, University of California, Berkeley, 1972.

[BW92] R. G. Brown and P. Y. C. Wang. *Introduction to Random Signals and Applied Kalman Filtering*. John Wiley, 2nd ed., 1992.

[BWC94] W. D. Blair, G. A. Watson, and T. A. Chmielewski. Benchmark problem for beam pointing control of phased array radar against maneuvering targets. *Proc. of the American Control Conference*, pp. 2071–2075, June 1994.

[CEF89] J. R. Cloutier, J. H. Evers, and J. J. Feeler. Assessment of air-to-air missile guidance and control technology. *IEEE Control Systems Magazine*, pp. 27–34, Oct. 1989.

[Cla70] J. M. C. Clark. The representation of functionals of Brownian motion by stochastic integrals. *Ann. Math. Statist.*, 41:1282–1295, 1970.

[COD91] E. Cotina, D. Otero, and C. E. D'Attellis. Maneuvering target tracking using extended Kalman filter. *IEEE Trans. on Aerospace and Electronic Systems*, 27(1):155–158, Jan. 1991.

[CS95] G. A. Clapp and D. D. Sworder. Modeling TAD/TMD decision maker dynamic response. *Proc. of the Fire Control Symposium*, pp. 529–547, Aug. 1995.

[CSB96a] G. A. Clapp, D. D. Sworder, and J. E. Boyd. Multisensor fusion in target tracking. *Proc. of the National Symposium on Sensor Fusion*, 1:305–319, March 1996.

[CSB96b] G. A. Clapp, D. D. Sworder, and J. E. Boyd. Multisensor fusion: Results and algorithms from the joint BMDO/AUSTRALIA tracking demonstration. *Proc. of the Fire Control Symposium*, pp. 56–70, Aug. 1996.

[CT84] C. B. Chang and J. A. Tabaczynski. Application of state estimation to target tracking. *IEEE Trans. Automatic Control*, 29:98–109, Feb. 1984.

[DB94] F. Dufour and P. Bertrand. The filtering problem for continuous-time linear systems with Markovian switching coefficients. *Systems and Control Letters*, 23:453–461, 1994.

[DB95] R. C. Dorf and R. H. Bishop. *Modern Control Systems*. Addison-Wesley, 7th ed., 1995.

[DB96] F. Dufour and P. Bertrand. An image-based filter for discrete time Markovian jump linear systems. *Automatica*, 32(2):241–247, 1996.

[DBE96] F. Dufour, P. Bertrand, and R. J. Elliott. Filtering for linear systems with jump parameters with high signal-to-noise ratio. *Proc. of the IFAC Conference*, pp. 445–450, June 1996.

[DE98] F. Dufour and R. J. Elliott. Adaptive control of linear systems with Markov perturbations. *IEEE Trans. Automatic Control*, 43(3):351–372, March 1998.

[DM82] C. Dellacherie and P. A. Meyer. *Probabilities and Potential B, Theory of Martingales*. North-Holland, 1982.

[DM91] F. Dufour and M. Mariton. Tracking 3D maneuvering target with passive sensors. *IEEE Trans. on Aerospace and Electronic Systems*, 27:725–739, July 1991.

[EAM95] R. J. Elliott, L. Aggoun, and J. B. Moore. *Hidden Markov Models: Estimation and Control*. Springer-Verlag, 1995.

[EB96] H. H. C. Everdji and H. A. P. Blom. Embedding adaptive JLQG into LQ Martingale control with a completely observable stochastic control matrix. *IEEE Trans. Automatic Control*, 41(3):424–430, 1996.

[EDS96] R. J. Elliott, F. Dufour, and D. D. Sworder. Exact hybrid filters in discrete time. *IEEE Trans. Automatic Control*, 41(12):1807–1810, 1996.

[EE99] J. S. Evans and R. J. Evans. Image-enhanced multiple model tracking. *Automatica*, 1999.

[Ell82] R. J. Elliott. *Stochastic Calculus and Applications*. Springer-Verlag, 1982.

[EvdH99] R. J. Elliott and J van der Hoek. A finite dimensional filter for hybrid observations. *IEEE Trans. Automatic Control*, Vol 43 (1998), 736–739.

[FKK72] M. Fujisaki, G. Kallianpur, and H. Kunita. Stochastic differential equations for the nonlinear filtering problem. *Osaka J. Math.*, 9:19–40, 1972.

[FLC96] Y. Fang, X. Li, and X. Chu. Almost sure stability of jump linear systems. *Proc. of the IFAC Conference*, pp. 457–462, June 1996.

[FLF95] Y. Fang, K. A. Loparo, and X. Feng. Stability of discrete time jump linear systems. *Journal of Mathematical Systems, Estimation and Control*, 5(3):275–321, 1995.

[FLJC92] X. Feng, K. A. Loparo, Y. Ji, and H. J. Chizeck. Stochastic stability properties of jump linear systems. *IEEE Trans. Automatic Control*, 37(1):38–52, Jan. 1992.

[Fri96] B. Friedland. *Advanced Control System Design*. Prentice-Hall, 1996.

[GA93] M. S. Grewal and A. P. Andrews. *Kalman Filtering: Theory and Practice*. Prentice-Hall, 1993.

[Gel84] A. Gelb. *Applied Optimal Estimation*. MIT Press, 1984.

[GM77] N. H. Gholson and R. L. Moose. Maneuvering target tracking using adaptive state estimation. *IEEE Trans. on Aerospace and Electronic Systems*, 13:310–316, May 1977.

[GW95] F. Gustafson and B. Wahlberg. Blind equalization by direct examination of the input sequences. *IEEE Trans. on Communications*, 43:2213–2222, July 1995.

[HBS89] A. Houles and Y. Bar-Shalom. Multisensor tracking of a maneuvering target in clutter. *IEEE Trans. on Aerospace and Electronic Systems*, 25:176–188, March 1989.

[HG90] S. A. R Hepner and H. P. Geering. Observability analysis for target maneuver estimation via bearing-only and bearing-rate-only measurements. *Journal of Guidance Control and Dynamics*, 13:977–983, Nov. 1990.

[HMZ87] G. A. Hewer, R. D. Martin, and J. Zeh. Robust processing for Kalman filtering of glint noise. *IEEE Trans. on Aerospace and Electronic Systems*, 23:120–128, Jan. 1987.

[HS92] R. G. Hutchins and D. D. Sworder. Image fusion algorithms for tracking maneuvering targets. *Journal of Guidance Control and Dynamics*, 15(1):175–184, Jan. 1992.

[Ilt90] R. A. Iltis. A Baysian channel and timing estimation algorithm for use with MLSE. *Proceedings of the Asilomar Conf. on Circuits, Systems, and Computers*, pp. 119–123, Oct. 1990.

[Itô51] K. Itô. Multiple Wiener integrals. *J. Math. Soc. Jpn.*, 3:157–169, 1951.

[Ken96] M. Kent. *Recursive HMM Probability Distribution Computation and Its Application to the Demodulation of CPM Signals*. PhD dissertation, University of California, San Diego, June 1996.

[Kle89] G. A. Klein. Recognition-primed decisions. In *Advances in Man Machine Research*, pp. 47–92, 1989. JAI Press.

[Kle91] G. A. Klein. Models of skilled decision making. *Proceedings of the 1991 Human Factors Society 35th Annual Meeting*, 1991.

[KMR81] J. D. Kendrick, P. S. Maybeck, and J. G. Reid. Estimation of aircraft target motion using orientation measurements. *IEEE Trans. on Aerospace and Electronic Systems*, 17:254–259, March 1981.

[Kor90] I. Korn. GMSK with differential phase detection in the satellite mobil channel. *IEEE Trans. on Communications*, 38:1980–1985, Nov. 1990.

[Kri84] V Krishnan. *Nonlinear Filtering and Smoothing*. John Wiley, 1984.

[KSE94] M. Kent, D. D. Sworder, and R. J. Elliott. GMSK for mobile communications. *Proceedings of the Asilomar Conf. on Circuits, Systems, and Computers*, pp. 455–459, Oct. 1994.

[Kuh92] F. P. Kuhl. Radar recognition of geometric classes in free space. *Remote Sensing Reviews*, 6:331–349, 1992.

[KW67] H. Kunita and S. Watanabe. On square integrable martingales. *Nagoya Math. J.*,30:209–245, 1967.

[LBS93] X. R. Li and Y. Bar-Shalom. Design of an interacting multiple model algorithm for air traffic control tracking. *IEEE Trans. on Control Systems Technology*, 1(3):186–194, Sept. 1993.

[LBV91] K. A. Loparo, M. R. Buchner, and K. S. Vasudeva. Leak detection in an experimental heat exchanger process: A multiple model approach. *IEEE Transactions on Automatic Control*, 36(2):167–177, Feb. 1991.

[LDB98] D Laneuville, F. Dufour, and P. Bertrand. Image based maneuvering target tracking. *Proc. of the American Control Conference*, pp. 2444–2449, June 1998.

[Lef84] C. C. Lefas. Using roll angle measurements to track aircraft maneuvers. *IEEE Trans. on Aerospace and Electronic Systems*, 20(6):671–681, Nov. 1984.

[LF91] S. S. Lim and M. Farooq. Maneuvering target tracking using jump processes. *Proc. of the Conf. on Decision and Control*, pp. 2049–2054, Dec. 1991.

[Lig88] P. L. Ligomenides. Real-time capture of experiential knowledge. *IEEE Trans. on Systems, Man, and Cybernetics*, SMC-18:542–551, July 1988.

[LKM85] M. L. Lee, W. J. Kolodzeij, and R. R. Moler. Stochastic dynamic system suboptimal control with uncertain parameters. *IEEE Trans. on Aerospace and Electronic Systems*, 21(5):594–599, Sept. 1985.

[LM90] J. H. Lodge and M. L. Moher. Maximum likelihood sequence estimation of CPM signals transmitted over Rayleigh flat-fading channels. *IEEE Trans. on Communications*, 38:787–794, June 1990.

[LRE86] K. A. Loparo, Z. Roth, and S. J. Eckert. Nonlinear filtering for systems with random structure. *IEEE Trans. Automatic Control*, 31(11):1064–1068, Nov. 1986.

[Mar90] M. Mariton. *Jump Linear Systems in Automatic Control*. Marcell Dekker, 1990.

[May79] P. S. Maybeck. *Stochastic Models, Estimation and Control*, Volume 1. Academic Press, 1979.

[May82a] P. S. Maybeck. *Stochastic Models, Estimation and Control*, Volume 2. Academic Press, 1982.

[May82b] P. S. Maybeck. *Stochastic Models, Estimation and Control*, Volume 3. Academic Press, 1982.

[MH81] K Murota and K. Hirade. GMSK modulation for digital mobile radio telephony. *IEEE Trans. on Communications Technology*, 29:1044–1050, July 1981.

[MS84] P. S. Maybeck and R. I. Suizu. Adaptive tracker field-of-view variation via multiple model filtering. *IEEE Trans. on Aerospace and Electronic Systems*, 21:529–539, July 1984.

[MS91] P. S. Maybeck and R. D. Stevens. Reconfigurable flight control via multiple model adaptive control methods. *IEEE Trans. on Aerospace and Electronic Systems*, 27:470–480, May 1991.

[MSG95] M. I. Miller, A. Srivastava, and U. Grenander. Conditional-mean estimation via jump-diffusion processes in multiple target tracking/recognition. *IEEE Trans. on Signal Processing*, 43:2678–2690, Nov. 1995.

[MVM79] R. L. Moose, V. F. Vanlandingham, and D. H. McCabe. Modeling and estimation for maneuvering targets. *IEEE Trans. on Aerospace and Electronic Systems*, 15:448–455, May 1979.

[Pap91] A. Papoulis. *Probability, Random Variables, and Stochastic Processes*. McGraw-Hill, 3rd ed., 1991.

[Pro95] J. G. Proakis. *Digital Communication*. McGraw-Hill, 1995.

[Rug91] W. J. Rugh. Analytical framework for gain scheduling. *IEEE Control Systems Magazine*, pp. 79–84, Jan. 1991.

[RW78] G. G. Ricker and J. R. Williams. Adaptive tracking filter for maneuvering targets. *IEEE Trans. on Aerospace and Electronic Systems*, 14(1):185–193, Jan. 1978.

[SA77] D. D. Sworder and S. M. Archer. Influence of sensor failures on LQG regulators. *Journal of Cybernetics*, 7:269–278, 1977.

[SB97a] D. D. Sworder and J. E. Boyd. Hybrid estimation algorithms II. *Journal of Optimization Theory and Applications*, 94(1), July 1997.

[SB97b] D. D. Sworder and J. E. Boyd. More on gain adaptive tracking. *Journal of Guidance Control and Dynamics*, 20(5):882–890, Sept. 1997.

[SBCV99] D. D. Sworder, J. E. Boyd, G. A. Clapp, and R. Vojak. Nonlinear trackers using image dependent gains. (Photogrammatic Engineering and Remote Sensing will appear in 1999), 1999.

[SC86] D. D. Sworder and D. S. Chou. Feedforward/feedback controls in a noisy environment. *IEEE Trans. on Systems, Man, and Cybernetics*, 16(4):522–531, July 1986.

[SC94] D. D. Sworder and G. A. Clapp. Decisionmaker styles in dynamic situation assessment. *Proc. of the Symposium on Command and Control Research Command and Control Research*, pp. 591–601, June 1994.

[Sch80] E. E. Schiring. Simplified linear dynamic model for first cut controller design of a solar-powered once-through boiler. *Instrumentation for the Power Industry*, 23:35–52, May 1980.

[SCK93] D. D. Sworder, G. A. Clapp, and T. W. Kidd. Model-based prediction of decision maker delays and uncertainty. *Cybernetics and Systems; An International Journal*, 24:51–68, 1993.

[Ses94] N. Seshardi. Joint data and channel estimation using blind trellis search techniques. In *Blind Deconvolution*, pp. 259–286, 1994. Prentice-Hall.

[SH89] D. D. Sworder and R. G. Hutchins. Image enhanced tracking. *IEEE Trans. on Aerospace and Electronic Systems*, 25(5):701–710, Sept. 1989.

[SH90] D. D. Sworder and R. G. Hutchins. Maneuver estimation using estimates of orientation. *IEEE Trans. on Aerospace and Electronic Systems*, 26(5):625–638, Sept. 1990.

[SH92] D. D. Sworder and R. G. Hutchins. Improved tracking of an agile target. *Journal of Guidance Control and Dynamics*, 15(5):1281–1284, Sept. 1992.

[Shi96] P. Shi. Robust control for linear systems with Markovian jumping parameters. *Proc. of the IFAC Conference*, pp. 433–438, June 1996.

[SHK93] D. D. Sworder, R. G. Hutchins, and M. Kent. Utility of imaging sensors in tracking systems. *Automatica*, 29(2):445–449, March 1993.

[Sin70] R. A. Singer. Estimating optimal tracking filter performance for manned maneuvering targers. *IEEE Trans. on Aerospace and Electronic Systems*, 10:473–482, July 1970.

[SKT90] J. L. Speyer, K. D. Kim, and M. Tahke. Passive homing missile guidance law based on new target maneuver models. *Journal of Guidance Control and Dynamics*, 13(5):803–812, Sept. 1990.

[SKVH95] D. D. Sworder, M. Kent, R. Vojak, and R. G. Hutchins. Renewal models for maneuvering targets. *IEEE Trans. on Aerospace and Electronic Systems*, 31(1):138–150, Jan. 1995.

[SR83] D. D. Sworder and R. O. Rogers. An LQ-solution to a control problem associated with a solar thermal central receiver. *IEEE Trans. Automatic Control*, 28(10):971–978, Oct. 1983.

[SV94] D. D. Sworder and R. Vojak. Tracking mobile vehicles using a non-Markovian maneuver model. *Journal of Guidance Control and Dynamics*, 17(4):870–873, July 1994.

[SVH93] D. D. Sworder, R. Vojak, and R. G. Hutchins. Gain adaptive tracking. *Journal of Guidance Control and Dynamics*, 16(5):865–873, Sept. 1993.

[SW83] M. K. Simon and C. C. Wang. Differential versus limiter-discriminator detection for narrow band FM. *IEEE Trans. on Communications*, 31(11):1227–1234, Nov. 1983.

[SW84] M. K. Simon and C. C. Wang. Differential detection of Gaussian MSK in a mobile radio environment. *IEEE Trans. on Vehicular Technology*, 33(4):307–320, Nov. 1984.

[Swo66a] D. D. Sworder. *Optimal Adaptive Control Systems*. Academic Press, 1966.

[Swo66b] D. D. Sworder. A study of the relationship between identification and control in adaptive control systems. *Journal of the Franklin Institute*, 281(3):198–231, 1966.

[Swo69a] D. D. Sworder. Feedback control of a class of linear systems with jump parameters. *IEEE Trans. Automatic Control*, 14(1):9–14, Feb. 1969.

[Swo69b] D. D. Sworder. On the control of stochastic systems: II. *International Journal of Control*, 10(3):271–277, 1969.

[Swo82] D. D. Sworder. Regulation of stochastic systems with wide-band transfer functions. *IEEE Trans. on Systems, Man, and Cybernetics*, 12(2):307–314, May 1982.

[Tin95] J. Tinsley. Foreign antiship cruise missiles – representative threat systems. *Proceedings of the Fire Control Symposium*, Vol. 2, pp. 61–67, Aug. 1995.

[TRK92] R. W. Taylor, A. P. Reeves, and F. P. Kuhl. Method for identifying object class, type and orientation in the presence of uncertainty. *Remote Sensing Reviews*, 6:183–206, 1992.

[Vau90] W. S. Vaughan. Toward a science of command and control: Challenges of distributed decision making. *Proc. of the Symposium on Command and Control Research Command and Control Research*, pp. 26–35, June 1990.

[WF88] D. E. Williams and B. Friedland. Target maneuver detection and estimation. *Proc. of the Conf. on Decision and Control*, pp. 851–855, Dec. 1988.

[WH85] E. Wong and B. Hajek. *Stochastic Processes in Engineering Systems*. Springer-Verlag, 1985.

[WKBS98] H. Wang, T. Kirubarajan, and Y. Bar-Shalom. Precision large scale air surveillance using IMM assignment estimators. *IEEE Trans. on Aerospace and Electronic Systems*, (35)1:255–266, Jan. 1999.

[Wol94] W. A. Wolovich. *Automatic Control Systems: Basic Analysis and Design*. Saunders College Publishing, 1994.

[YMF88] A. Yougacogla, D. Makrakis, and M. Feher. Differential detection of GMSK using decision feedback. *IEEE Trans. on Communications*, 36:641–649, June 1988.

INDEX

GLOSSARY

$A(.,j)$, the jth column of matrix A, 73

$A(i,.)$, the ith row of matrix A, 73

$[\eta,\eta]_t$, optional quadratic variation, xviii

$\approx$, approximately equal to, xviii

$\int_\Omega$, an integral over the entire applicable space or domain Ω, xv

$\mapsto$, a modal state transition, e.g., $\mathbf{e}_i \mapsto \mathbf{e}_j$, 6

$*$, Hadamard (element by element) vector product, xv

1, a vector of all ones, dimension clear from the context, xv

$A(r:s,t:v)$, the submatrix of A containing rows r through s of columns t through v, xv

$\{\nu[k](z)\}$, a (possibly vector) discrete sequence of functions of z, xv

$\{\nu[k]\}$, a discrete sequence indexed by k, xv

$\{\nu_1\}$, the first component of the vector sequence $\{\nu_t\}$, xv

$\{\nu_x\}$, a subvector process in $\{\nu_t\}$ associated with the process $\{x_t\}$, xv

$\{\nu_i[k](z)\}$, the ith component of $\{\nu[k](z)\}$, xv

$\{\nu_t\}$, a continuous-time process indexed by t, xv

$\mathbf{A}_i$, a composite plant dynamics matrix, $A_i + \sum_l Q_{il} M(i,l)$, 9

Δm_t, $m_t - m_{t-}$, xviii

$D^j_{yy}[k+1]$, the inverse of $R^j_{yy}[k+1]$, 30

$d[k]$, $D_{xx}[k]\hat{x}[k]$, 29

$\Delta\vartheta_t$ scaled modal observation, $\Delta\vartheta_t = h(\boldsymbol{\lambda}_t^{-1} * \Delta z_t)$, 49

dw, or dw_t, for a stochastic process $\{w_t\}$ the *forward* differential increment $w_{t+dt} - w_t$, where $dt > 0$, xvii

D_x, the inverse of observation noise covariance, $D_x = R_x^{-1}$, 29

$D_{xx}[k]$, the inverse of $P_{xx}[k]$, 29

$D_{yy}[k]$, the matrix inverse of $R_{yy}[k]$, 12

$\mathbf{e}$, canonical unit vector, xv

η^c, the continuous part of the martingale process η, xviii

η^d, the purely discontinuous part of the martingale process η, xviii
$\mathbf{e}_i$, the ith canonical unit vector, dimension understood from context, xv
$\mathbf{E}_i$, a square matrix of all zeros except for a 1 at position (i, i), xv
$\mathbf{E}_{ij}$, a square matrix of all zeros except for a 1 at position (i, j), xv
$E[x \mid \mathcal{Y}]$, the conditional expectation of x given the σ-field $\mathcal{Y}$, xvi
$\mathcal{F}$, a σ-field, a set of events (subsets) on an outcome set Ω, xvi
$\phi(k - s : k - 1)$, the s-length sequence of modal states taken by ϕ on the time-discrete interval $[k - s, k - 1]$, 28
$\hat{\phi}([\kappa]; [k - s : k - 1])$, the probability that ϕ on the time-discrete interval $[k - s, k - 1]$ is the κth such sequence, 28
$\{\phi_t\}$, a process (usually Markov) on the canonical unit vectors, $\mathbf{e}_i, i \in S$, xvii
$\{\mathcal{F}_t\}$, a (fundamental) filtration on Ω contained in $\mathcal{F}$, 2
$\{\mathcal{F}_t\}$, the filtration generated by $\{x_t, \phi_t, y_t, z_t\}$, 156
$\{\mathcal{G}_t^{\phi}\}$, the filtration generated by $\{y_t, \phi_t\}$, 11, 156
$\{\mathcal{G}_t^{\phi T}\}$, the filtration generated by $\{y_t\}$ and $\{\phi_\tau; \tau \in [0,\mathrm{T}]\}$, 156
$\{\mathcal{G}_t\}$, the filtration generated by $\{g_t\}$, 3
$\{\mathcal{G}_t\}$, the filtration generated by $\{y_t, z_t\}$, 156
$\mathbf{I}$, the identity matrix, size clear from context, xv
λ^{-1}, where λ is a vector, the vector of reciprocals, xv
$\boldsymbol{\lambda}_t$, the $\mathcal{F}_t$-conditional rate of modal-state observations, 39
$M(i, l)$, a base-state rotation or scaling associated with a modal-state transition $i \mapsto l$, 6
$\mathbf{N}(m, P)$, the Gaussian (normal) distribution with mean m and covariance P, xv
$\{\mathcal{O}_t\}$, 48
Π, as a variable, a modal transition matrix, 23
$P_{\phi\phi}$, the $\mathcal{G}_t$-covariance of ϕ_t, 46
$P_{\phi\chi}$, the $\mathcal{G}_t$-covariance of ϕ and χ, 46
PME(0.55,10), a PME with target type measurement fidelity 0.55 and measurement rate 10 /s, 177
$\{\boldsymbol{\Phi}_t\}$, a modal-state process, 2
$P_{x\phi\phi_i}(t)$ the third central moment, $E[\tilde{x}_t \widetilde{\phi_t' \tilde{\phi}_i} \mid \mathcal{G}_t]$, 50
$P_{(x\tilde{\phi}_i)x\phi_m}$, the compound central moment $E[(\widetilde{x_t \tilde{\phi}_i}) \tilde{x}_t' \tilde{\phi}_m$, 50
$P_{x_i x}$, the ith row of P_{xx}, usually associated with the ith component of the vector x, xv
$P_{x\chi}$, covariance process for x_t and χ_t, 54
$P_{xx}(0)$, P_{xx} at $t = 0$, 3
$P_{xx\phi_i}$, the third central moment $E[\tilde{x}_t \tilde{x}_t \hat{\phi}_i \mid \mathcal{G}_t]$, 50
$P_{xx\phi_i\phi_m}$, the fourth central moment $E[\tilde{x}_t \tilde{x}_t' \tilde{\phi}_r \underline{\tilde{\phi}_m \mid \mathcal{G}_t}]$, 50
$P_{(xx\phi_i)\phi_m}$, the compound central moment $E[(x_t x_t' \phi_i) \tilde{\phi}_m \mid \mathcal{G}_t]$, 50

P_{xx_i}, the ith column of P_{xx}, usually associated with the ith component of x, xvi
$P^j_{xx}[k+1]$, the x-covariance for the jth model in a multiple model estimator, 29
$P_{\chi\chi}$, covariance process for χ_t, 54
Q, a Markov transition matrix, xvii
r, radians, 14
$\mathbb{R}$, the field of real numbers, xv
$R_{\phi\phi}$, $E[\phi_t\phi_t' \mid \mathcal{G}_t$, 46
$R_{\phi\chi}$, $E[\phi\chi' \mid \mathcal{G}_t]$, the $\mathcal{G}_t$-cross correlation of ϕ and χ, 46
$\rho(i,l)$, a translational discontinuity in plant base-state associated with a modal-state transition from $i \mapsto l$, 6
$R^j_{yy}[k+1]$, the jth model covariance of y at time $k+1$, 30
$\mathbb{R}^k$, the k-dimensional space of vectors over $\mathbb{R}$, xv
r/s, radians per second, 14
$R_{x\phi\phi_i}(t)$, the third moment $E[x_t\phi_t'\phi_i \mid \mathcal{G}_t]$, 50
$R_{xx\phi_i}$, the third moment $E[x_t x_t'\phi_i \mid \mathcal{G}_t]$, 50
R_{xy}, the correlation of x and y, $E[xy']$, also written with explicit reference to time dependence, $R_{xy}(t)$, or for discrete time, $R_{xy}[k]$, xv
$R_{yy}[k]$, the covariance of $y[k]$, 12
S, integer index set, xv
$S^j_{yy}[k+1]$, a matrix square root of $D^j_{yy}[k+1]$, 30
$S_{yy}(t)$, a matrix square root of the positive definite symmetric matrix P_{yy}, $S_{yy}(t)'S_{yy}(t) = P_{yy}(t)$, xvi
$S_{yy}[k]$, a matrix square root of $R_{yy}[k]$, 12
T, final time of interest, as in $[0,\mathsf{T}]$, xv
$\mathrm{Tr}(A)$, the trace of the matrix A, 72
υ_n, a nominal value of the actuating signal υ_t associated with the nominal operating point χ_n, 3
υ, a block matrix array of the nominal actuating inputs $\upsilon_i : \upsilon = [\upsilon_1|\upsilon_2|\cdots|\upsilon_n|$, 4
$\{u_t\}$, an exogenous plant input process and, in particular, a perturbation input given by $u_t = \upsilon_t - \upsilon_n$, 3
W, the intensity of the Brownian motion $\{w_t\}$, xvii
$\{w_t\}$, a Brownian motion process, xvii
$\{\chi_t\}$, a plant state process, 2
χ, a block matrix array of the nominal operating points $\chi_i : \chi = [\chi_1|\chi_2|\cdots|\chi_n]$, 4
$\{\mathcal{X}^{\phi\mathsf{T}}_t\}$, the filtration generated by $\{x_t\}$ and $\{\phi_\tau; \tau \in [0,\mathsf{T}]\}$, 156
$\{\mathcal{X}^{\phi}_t\}$, the filtration generated by $\{x_t, \phi_t\}$, 156
$\hat{x}^j[k+1]$, the jth estimate of $x[k+1]$ in a multiple model context, 29
$\hat{x}_t$, the conditional mean of x_t, xvii

$\hat{x}_t; \mathcal{Y}_t$, the conditional mean of $\hat{x}_t$ with explicit reference to the conditioning σ-field $\mathcal{Y}_t$, xvii

χ_n, a nominal operating point for the state process χ_t, 3

$\tilde{x}_t$, the estimation error in $\hat{x}_t$, $x_t - \tilde{x}_t$, xvii

$\{x_t\}$, a base-state process and, in particular, a perturbation process given by $x_t = \chi_t = \chi_n$, 3

$\{\mathcal{X}_t^z\}$, the filtration generated by $\{x_t, z_t\}$, 156

$\mathcal{Y}$, a σ-field on an outcome set Ω, xvi

$\mathcal{Y}_s$, the coarsest σ-field with respect to which all random variables $\{y(\tau); 0 \leq \tau \leq s\}$ are measurable, where $\{y_t\}$ is a random process on [0,T] and $0 \leq s \leq$T, xvi

$\{y_t\}$, a stochastic process; in particular, an observation process, 41

$\{\mathcal{Y}_t\}$, the filtration generated by $\{y_t\}$, 4

$\{\mathcal{Y}_{t-}\}$, the left continuous version of $\{\mathcal{Y}_t\}$, xvi

y_{t+}, $\lim_{\tau \to t+} y_\tau$, xvi

$(\Omega, \mathcal{F}, \mathcal{P})$, a probability space on the set Ω with σ-algebra $\mathcal{F}$ and probability measure $\mathcal{P}$, xvi

$(\Omega, \mathcal{F}, \mathcal{P}; \mathcal{F}_t)$, a probability triple together with a filtration, $\{\mathcal{F}_t\}$, xvii

$\{z_t\}$, a stochastic process; in particular, the modal-state observation process, 43

$\mathcal{Z}_t$, the filtration generated by $\{z_t\}$, 38